Laboratory M

# Essentials
# of Biology

**Seventh Edition**

## Sylvia S. Mader

**Contributor**

## Dave Cox

*Lincoln Land Community College*

McGraw Hill

LABORATORY MANUAL TO ACCOMPANY ESSENTIALS OF BIOLOGY, SEVENTH EDITION

Published by McGraw Hill LLC, 1325 Avenue of the Americas, New York, NY 10019. Copyright ©2024 by McGraw Hill LLC. All rights reserved. Printed in the United States of America. Previous editions ©2021, 2018, and 2015. No part of this publication may be reproduced or distributed in any form or by any means, or stored in a database or retrieval system, without the prior written consent of McGraw Hill LLC, including, but not limited to, in any network or other electronic storage or transmission, or broadcast for distance learning.

Some ancillaries, including electronic and print components, may not be available to customers outside the United States.

This book is printed on acid-free paper.

1 2 3 4 5 6 7 8 9 LMN 28 27 26 25 24 23

ISBN 978-1-266-09137-7
MHID 1-266-09137-8

Portfolio Manager: *Lora Neyens*
Product Developer: *Anne Winch*
Marketing Manager: *Erin Martin*
Content Project Manager: *Lisa Bruflodt*
Buyer: *Sandy Ludovissy*
Content Licensing Specialist: *Lori Hancock*
Cover Image: *Eric Weber/Inspire Edventures LLC*
Compositor: *Aptara®, Inc.*

All credits appearing on page or at the end of the book are considered to be an extension of the copyright page.

The Internet addresses listed in the text were accurate at the time of publication. The inclusion of a website does not indicate an endorsement by the authors or McGraw Hill LLC, and McGraw Hill LLC does not guarantee the accuracy of the information presented at these sites.

# Contents

# Preface

The 25 laboratory sessions in this manual have been designed to introduce beginning students to the major concepts of biology, while keeping in mind minimal preparation for sequential laboratory use. The laboratories are coordinated with *Essentials of Biology*, a general biology text that covers all fields of biology. In addition, this Laboratory Manual can be adapted to a variety of course orientations and designs. There are a sufficient number of laboratories and exercises within each lab to tailor the laboratory experience as desired. Then, too, many exercises may be performed as demonstrations rather than as student activities, thereby shortening the time required to cover a particular concept.

## Laboratory Resource Guide

The Laboratory Resource Guide, an essential aid for instructors and laboratory assistants, free to adopters of the *Essentials of Biology*, is available in the Instructor Resources area of Connect. The answers to the Laboratory Review questions are in the Laboratory Resource Guide.

## Customized Editions

With McGraw-Hill Create™ you can easily rearrange the labs in this manual and combine them with content from other sources, including your own syllabus or teaching notes. Go to create.mheducation.com to learn how to create your own version of the *Essentials of Biology Laboratory Manual.*

## The Exercises

All exercises have been tested for student interest, preparation time, estimated time of completion, and feasibility. The following features are particularly appreciated by adopters:

**Integrated opening:** Each laboratory begins with a list of Learning Outcomes organized according to the major sections of the laboratory. The major sections of the laboratory are numbered on the opening page and in the laboratory text material. This organization will help students better understand the goals of each laboratory session.

**Pre-Lab Questions:** Each lab now contains a set of pre-lab questions designed to familiarize students with the lab before coming to class. The pre-lab questions are presented at the beginning of each section in the lab and include questions about the lab's purpose, procedure, and concepts. Students will need to carefully read each section to answer the questions.

**Self-contained content:** Each laboratory contains all the background information necessary to understand the concepts being studied and to answer the questions asked. This feature will reduce student frustration and increase learning.

**Scientific process:** All laboratories stress the scientific process, and many opportunities are given for students to gain an appreciation of the scientific method. The first laboratory of this edition explicitly explains the steps of the scientific method and gives students an opportunity to use them.

**Student activities:** Sequentially numbered steps guide students as they perform each activity. Some student exercises are Observations (designated by a tan bar) and some are Experimental Procedures (designated by a blue bar). A time icon appears whenever a procedure requires a period of time before results can be viewed.

**Live materials:** Although students work with living material during some part of almost all laboratories, the exercises are designed to be completed within one laboratory session. This facilitates the use of the manual in multiple-session courses.

**Laboratory safety:** Laboratory safety, a listing on the website www.mhhe.com, will assist instructors in making the laboratory experience a safe one. Throughout the laboratories, safety precautions specific to an activity are highlighted and identified by a caution symbol.

# Contents

# Preface

The 25 laboratory sessions in this manual have been designed to introduce beginning students to the major concepts of biology, while keeping in mind minimal preparation for sequential laboratory use. The laboratories are coordinated with *Essentials of Biology*, a general biology text that covers all fields of biology. In addition, this Laboratory Manual can be adapted to a variety of course orientations and designs. There are a sufficient number of laboratories and exercises within each lab to tailor the laboratory experience as desired. Then, too, many exercises may be performed as demonstrations rather than as student activities, thereby shortening the time required to cover a particular concept.

## Laboratory Resource Guide

The Laboratory Resource Guide, an essential aid for instructors and laboratory assistants, free to adopters of the *Essentials of Biology*, is available in the Instructor Resources area of Connect. The answers to the Laboratory Review questions are in the Laboratory Resource Guide.

## Customized Editions

With McGraw-Hill Create™ you can easily rearrange the labs in this manual and combine them with content from other sources, including your own syllabus or teaching notes. Go to create.mheducation.com to learn how to create your own version of the *Essentials of Biology Laboratory Manual.*

## The Exercises

All exercises have been tested for student interest, preparation time, estimated time of completion, and feasibility. The following features are particularly appreciated by adopters:

**Integrated opening:** Each laboratory begins with a list of Learning Outcomes organized according to the major sections of the laboratory. The major sections of the laboratory are numbered on the opening page and in the laboratory text material. This organization will help students better understand the goals of each laboratory session.

**Pre-Lab Questions:** Each lab now contains a set of pre-lab questions designed to familiarize students with the lab before coming to class. The pre-lab questions are presented at the beginning of each section in the lab and include questions about the lab's purpose, procedure, and concepts. Students will need to carefully read each section to answer the questions.

**Self-contained content:** Each laboratory contains all the background information necessary to understand the concepts being studied and to answer the questions asked. This feature will reduce student frustration and increase learning.

**Scientific process:** All laboratories stress the scientific process, and many opportunities are given for students to gain an appreciation of the scientific method. The first laboratory of this edition explicitly explains the steps of the scientific method and gives students an opportunity to use them.

**Student activities:** Sequentially numbered steps guide students as they perform each activity. Some student exercises are Observations (designated by a tan bar) and some are Experimental Procedures (designated by a blue bar). A time icon appears whenever a procedure requires a period of time before results can be viewed.

**Live materials:** Although students work with living material during some part of almost all laboratories, the exercises are designed to be completed within one laboratory session. This facilitates the use of the manual in multiple-session courses.

**Laboratory safety:** Laboratory safety, a listing on the website www.mhhe.com, will assist instructors in making the laboratory experience a safe one. Throughout the laboratories, safety precautions specific to an activity are highlighted and identified by a caution symbol.

## Improvements This Edition

The Laboratory Review section has been updated to better reflect the Learning Outcomes of each particular lab. In addition, many labeling exercises have been added to engage students with the lab diagrams.

For this edition, the laboratory manual author worked with a diversity, equity, and inclusion reviewer to identify areas in which the content could be made more sensitive to, and representative of, the broad spectrum of learners taking biology courses today. The narrative has been revised to be gender-neutral where possible, though the terms *male* and *female* are used where needed to indicate biological sex. Improvements have also been made in the accessibility of the artwork in the manual.

**Lab 2 Measuring with Metric.** Table 2.1 Metric Units of Length Measurement was expanded to include the kilometer unit. This will help students apply metrics to practical life situations.

**Lab 8 Meiosis: Sexual Reproduction.** Several figures have been updated so they match the figures in the textbook. This will help students make a stronger connection between the content in the text and the content in the lab.

**Lab 15 Seed Plants.** Table 15.1 Monocots and Eudicots was updated to reflect additional features used to identify a Monocot versus a Eudicot.

**Lab 16 Organization of Flowering Plants.** Figure 16.5 was updated to reflect additional features used to identify a Monocot versus a Eudicot.

# 1

# Scientific Method

## Learning Outcomes

**Introduction**
- In general, describe pillbug external anatomy and lifestyle.
- Outline the steps of the scientific method.

**1.1 Observing a Pillbug**
- Observe and describe the external anatomy of a pillbug, *Armadillidium vulgare*.
- Observe and describe how a pillbug moves.

**1.2 Formulating Hypotheses**
- Formulate a hypothesis based on appropriate observations.

**1.3 Performing the Experiment and Coming to a Conclusion**
- Reach a conclusion based on observation and experimentation.

## Introduction

### Pre-Lab

1. What is the purpose of this lab? _____
   _____

2. What is the purpose of the scientific method? _____
   _____

This laboratory will provide you with an opportunity to practice the scientific method using a type of crustacean called a pillbug (*Armadillidium vulgare*) as a subject.

Pillbugs have an exoskeleton consisting of overlapping "armored" plates that make them look like little armadillos. As pillbugs grow, they molt (shed the exoskeleton) four or five times during a lifetime. A pillbug can roll up into such a tight ball that its legs and head are no longer visible, earning it the nickname "roly-poly." Pillbugs have three body parts: head, thorax, and abdomen. The head bears compound eyes and two pairs of antennae. The thorax bears pairs of walking legs; gills for gas exchange are located at the top of the first five pairs. The gills must be kept slightly moist, which explains why pillbugs are usually found in damp places. The final pair of appendages, the uropods, which are sensory and defensive in function, project from the abdomen of the animal.

Pillbugs are commonly found in damp leaf litter, under rocks, and in basements or crawl spaces under houses. Following an inactive winter, pillbugs mate in the spring. Several weeks later, the eggs hatch and remain for six weeks in a brood pouch on the underside of the female's body. Once they leave the pouch, they eat primarily dead organic matter, including decaying leaves.

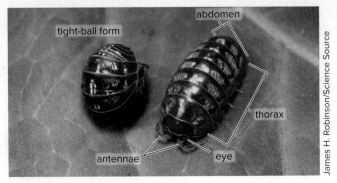
Pillbugs on leaf

## The Scientific Method

To gain knowledge about the natural world, scientists follow a series of steps when doing research called the **scientific method.** However, it's important to realize the scientific method is not a rigid procedure that must be followed, but a guideline that can be adapted and modified.

Most often, the scientific method follows these steps, as shown in Figure 1.1:

**Figure 1.1  Flow diagram for the scientific method.**
Scientists can predict the results by formulating hypotheses that are based on prior observations. Experiments are conducted to test the hypotheses that are supported or not supported by the analysis of data collected. Modification of the hypotheses can lead to more experiments. The conclusions of many related experiments can lead to the development of a theory.

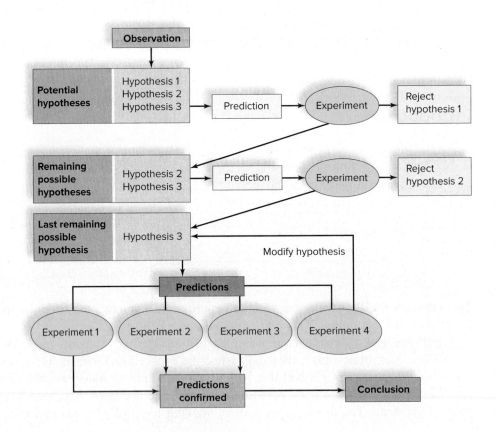

**Making observations.** An observation of the natural world typically leads to questions about the natural world. For example, watching leaves change color in autumn might lead you to use the scientific method to learn why they change color. A critical part of the scientific method is researching what other scientists have learned about the topic. This provides background knowledge to perform the next step of formulating a hypothesis.

**Hypothesis.** A **hypothesis** is an informed statement that can be tested with the scientific method. To formulate a hypothesis, scientists need both an observation to hypothesize about and background knowledge to create a hypothesis that is plausible. Multiple hypotheses may even be created, for example:

> Leaves change color because of cold temperatures.

> Leaves change color because of decreasing day length.

**Testing the hypothesis.** Scientists use a variety of different experiments to test a hypothesis. The results can tell a scientist whether the hypothesis is supported or not supported. Many experiments, like yours today with the pillbug, are controlled experiments that follow a certain testing procedure. Controlled experiments eliminate all other variables except those stated in the hypothesis and also use a control. A **control** is a sample or event in the experiment that is not exposed to the testing procedure.

For example, if you are testing the hypothesis that leaves change color because of cold temperatures, you would compare the color change of plants placed in cold temperatures to a control of plants not placed in cold temperatures.

A scientist must record detailed information during an experiment called **data.** Accurate data allow other scientists to retest and validate the experiment.

**Coming to a conclusion.** Scientists conclude whether their data from the experiment *support* or *do not support* the hypothesis. A scientist cannot say a hypothesis is proven true or false because some future knowledge or alternate experiment may change the conclusion.

**Developing a scientific theory.** If many related experiments give similar conclusions and have similar supported hypotheses, it can lead to a scientific theory. A **scientific theory** is an encompassing conclusion based on many individual conclusions. Theories develop over many years and are supported by overwhelming evidence, for example, gene theory and the theory of evolution.

## 1.1   Observing a Pillbug

**Pre-Lab**

3. Considering you will be testing how pillbugs move toward or away from particular foods or repellents, how are the observations from this section important? _____

_____

Wash your hands before and after handling pillbugs. Please handle them carefully so they are not crushed. When touched, they roll up into a ball or "pill" shape as a defense mechanism. They will soon recover if left alone.

### *Observation: Pillbug's External Anatomy*

Place a pillbug in a small dish to keep it contained. Examine the pillbug with your unaided eye and under magnification. In the following space, draw an outline of your pillbug (at least 7 centimeters [cm] long). *Label the head, thorax, abdomen, antennae, eyes, legs, and uropods (leaflike growths at the base of some legs of females where developing eggs and embryos are held in pouches).*

1. Observe how the pillbug moves and behaves for a few moments. You may hold it in your hand or watch it from underneath the dish. Watch it under magnification to see details of its legs or mouthparts. Describe in detail what you observed.

   _____

   _____

   _____

   _____

   _____

   _____

2. Measure the speed of three pillbugs.
   a. Place each pillbug on a metric ruler, and use a stopwatch to measure the number of seconds (sec) it takes for the pillbug to move several centimeters. Record the number of centimeters moved and time in seconds for each pillbug in Table 1.1.
   b. In the table, record the speed of each pillbug in centimeters per second (cm/sec) by dividing the number of centimeters traveled by the time in seconds.
   c. Add the three speeds together and divide by 3 to determine the average speed for the three pillbugs.

| Table 1.1 Pillbug Speed | | | |
|---|---|---|---|
| Pillbug | Centimeters (cm) Traveled | Time (sec) | Speed (cm/sec) |
| 1 | | | |
| 2 | | | |
| 3 | | | |
| | | Average speed: | |

## 1.2 Formulating Hypotheses

### Pre-Lab

4. Why is it important to have sand and water as controls in this experiment? _____

   _____

You will be testing whether pillbugs are attracted to (move toward and eat), repelled by (move away from), or unresponsive to (don't move away from and do not move toward and eat) the particular substances, which are potential foods. If a pillbug simply rolls into a ball, nothing can be concluded, and you may wish to choose another pillbug or wait a minute or two to check for further response.

1. Choose
   a. two dry substances, such as flour, cornstarch, coffee creamer, or baking soda. Fine sand will serve as a control for dry substances. Record your "dry" choices as 1, 2, and 3 in the first column of Table 1.2.

**b.** two liquids, such as milk, orange juice, ketchup, applesauce, or carbonated beverage. Water will serve as a control for liquid substances. Record your "wet" choices as 4, 5, and 6 in the first column of Table 1.2.

**2.** In the second column of Table 1.2, hypothesize how you expect the pillbug to respond to each substance. Use a plus (+) sign if you hypothesize that the pillbug will move toward and eat the substance; a minus (−) sign if you hypothesize that the pillbug will be repelled by the substance; and a zero (0) if you expect the pillbug to show neither behavior.

**3.** In the third column of Table 1.2, offer a reason for your hypothesis based on your knowledge of pillbugs from the introduction and your examination of the animal.

| Table 1.2 Hypotheses About Pillbug's Response to Potential Foods | | |
|---|---|---|
| **Substance** | **Hypothesis About Pillbug's Response** | **Reason for Hypothesis** |
| 1 | | |
| 2 | | |
| 3 | (control) | |
| 4 | | |
| 5 | | |
| 6 | (control) | |

## 1.3 Performing the Experiment and Coming to a Conclusion

### Pre-Lab

**5.** Briefly list the steps of this Experimental Procedure. _____

_____

**6.** If a pillbug shows the same attraction toward a control and food, can you conclude the pillbug is attracted to the food? _____

_____

A good experimental design would be to keep your pillbug in a petri dish to test its reaction to the chosen substances. During your experiment, no substance must be put directly on the pillbug, nor can the pillbug be placed directly onto the substances.

### *Experimental Procedure: Pillbug's Response to Potential Foods*

**1.** Before testing the pillbug's reaction, fill in the first column of Table 1.3. It will look exactly like the first column of Table 1.2.

**2.** Since pillbugs tend to walk around the edge of a petri dish, you could put the wet or dry substance there; or for the wet substance, you could put liquid-soaked cotton in the pillbug's path.

3. Rinse your pillbug between procedures by spritzing it with distilled water from a spray bottle. Then put it on a paper towel to dry it off.
4. Watch the pillbug's response to each substance, and record it in Table 1.3, using +, −, or 0 as before.

| Table 1.3 Pillbug's Response to Potential Foods | | |
|---|---|---|
| **Substance** | **Pillbug's Response** | **Hypothesis Supported?** |
| 1 | | |
| 2 | | |
| 3 (control) | | |
| 4 | | |
| 5 | | |
| 6 (control) | | |

## Conclusion

5. Do your results support your hypotheses? Answer yes or no in the last column of Table 1.3.
6. **Class Results.** Compare your results with those of other students who tested the same substance. Calculate the proportional response to each potential food (%+, %−, %0) and record your calculations in Table 1.4. As a group, your class can decide what proportion is needed to designate this response as typical. For example, if the pillbugs as a whole were attracted to a substance 70% or more of the time, you can call that response the "typical response."
7. Scientists prefer to come to conclusions on the basis of many trials. Why is this the best methodology?

_____

8. Comparing the results from the controls to the other substances, did the pillbugs behave differently? How do the results of the control affect your conclusions about the tested substances? _____

_____

| Table 1.4 Pillbug's Response to Potential Foods: Class Results | | | | |
|---|---|---|---|---|
| **Substance** | **Pillbug's Response** | | | **Hypothesis Supported?** |
| 1 | %+ | %− | %0 | |
| 2 | %+ | %− | %0 | |
| 3 (control) | %+ | %− | %0 | |
| 4 | %+ | %− | %0 | |
| 5 | %+ | %− | %0 | |
| 6 (control) | %+ | %− | %0 | |

_____ **1.** What kind of animal is a pillbug?

_____ **2.** What kind of skeleton does a pillbug have?

_____ **3.** What structures do pillbugs use for gas exchange, and what condition must these structures be in to function properly?

_____ **4.** What do scientists do first when they begin to study a specific topic?

_____ **5.** What is a tentative answer about the outcome of an experiment?

_____ **6.** What do we call the sample that lacks the factor being tested and goes through all the experimental steps?

_____ **7.** What do scientists call the information they collect while doing experiments or making observations?

_____ **8.** Which is made by a scientist following experiments and observations, a theory or a conclusion?

_____ **9.** What do scientists develop after many years of experimentation and a lot of similar individual conclusions?

_____ **10.** If your hypothesis was that your pillbug would be attracted to applesauce and your pillbug moved toward the applesauce, what would you say about your hypothesis?

_____ **11.** If a pillbug travels 3 millimeters (mm) in 30 seconds, what is its rate of speed?

_____ **12.** If a pillbug moves toward a substance, is it attracted to or repelled by that substance?

For 13 and 14, indicate whether the statements are hypotheses, conclusions, or scientific theories.

_____ **13.** All organisms are made of cells.

_____ **14.** The data show that trans fat intake raises cholesterol and contributes to heart disease.

## Thought Questions

**15.** Why is a theory more comprehensive than a conclusion?

**16.** Why is it important to have a control substance for an experiment?

**17.** Why is it important to test a pillbug's response using one substance at a time?

# 2

# Measuring with Metric

## Learning Outcomes

**2.1 Length**
- Perform measurements of length.
- Convert the metric units for length from one metric unit to another.

**2.2 Weight**
- Perform measurements of weight.
- Convert the metric units for weight from one metric unit to another.

**2.3 Volume**
- Perform measurements of volume.
- Determine strategies for measuring volume in various circumstances.
- Convert the metric units for volume from one metric unit to another.

**2.4 Temperature**
- Compare and contrast the Fahrenheit (F) and Celsius (C) temperature scales.
- Perform measurements of temperature.
- Convert one unit of temperature into another using a provided equation.

## Introduction

### Pre-Lab

1. Why is a standard system of measurement useful across the sciences? _____

_____

2. Which types of measurements will you perform in this lab? _____

The metric system is the standard system of measurement in the sciences, including biology, chemistry, and physics (Fig. 2.1). It has tremendous advantages because all conversions, whether for volume, mass (weight), or length, are in units of ten.

**Figure 2.1**
The metric system is the system of measurement used in scientific laboratories.

Harmik Nazarian/Hill Street Studios/Blend Images/Getty Images

This base-ten system is similar to our monetary system, in which 10 cents equals a dime, 10 dimes equals a dollar, and so on. In this laboratory, you will gain experience making measurements of length, volume, mass, and temperature.

## 2.1 Length

### Pre-Lab

**3.** What is the base unit for length? _____

**4.** Which metric units of measurement are on a typical ruler? _____

Metric units of length measurement include the **kilometer (km), meter (m), centimeter (cm), millimeter (mm), micrometer (µm),** and **nanometer (nm)** (Table 2.1). The prefixes *milli* ($10^{-3}$), *micro* ($10^{-6}$), and *nano* ($10^{-9}$) are used with length, weight, and volume.

**Table 2.1  Metric Units of Length Measurement**

| Unit | Meters | Centimeters | Millimeters | Relative Size |
|---|---|---|---|---|
| Kilometer (km) | 1,000 ($10^3$) m | 100,000 cm | 1,000,000 mm | Largest |
| Meter (m) | 1 m | 100 cm | 1,000 mm | |
| Centimeter (cm) | 0.01 ($10^{-2}$) m | 1 cm | 10 mm | |
| Millimeter (mm) | 0.001 ($10^{-3}$) m | 0.1 cm | 1.0 mm | |
| Micrometer (µm) | 0.000001 ($10^{-6}$) m | 0.0001 ($10^{-4}$) cm | 0.001 ($10^{-3}$) mm | |
| Nanometer (nm) | 0.000000001 ($10^{-9}$) m | 0.0000001 ($10^{-7}$) cm | 0.000001 ($10^{-6}$) mm | Smallest |

### Experimental Procedure: Length

**1.** Obtain a small ruler marked in centimeters and millimeters. One centimeter equals how many millimeters? _____ To express the size of small objects, such as cell contents, biologists use even smaller units of the metric system than those on the ruler. These units are the micrometer (µm) and the nanometer (nm).

According to Table 2.1, 1 µm = _____ mm, and 1 nm = _____ mm.

**2.** Measure the diameter of the circle shown to the nearest millimeter. This circle's diameter is _____ mm = _____ µm = _____ nm.

**3.** Obtain a meter stick. On one side, find the numbers 1 through 39, which denote inches. One meter equals 39.37 inches (in.); therefore, 1 meter is roughly equivalent to 1 yard (yd). Turn the meter stick over, and observe the metric subdivisions. How many centimeters are in a meter? _____ How many millimeters are in a meter? _____ How many meters are in a kilometer? _____.

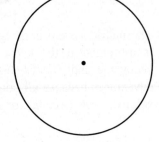

**4.** Use the meter stick and the method shown in Figure 2.2 to measure the length of two long bones from a disarticulated human skeleton. Lay the meter stick flat on the lab table. Place a long bone next to the meter stick between two pieces of cardboard (each about 10 cm × 30 cm), held upright at right angles to the stick. The narrow end of each piece of cardboard should touch the meter stick. The length between the cards is the length of the bone in centimeters. For example, if the bone measures from the 22-cm mark to the 50-cm mark, the length of the bone is _____ cm. If the bone measures from the 22-cm mark to midway between the 50-cm and 51-cm marks, its length is _____ mm, or _____ cm.

5. Record the length of two bones. First bone: _____ cm = _____ mm.

Second bone: _____ cm = _____ mm.

**Figure 2.2  Measurement of a long bone.**
How to measure a long bone using a meter stick.

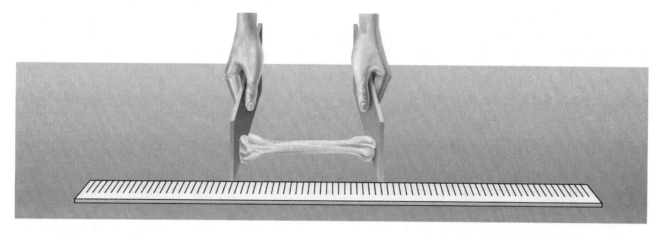

## 2.2  Weight

### Pre-Lab

5. What is the base metric unit for weight? _____

6. Which metric unit is typically used for larger weights such as human body weight?

_____

Two metric units of weight are the **gram (g)** and the **milligram (mg)**. A paper clip weighs about 1 g, which equals 1,000 mg. 2 g = _____ mg; 0.2 g = _____ mg; and 2 mg = _____ g.

### Experimental Procedure: Weight

1. Use a balance scale to measure the weight of a small wooden block.

2. Measure the weight of the block to the tenth of a gram. _____ g = _____ mg.

3. Measure the weight of an item small enough to fit inside the opening of a 50-milliliter (ml) graduated cylinder. The item, a(n) _____, is _____ g = _____ mg.

## 2.3 Volume

### Pre-Lab

**7.** What is the base metric unit for volume? _____

**8.** What is another way 1 ml can be expressed? _____

Two metric units of volume are the **liter (l)** and the **milliliter (ml).** One liter = 1,000 ml.

### Experimental Procedure: Volume

1. Graduated cylinders are used to make accurate measures of liquid volume. To compare the accuracy of graduated cylinders to a beaker, fill a 100-ml beaker to the 40-ml mark. Now, pour the contents of the beaker into a 50-ml graduated cylinder. Read the volume of water at the meniscus (Fig. 2.3), or lowest margin of the liquid level. What is the actual volume of water? _____

2. It's useful to know about how many drops of water equal 1 ml. Fill a graduated cylinder to exactly 20 ml. A dropper bottle is necessary to obtain an exact amount. Now, add drops of water from the bottle or pipette until the meniscus reaches exactly 21 ml. About how many drops equal 1 ml of water? _____

3. The volume of solid objects can be measured using liquid. Your instructor will provide a small object. Hypothesize how you can measure the volume of the object with the water in the graduated cylinder.

_____

_____

Use your method to find the volume of the object. _____ ml

Instead of ml, the volume of solid objects is typically represented as the cube of a length measurement, such as cubic centimeter ($cm^3$). Conveniently, 1 ml = 1 $cm^3$. What is the volume of your object in cubic centimeters? _____

Why is it easier to use this method to measure volume of the object compared to using length measurements to calculate the volume? _____

_____

_____

**Figure 2.3  Meniscus.**
The proper way to view the meniscus.

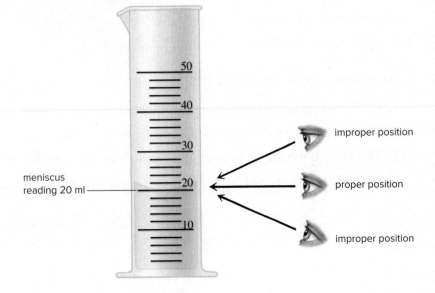

## 2.4 Temperature

### Pre-Lab

**9.** Convert 98.6°F to Celsius. _____

There are two temperature scales: the **Fahrenheit (F)** and **Celsius (centigrade, C)** scales (Fig. 2.4). Scientists use the Celsius scale.

### Experimental Procedure: Temperature

1. Study the two scales in Figure 2.4, and complete the following information:

   **a.** Water freezes at either _____ °F or _____ °C.

   **b.** Water boils at either _____ °F or _____ °C.

2. To convert from the Fahrenheit to the Celsius scale, use the following equation:

$$°C = (°F - 32°)/1.8$$
$$\text{or}$$
$$°F = (1.8°C) + 32$$

   Human body temperature of 98°F is what temperature on the Celsius

   scale? _____

3. Record any two of the following temperatures in your lab environment. In each case, allow the end bulb of the Celsius thermometer to remain in or on the sample for 1 minute.

   Room temperature = _____ °C

   Surface of your skin = _____ °C

   Cold tap water in a 50-ml beaker = _____ °C

   Hot tap water in a 50-ml beaker = _____ °C

   Ice water = _____ °C

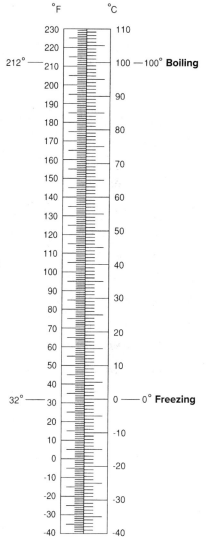

**Figure 2.4   Temperature scales.**
The Fahrenheit (°F) scale is on the left, and the Celsius (°C) scale is on the right.

## Laboratory Review 2

_____ 1. What system of measurement is used in science?

_____ 2. Which unit of measurements were examined in this lab?

_____ 3. What is the base unit for length?

_____ 4. 19 mm equals how many centimeters?

_____ 5. 880 mm equals how many meters?

_____ 6. What instruments are used to observe objects smaller than a millimeter?

_____ 7. What is the base unit for mass?

_____ 8. What instrument is used to measure mass?

_____ 9. 2,700 mg equals how many grams?

_____ 10. 1 ml is equal to how many cubic centimeters?

_____ 11. 3.4 l equals how many milliliters?

_____ 12. To properly measure 20 ml of water, what must be at the 20-ml mark of the graduated cylinder?

_____ 13. Which temperature scale is used in science?

_____ 14. 22°C equals how many degrees F?

## Thought Questions

15. Describe the advantages of using the metric system you discovered during this lab.

16. Explain how you would measure the volume of a solid object, such as a rock.

17. Why is it advantageous to use a standard measurement system in all sciences?

# 3
# Microscopy

## Learning Outcomes

**Introduction**
- Describe similarities and differences between the stereomicroscope (dissecting microscope), the compound light microscope, and the electron microscope.

**3.1 Stereomicroscope (Dissecting Microscope)**
- Identify the parts and explain how to focus the stereomicroscope.

**3.2 Use of the Compound Light Microscope**
- Identify and give the function of the basic parts of the compound light microscope.
- List, in proper order, the steps for bringing an object into focus with the compound light microscope.
- Examine microscope characteristics, including inversion, total magnification, diameter of field, and depth of field.

**3.3 Microscopic Observations**
- State two differences between human epithelial cells and onion epidermal cells.
- Examine a wet mount of pond water and describe the organisms observed.

## Introduction

### Pre-Lab

1. What is the purpose of this lab and what specific tool will you be using? _____
   _____

2. List several differences between a light and electron microscope. _____
   _____

3. Looking at Figure 3.1, what specimens or structures might you see today? _____

Cells are the basic unit of life, but they are extremely small and weren't discovered until the 1600s when the first microscopes were invented. Today, microscopes are an essential tool in biology. This laboratory examines the features, functions, and use of microscopes.

Light microscopes and electron microscopes are both used to view structures and details invisible to the human eye (Fig. 3.1). Light microscopes, which you will be using in the lab today, are the most common. Electron microscopes were invented in the 1930s and the most powerful types are able to view single atoms!

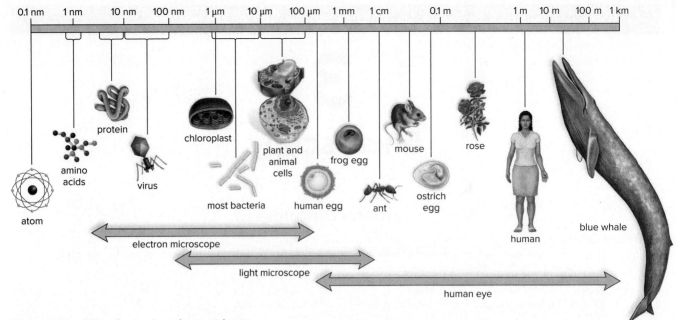

**Figure 3.1  The sizes of various objects.**
It takes a microscope to see most cells and lower levels of biological organization. Cells are visible with the light microscope, but not in much detail. An electron microscope is necessary to see organelles in detail and to observe viruses and molecules.

## Light Microscopes

Light microscopes use light rays passing through lenses to magnify an object. The **stereomicroscope (dissecting microscope)** is designed to study entire objects in three dimensions at low magnification. The **compound light microscope** is used for examining small or thinly sliced sections of objects under higher magnification than that of the stereomicroscope (Fig. 3.2a). The term **compound** refers to the use of two sets of lenses: the ocular lenses located near the eyes and the objective lenses located near the object. Illumination is from below, and visible light passes through clear portions but does not pass through opaque portions. To improve contrast, stains or dyes that bind to cellular structures and absorb light are often used.

## Electron Microscopes

Electron microscopes use beams of electrons to magnify the object. The beams are focused on a photographic plate by means of electromagnets. The **transmission electron microscope** is analogous to the compound light microscope. The object is ultra-thinly sliced and treated with heavy metal salts to improve contrast. The **scanning electron microscope** is analogous to the dissecting light microscope. It gives an image of the surface and dimensions of an object, as is apparent from the scanning electron micrograph in Figure 3.2b.

The electron microscope has greater resolving power. **Resolution** is the minimum distance between two objects at which they can still be seen, or resolved, as two separate objects. A compound light microscope can resolve objects about 200 nm apart, but electron microscopes are 2,000 times more powerful and can resolve objects at less than 0.1 nm!

**Figure 3.2  Comparative images of a lymphocyte.**
A lymphocyte is a type of white blood cell. A compound light microscope (a) shows less detail in the lymphocyte than a scanning electron microscope (b).

(a) Michael Ross/Science Source; (b) Steve Gschmeissner/Science Source

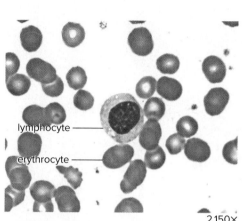

2,150×

**a.** Photomicrograph or light micrograph (LM)

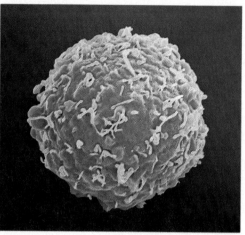

5,000×

**b.** Scanning electron micrograph (SEM)

# 3.1   Stereomicroscope (Dissecting Microscope)

## Pre-Lab

**4.** *Label Figure 3.3 with the help of your textbook.*

**5.** List several unique objects or specimens you could view with a stereomicroscope. _____

_____

The **stereomicroscope (dissecting microscope,** Fig. 3.3) allows you to view objects in three dimensions at low magnifications. It is used to study entire small organisms, any object requiring lower magnification, and opaque objects that can be viewed only by reflected light. It is called a stereomicroscope because it produces a three-dimensional image.

## Identifying the Parts

After your instructor has explained how to carry a microscope, obtain a stereomicroscope and a separate illuminator, if necessary, from the storage area. Place it securely on the table. Plug in the power cord, and turn on the illuminator. There is a wide variety of stereomicroscope styles, and your instructor will discuss the specific style(s) available to you. Regardless of style, the following features should be present:

1. **Binocular head:** holds two eyepiece lenses that move to accommodate for the various distances between different individuals' eyes.
2. **Eyepiece lenses:** the two lenses located on the binocular head. What is the magnification of your

   eyepieces? _____ Some models have one **independent focusing eyepiece** with a knurled knob to allow independent adjustment of each eye. The nonadjustable eyepiece is called the **fixed eyepiece.**

# Figure 3.3 Binocular dissecting microscope (stereomicroscope).

*Label this stereomicroscope with the help of the text material.*

©Leica Microsystems, GmbH

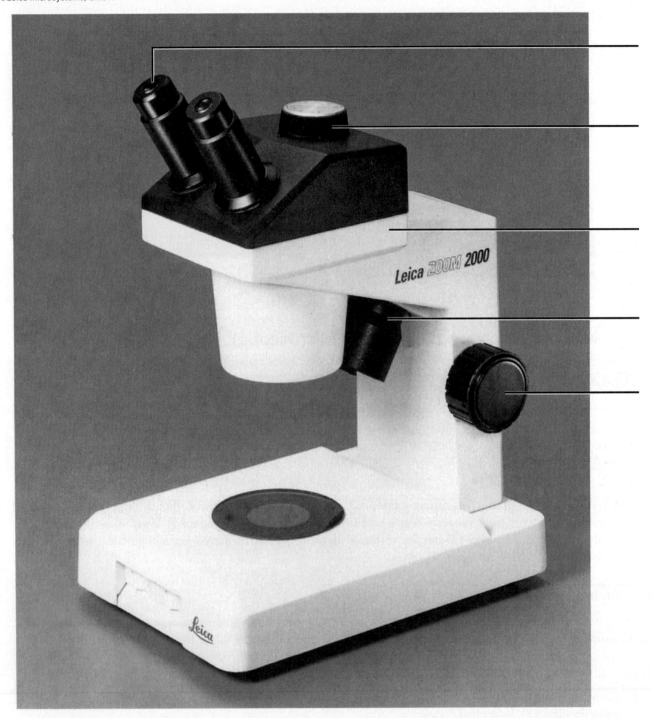

3. **Focusing knob:** a large, black or gray knob located on the arm; used for changing the focus of both eyepieces together.

4. **Magnification changing knob:** a knob, often built into the binocular head, used to change magnification in both eyepieces simultaneously. This may be a **zoom** mechanism or a **rotating lens** mechanism of different powers that clicks into place.

5. **Illuminator:** used to illuminate an object from above; may be built into the microscope or separate.

## Focusing the Stereomicroscope

1. In the center of the stage, place the specimen you are going to be observing.
2. Adjust the distance between the eyepieces on the binocular head so that they comfortably fit the distance between your eyes. You should be able to see the object with both eyes as one three-dimensional image.
3. Use the focusing knob to bring the object into focus.
4. Does your microscope have an independent focusing eyepiece? _____ If so, use the focusing knob to bring the image in the fixed eyepiece into focus, while keeping the eye at the independent focusing eyepiece closed. Then adjust the independent focusing eyepiece so that the image is clear, while keeping the other eye closed. Is the image inverted? _____
5. Turn the magnification changing knob, and determine the kind of mechanism on your microscope. A zoom mechanism allows continuous viewing while changing the magnification. A rotating lens mechanism blocks the view of the object as the new lenses are rotated. Be sure to click each lens firmly into place. If you do not, the field will be only partially visible. What kind of mechanism is on your microscope? _____
6. Set the magnification changing knob on the lowest magnification. Sketch the object in the following circle as though this represents your entire field of view:

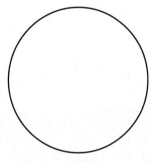

7. Rotate the magnification changing knob to the highest magnification. Draw another circle within the one provided to indicate the reduction of the field of view.
8. Experiment with various objects at various magnifications until you are comfortable with using the stereomicroscope.

## 3.2 Use of the Compound Light Microscope

### Pre-Lab

6. *Label Figure 3.4 with the help of your textbook.*
7. Compound light microscopes are parfocal and parcentric. In terms of using the multiple objective lenses, why are these features beneficial? _____
   _____
   _____

As mentioned, the name **compound light microscope** indicates that it uses two sets of lenses and light to view an object. The two sets of lenses are the ocular lenses located near the eyes and the objective lenses located near the object. Illumination is from below, and the light passes through clear portions but does not pass through opaque portions. This microscope is used to examine small or thinly sliced sections of objects under higher magnification than would be possible with the stereomicroscope.

## Identifying the Parts

Obtain a compound light microscope from the storage area, and place it securely on the table.

**Figure 3.4   Compound light microscope.**
Compound light microscope with binocular head and mechanical stage. *Label this microscope with the help of the text material.*
©Leica Microsystems GmbH

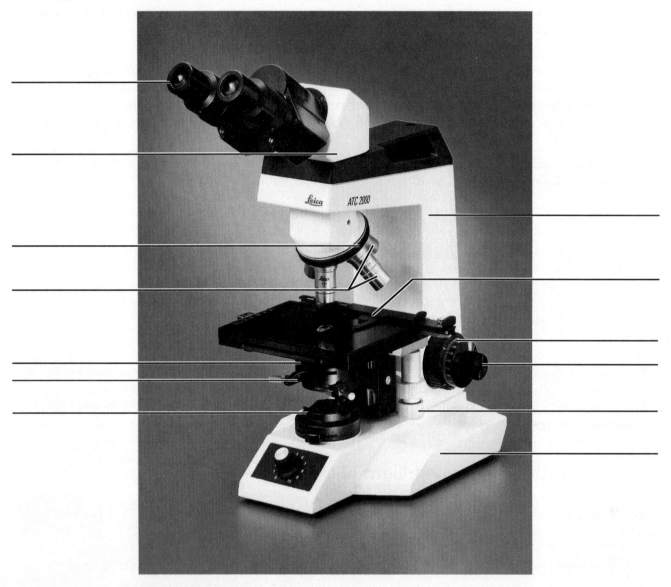

1. **Eyepieces** (ocular lenses): What is the magnifying power of the ocular lenses on your microscope? _____
2. **Viewing head:** holds the ocular lenses.
3. **Arm:** supports upper parts and provides carrying handle.
4. **Nosepiece:** revolving device that holds objectives.
5. **Objectives** (objective lenses):
   a. **Scanning objective:** This is the shortest of the objective lenses and is used to scan the whole slide. The magnification is stamped on the housing of the lens. It is a number followed by an ×. What is

   the magnifying power of the scanning objective lens on your microscope? _____

b. **Low-power objective:** This lens is longer than the scanning objective lens and is used to view objects in greater detail. What is the magnifying power of the low-power objective lens on your microscope? _____

c. **Power objective:** If your microscope has three objective lenses, this lens will be the longest. It is used to view an object in even greater detail. What is the magnifying power of the high-power objective lens on your microscope? _____

d. **Oil immersion objective** (on microscopes with four objective lenses): holds a 95× (to 100×) lens and is used in conjunction with immersion oil to view objects with the greatest magnification. Does your microscope have an oil immersion objective? _____ If this lens is available, your instructor will discuss its use when the lens is needed.

6. **Stage:** platform that holds and supports microscope slides. A mechanical stage is a movable stage that aids in the accurate positioning of the slide. Does your microscope have a mechanical stage? _____

   a. **Stage clips:** clips that hold a slide in place on the stage.
   b. **Mechanical stage control knobs:** two knobs that control forward/reverse movement and right/left movement, respectively.

7. **Coarse-adjustment knob:** knob used to bring object into approximate focus; used only with low-power objective.
8. **Fine-adjustment knob:** knob used to bring object into final focus.
9. **Condenser:** lens system below the stage used to focus the beam of light on the object being viewed.

   a. **Diaphragm or diaphragm control lever:** lever that controls the amount of light passing through the condenser.

10. **Light source:** an attached lamp that directs a beam of light up through the object.
11. **Base:** the flat surface of the microscope that rests on the table.

## Rules for Microscope Use

**Observe the following rules for using a microscope:**

1. The lowest-power objective (scanning or low) should be in position at both the beginning and the end of microscope use.
2. Use only lens paper for cleaning lenses.
3. Do not tilt the microscope because the eyepieces could fall out, or wet mounts could be ruined.
4. Keep the stage clean and dry to prevent rust and corrosion.
5. Do not remove parts of the microscope.
6. Keep the microscope dust-free by covering it after use.
7. Report any malfunctions.

## Focusing the Compound Light Microscope—Lowest Power

1. Turn the nosepiece so that the *lowest*-power objective on your microscope is in straight alignment over the stage.
2. Always begin focusing with the *lowest*-power objective on your microscope (4× [scanning] or 10× [low power]).
3. With the coarse-adjustment knob, lower the stage (or raise the objectives) until it stops.

4. Place a slide of the letter *e* on the stage, and stabilize it with the clips. (If your microscope has a mechanical stage, pinch the spring of the slide arms on the stage, and insert the slide.) Center the *e* as best you can on the stage or use the two control knobs located below the stage (if your microscope has a mechanical stage) to center the *e*.

5. Again, be sure that the lowest-power objective is in place. Then, as you look from the side, decrease the distance between the stage and the tip of the objective lens until the lens comes to an automatic stop or is no closer than 3 mm above the slide.

6. While looking into the eyepiece, rotate the diaphragm (or diaphragm control lever) to give the appropriate amount of light.

7. Using the coarse-adjustment knob, slowly increase the distance between the stage and the objective lens until the object—in this case, the letter *e*—comes into view, or focus.

8. Once the object is seen, you may need to adjust the amount of light. To increase or decrease the contrast, rotate the diaphragm slightly.

9. Use the fine-adjustment knob to sharpen the focus if necessary.

10. Practice having both eyes open when looking through the eyepiece, as this greatly reduces eyestrain.

## Inversion

Inversion refers to the fact that a microscopic image is upside down and reversed.

### Observation: Inversion

1. Draw the letter *e* as it appears on the slide (with the unaided eye, not looking through the eyepiece). _____

2. Draw the letter *e* as it appears when you look through the eyepiece. _____

3. What differences do you notice? _____

4. Move the slide to the right. Which way does the image appear to move? _____

5. Move the slide toward you. Which way does the image appear to move? _____

## Focusing the Compound Light Microscope—Higher Powers

Compound light microscopes are **parfocal**; that is, once the object is in focus with the lowest power, it should also be almost in focus with the higher power.

1. Bring the object into focus under the scanning power by following the instructions in the previous section.

2. Make sure that the letter *e* is centered in the field of the lowest objective.

3. Move to the next higher objective (low power [10×] or high power [40×]) by turning the nosepiece until you hear it click into place. Do not change the focus; parfocal microscope objectives will not "hit" normal slides when changing the focus if the lowest objective is initially in focus. (If you are on low power [10×], proceed to high power [40×] before going on to step 4.)

4. If any adjustment is needed, use only the *fine*-adjustment knob.
   (*Note:* Always use only the fine-adjustment knob with high power, and do not use the coarse-adjustment knob.)

5. On a drawing of the letter *e* to the right, *draw a circle around the portion of the letter that you are now seeing with high-power magnification.* The letter *e* will not disappear because your microscope is bold (the focus remains near the center).

6. When you have finished your observations of this slide (or any slide), rotate the nosepiece until the lowest-power objective clicks into place, and then remove the slide.

# Total Magnification

Total magnification is calculated by multiplying the magnification of the ocular lens (eyepiece) by the magnification of the objective lens. The magnification of a lens is imprinted on the lens casing.

## Observation: Total Magnification

Calculate total magnification figures for your microscope, and record your findings in Table 3.1.

| Table 3.1 Total Magnification | | | |
|---|---|---|---|
| **Objective** | **Ocular Lens** | **Objective Lens** | **Total Magnification** |
| Scanning power (if present) | | | |
| Low power | | | |
| High power | | | |
| Oil immersion (if present) | | | |

# Field of View

A microscope's **field of view** is the circle visible through the lenses. The **diameter of field** is the length of the field from one edge to the other.

## Observation: Field of View

### Low-Power (10×) Diameter of Field

1. Place a clear plastic ruler across the stage so that the edge of the ruler is visible as a horizontal line along the diameter of the low-power (not scanning) field. Be sure that you are looking at the millimeter side of the ruler.

2. Estimate the number of millimeters, to tenths, that you see along the field: _____ mm. (*Hint:* Start by placing any millimeter marker at the edge of the field.) Convert the observed number of millimeters to micrometers: _____ μm. This is the **low-power diameter of field (LPD)** for your microscope in micrometers.

### High-Power (40×) Diameter of Field

1. To compute the **high-power diameter of field (HPD),** substitute these data into the formula given:

   a. LPD = low-power diameter of field (in micrometers) = _____

   b. LPM = low-power total magnification (from Table 3.1) = _____

   c. HPM = high-power total magnification (from Table 3.2) = _____

Example: If the diameter of field is about 2 mm, then the LPD is 2,000 μm. Using the LPM and HPM values from Table 3.1, the HPD would be 500 μm.

$$HPD = LPD \times \frac{LPM}{HPM}$$

$$HPD = (\quad\quad) \times \frac{(\quad\quad\quad)}{(\quad\quad\quad)} = \text{_____}$$

## Depth of Field

When viewing an object on a slide under high power, the **depth of field** (Fig. 3.5) is the area—from top to bottom—that comes into focus while slowly focusing up and down with the microscope's fine-adjustment knob.

### Observation: Depth of Field

1. Obtain a prepared slide with three or four colored threads mounted together, or prepare a wet-mount slide with three or four crossing threads or hairs of different colors. (Directions for preparing a wet mount are given in Section 3.3.)

2. With low power, find a point where the threads or hairs cross. Slowly focus up and down. Notice that when one thread or hair is in focus, the others seem blurred. Remember, as the stage moves upward (or the objectives move downward), objects on top come into focus first. Determine the order of the threads or hairs, and complete Table 3.2.

3. Switch to high power, and notice that the depth of field is more shallow with high power than with low power. Focusing up and down with the fine-adjustment knob when viewing a slide with high power will give you an idea of the specimen's three-dimensional form. For example, viewing a number of sections from bottom to top allows reconstruction of the three-dimensional structure, as demonstrated in Figure 3.5.

**Figure 3.5  Depth of field.**
A demonstration of how focusing at depths 1, 2, and 3 would produce three different images (views) that could be used to reconstruct the original three-dimensional structure of the object.

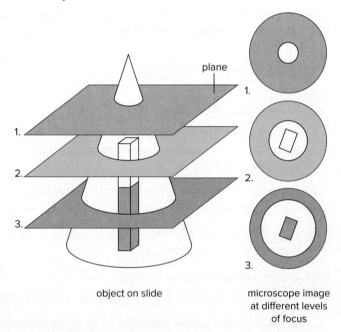

object on slide

microscope image at different levels of focus

| Table 3.2   Order of Threads (or Hairs) | |
|---|---|
| **Depth** | **Thread (or Hair) Color** |
| Top | |
| Middle | |
| Bottom | |

# 3.3 Microscopic Observations

**Pre-Lab**

8. Review the Observation exercises below for Onion Epidermal Cells, Human Epithelial Cells, and Pond Water. Which specific cell structures do you predict will be most visible? _____

_____

When a specimen is prepared for observation, the object should always be viewed as a **wet mount.** A wet mount is prepared by placing a drop of liquid on a slide or, if the material is dry, by placing it directly on the slide and adding a drop of water or stain. The mount is then covered with a coverslip, as illustrated in Figure 3.6. Dry the bottom of your slide before placing it on the stage.

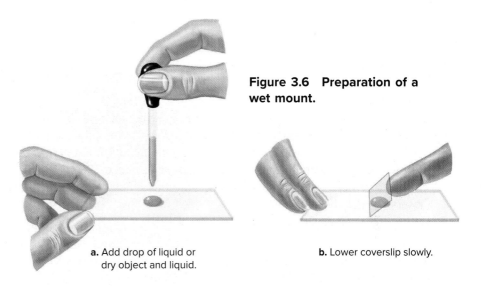

**Figure 3.6   Preparation of a wet mount.**

**a.** Add drop of liquid or dry object and liquid.

**b.** Lower coverslip slowly.

## Onion Epidermal Cells

Epidermal cells cover the surfaces of plant organs, such as leaves. The bulb of an onion is made up of fleshy leaves.

### *Observation: Onion Epidermal Cells*

1. With a scalpel, strip a small, thin, transparent layer of cells from the inside of a fresh onion leaf.

**Scalpel** Exercise care when using a scalpel.

**Figure 3.7   Onion epidermal cells.**
*Label the cell wall and the nucleus.*
Ted Kinsman/Science Source

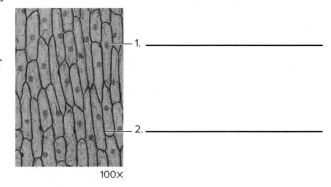

2. Place it gently on a clean, dry slide, and add a drop of iodine solution (or methylene blue). Cover with a coverslip.
3. Observe under the microscope.
4. Locate the cell wall and the nucleus. *Label Figure 3.7.*
5. Count the number of onion cells that line up end to end in a single line across the diameter of the high-power (40×) field. _____

1. _____

2. _____

100×

Based on what you learned about measuring diameter of field, what is the high-power diameter of field (HPD) in micrometers? _____ µm

Calculate the length of each onion cell:

(HPD ÷ number of cells): _____ µm

6. Wash and reuse this slide for the next exercise (Human Epithelial Cells).

## Human Epithelial Cells

Epithelial cells (Fig. 3.8) cover the body's surface and line its cavities.

### Observation: Human Epithelial Cells

1. Obtain a prepared slide, or make your own as follows:
   a. Obtain a prepackaged flat toothpick (or sanitize one with alcohol or alcohol swabs).
   b. Gently scrape the inside of your cheek with the toothpick, and place the scrapings on a clean, dry slide. Smear your cheek scraping approximately 2 cm across the center of the slide. Discard used toothpicks in the biohazard waste container provided.
   c. Add a drop of very weak *methylene blue* or *iodine solution,* and cover with a coverslip.
2. Observe under the microscope.
3. Locate the nucleus (the central, round body), the cytoplasm, and the plasma membrane (outer cell boundary). *Label Figure 3.8.*
4. Because your epithelial slides are biohazardous, they must be disposed of as indicated by your instructor.
5. Note some obvious differences between the onion cells and the human cheek cells, and list them in Table 3.3.

> ⚠ **Methylene blue** Avoid ingestion, inhalation, and contact with skin, eyes, and mucous membranes. If any should spill on your skin, wash the area with mild soap and water. Methylene blue will also stain clothing.

**Figure 3.8   Cheek epithelial cells.**
*Label the nucleus, the cytoplasm, and the plasma membrane.*
Dr. Gopal Murti/Science Source

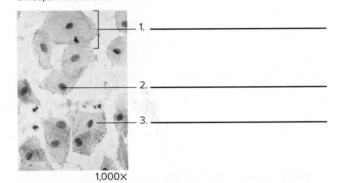

1. _____

2. _____

3. _____

1,000×

| Table 3.3   Differences Between Onion Epidermal and Human Epithelial Cells | | |
|---|---|---|
| **Differences** | **Onion Epidermal Cells** | **Human Epithelial Cells (Cheek)** |
| Shape | | |
| Orientation | | |
| Boundary | | |

## Pond Water

Examination of pond water will test your ability to observe objects with the microscope, to utilize depth of field, and to control illumination to heighten contrast.

1. Make a wet mount by using a drop of pond water and adding a drop of Protoslo (methyl cellulose solution) onto a slide. The Protoslo slows the organism's swimming.
2. Mix thoroughly with a toothpick, and add a coverslip.
3. Scan the slide for living organisms. Use Figure 3.9 to help identify what you find.

**Figure 3.9   Microorganisms found in pond water (not actual size).**

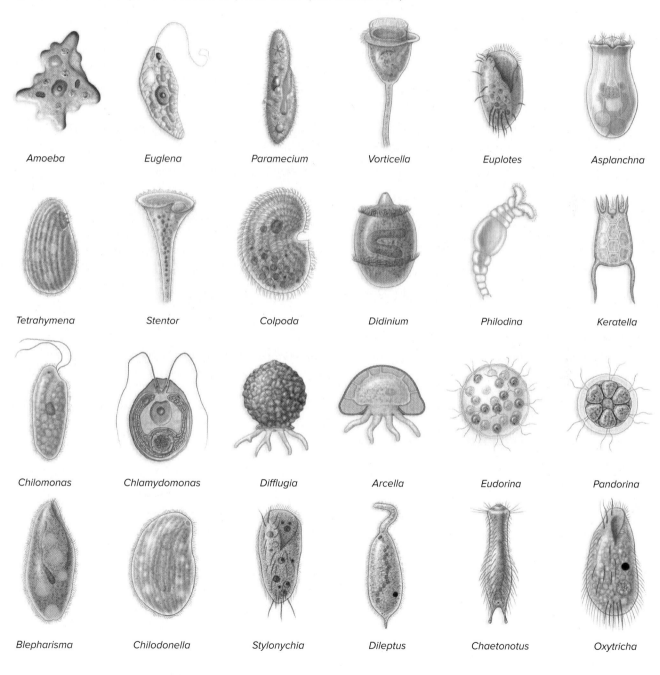

|  |  |  |  |  |  |
|---|---|---|---|---|---|
| Amoeba | Euglena | Paramecium | Vorticella | Euplotes | Asplanchna |
| Tetrahymena | Stentor | Colpoda | Didinium | Philodina | Keratella |
| Chilomonas | Chlamydomonas | Difflugia | Arcella | Eudorina | Pandorina |
| Blepharisma | Chilodonella | Stylonychia | Dileptus | Chaetonotus | Oxytricha |

_____ **1.** List several types of microscopes.

_____ **2.** What is the difference between a longitudinal section and a cross section of an object?

_____ **3.** What are the differences between a stereomicroscope and a compound light microscope?

_____ **4.** What does the term *compound* mean?

_____ **5.** What is used to improve contrast when viewing clear portions of cells?

_____ **6.** What kind of microscope would be used to study a whole or opaque object?

_____ **7.** What is the name for the lenses located near the eye?

_____ **8.** Which term describes the minimum distance between two objects required to distinguish them as two separate objects?

_____ **9.** If the total magnification of a slide is 400× and the ocular lenses are 10×, what is the magnifying power of the objective being used?

_____ **10.** If the amount of light passing through the condenser needs to be decreased, what microscope part should be adjusted?

_____ **11.** What word describes a microscope that remains in focus when the objective lenses are changed?

_____ **12.** If a slide is being viewed with the high-power objective, which adjustment knob should be used to sharpen the focus?

_____ **13.** Which objective should be in place when the microscope is put away?

_____ **14.** If threads are layered from top to bottom, brown, green, red, which layer will come into focus first if you are using the microscope properly?

## Thought Questions

**15.** Explain the type of microscope you should use to view a virus that is 50 nm in size?

**16.** While studying the pond water wet mount the organisms were swimming to the left, which way should you move your slide to keep them in view?

**17.** Explain the problems you will encounter if you use the course adjustment knob to focus on an object with the higher-power objective.

# 4

# Cell Structure and Function

## Learning Outcomes

**4.1 Prokaryotic Versus Eukaryotic Cells**
- Distinguish between prokaryotic and eukaryotic cells by description and examples.

**4.2 Animal Cell and Plant Cell Structure**
- Describe the structure and function of plant and animal cells.
- Use microscopic techniques to observe plant cell structure.

**4.3 Diffusion**
- Use Experimental Procedures to observe the process of diffusion.

**4.4 Osmosis: Diffusion of Water Across Plasma Membrane**
- Define osmosis, and explain the movement of water across a membrane.
- Predict the effect of different tonicities on animal (e.g., red blood) cells and on plant (e.g., *Elodea*) cells.

**4.5 pH and Cells**
- Predict the change in pH before and after the addition of an acid to nonbuffered and buffered solutions.
- Identify the effectiveness of antacid medications.

## Introduction

### Pre-Lab

1. This lab helps you investigate cell structure and cell function. What do you learn studying structure, and what do you learn studying function? _____

_____

The basic units of life are cells. The **cell theory** states that all organisms are composed of cells and that cells come only from other cells. While we are accustomed to considering the heart, the liver, or the intestines as enabling the human body to function, it is actually cells that do the work of these organs.

Figure 3.8 shows human cheek epithelial cells as viewed by an ordinary compound light microscope available in general biology laboratories. It shows that the content of a cell, called the **cytoplasm,** is enclosed by a **plasma membrane.** The plasma membrane regulates the movement of molecules into and out of the cytoplasm. In this

> **Planning Ahead** To save time, your instructor may have you start a boiling water bath and the potato strip experiment at the beginning of the laboratory.

lab, we will study how the passage of water into a cell depends on the difference in concentration of solutes (particles) between the cytoplasm and the surrounding medium or solution. The well-being of cells also depends on the pH of the solution surrounding them. We will see how a buffer can maintain the pH within a narrow range and how buffers within cells can protect them against damaging pH changes.

# 4.1 Prokaryotic Versus Eukaryotic Cells

### Pre-Lab

**2.** Based on their structure, is it likely that prokaryotic or eukaryotic cells evolved first? Why?

_____

_____

All living cells are classified as either prokaryotic or eukaryotic. One of the basic differences between the two types is that **prokaryotic cells** do not contain nuclei (*pro* means "before"; *karyote* means "nucleus"), while eukaryotic cells do contain nuclei (*eu* means "true"; *karyote* means "nucleus"). Bacteria (including cyanobacteria) and archaea are classified as prokaryotes; all other organisms are eukaryotes.

Prokaryotes also don't have the organelles found in **eukaryotic cells** (Fig. 4.1). **Organelles** are small, membranous bodies, each with a specific structure and function. Prokaryotes do have **cytoplasm,** the material enclosed by a plasma membrane and cell wall. The cytoplasm contains ribosomes, small granules that

**Figure 4.1 Prokaryotic cell.**
Prokaryotic cells lack membrane-bound organelles, as well as a nucleus. Their DNA is in a nucleoid region.
Sercomi/Science Source

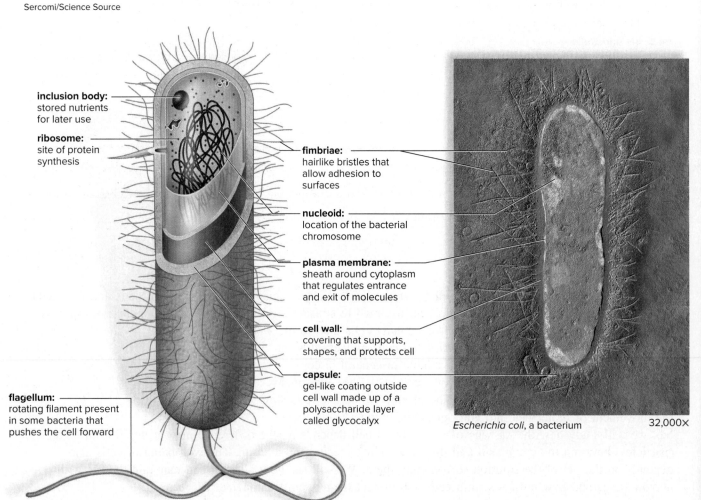

**inclusion body:**
stored nutrients
for later use

**ribosome:**
site of protein
synthesis

**fimbriae:**
hairlike bristles that
allow adhesion to
surfaces

**nucleoid:**
location of the bacterial
chromosome

**plasma membrane:**
sheath around cytoplasm
that regulates entrance
and exit of molecules

**cell wall:**
covering that supports,
shapes, and protects cell

**capsule:**
gel-like coating outside
cell wall made up of a
polysaccharide layer
called glycocalyx

**flagellum:**
rotating filament present
in some bacteria that
pushes the cell forward

*Escherichia coli*, a bacterium          32,000×

coordinate the synthesis of proteins; thylakoids (only in cyanobacteria) that participate in photosynthesis; and innumerable enzymes. Prokaryotes also have a nucleoid, a region in the bacterial cell interior in which the DNA is physically organized but not enclosed by a membrane.

### Observation: Prokaryotic/Eukaryotic Cells

Two microscope slides on display will show you the main difference between prokaryotic and eukaryotic cells.

Describe the differences you notice: _____

_____

# 4.2   Animal Cell and Plant Cell Structure

## Pre-Lab

**3.** Complete the Animal Cell Structure and Plant Cell Structure activities. Both activities will require the help of your textbook. Be sure to complete the structure/function descriptions and cell labeling.

## Animal Cell Structure

*Label Figure 4.2.* With the help of your textbook, give a function for each labeled structure.

| Structure | Function |
|---|---|
| Plasma membrane | _____ |
| Nucleus | _____ |
| Nucleolus | _____ |
| Ribosome | _____ |
| Endoplasmic reticulum | _____ |
|     Rough ER | _____ |
|     Smooth ER | _____ |
| Golgi apparatus | _____ |
| Vesicles | _____ |
| Lysosome | _____ |
| Mitochondrion | _____ |
| Centrioles in centrosome | _____ |
| Cytoskeleton | _____ |

# Figure 4.2 Animal cell structure.

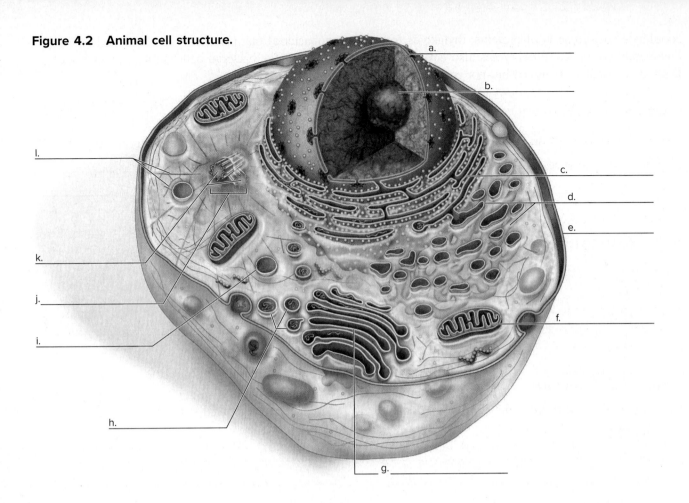

## Plant Cell Structure

*Label Figure 4.3.* With the help of your textbook, give a function for each labeled structure unique to plant cells.

| Structure | Function |
|---|---|
| Cell wall | _____ |
| Central vacuole, large | _____ |
| Chloroplast | _____ |

**Figure 4.3   Plant cell structure.**

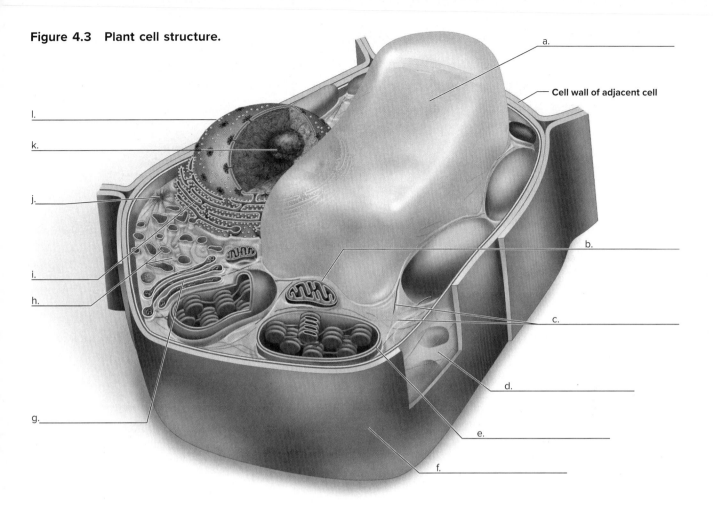

a. _____

Cell wall of adjacent cell

l. _____

k. _____

j. _____

i. _____

h. _____

b. _____

c. _____

d. _____

g. _____

e. _____

f. _____

## Observation: Microscopic Study of Plant Cells

1. Prepare a wet mount of a small piece of young *Elodea* leaf in fresh water. *Elodea* is a multicellular, eukaryotic plant found in freshwater ponds and lakes.
2. Have the drop of *water* ready on your slide so that the leaf does not dry out, even for a few seconds. Take care that the leaf is mounted with its top side up.
3. Examine the slide using low power, focusing sharply on the leaf surface.
4. Select a cell with numerous chloroplasts for further study, and switch to high power.
5. Carefully focus on the side and end walls of the cell. The chloroplasts appear to be only along the sides of the cell because the large, fluid-filled, membrane-bound central vacuole pushes the cytoplasm against the cell walls (Fig. 4.4*a*). Then focus on the surface and notice an even distribution of chloroplasts (Fig. 4.4*b*).

## Figure 4.4 *Elodea* cell structure.

(a–b) Ray F. Evert

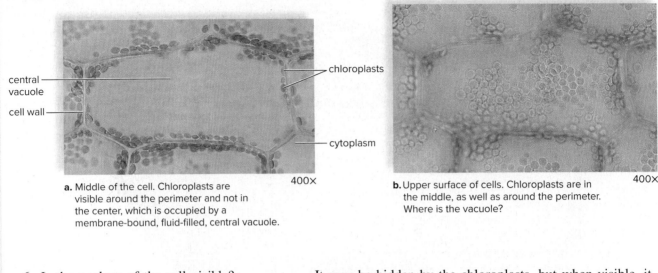

a. Middle of the cell. Chloroplasts are visible around the perimeter and not in the center, which is occupied by a membrane-bound, fluid-filled, central vacuole.

b. Upper surface of cells. Chloroplasts are in the middle, as well as around the perimeter. Where is the vacuole?

6. Is the nucleus of the cell visible? _____ It may be hidden by the chloroplasts, but when visible, it appears as a faint, gray lump on one side of the cell.

7. Why can't you see the other organelles featured in Figure 4.3? _____

_____

8. Can you detect movement of chloroplasts in this cell or any other cell? _____ The chloroplasts are not moving under their own power but are being carried by a streaming of the nearly invisible cytoplasm.

9. Save your slide for use later in this laboratory.

## 4.3  Diffusion

### Pre-Lab

4. How would you expect an increased temperature to affect the speed of diffusion? _____

_____

5. Review the Solute Diffusion Across the Plasma Membrane experiment and Figure 4.6. Where do you expect each solute—iodine, glucose, and starch—to be located after 5 minutes (beaker, dialysis bag, or both). Why? _____

_____

**Diffusion** is the movement of molecules from a higher to a lower concentration until equilibrium is achieved and the molecules are distributed equally (Fig. 4.5). At equilibrium, molecules may still be moving back and forth, but there is no net movement in any one direction.

Diffusion is a general phenomenon in the environment. The speed of diffusion is dependent on such factors as the temperature, the size of the molecule, and the type of medium through which the molecules move.

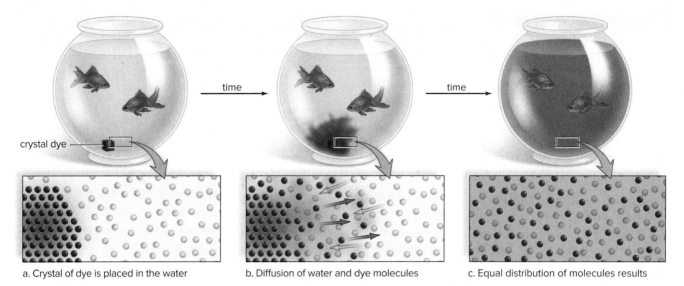

a. Crystal of dye is placed in the water     b. Diffusion of water and dye molecules     c. Equal distribution of molecules results

**Figure 4.5 Process of diffusion.**
Diffusion is apparent when dye molecules have equally dispersed.

## Experimental Procedure: Speed of Diffusion

### Solute Diffusion Through a Semisolid

1. Observe a petri dish containing 1.5% gelatin (or agar) to which potassium permanganate ($KMnO_4$) was added in the center depression at the beginning of the lab.

> ⚠ **Potassium permanganate ($KMnO_4$)** $KMnO_4$ is highly poisonous and is a strong oxidizer. Avoid contact with skin and eyes and with combustible materials. If spillage occurs, wash all surfaces thoroughly. $KMnO_4$ will also stain clothing.

2. Obtain the start time from your instructor, and record the start and ending time (now) in Table 4.1. What is the length of time in hours and minutes? _____ What is the time in hours only? _____ hr.

3. Using a ruler placed over the petri dish, measure (in mm) the movement of potassium permanganate ($KMnO_4$) from the center of the depression outward in one direction: _____ mm.

4. Calculate the speed of diffusion: _____ mm/60 min = mm/hr.

5. Record all data in Table 4.1.

### Solute Diffusion Through a Liquid

1. Add enough water to cover the bottom of a glass petri dish.
2. Place the petri dish over a thin, flat ruler. Position the petri dish directly over the mm measurement line.
3. With tweezers, add a crystal of potassium permanganate ($KMnO_4$) directly over the mm measurement line. Note the starting time in Table 4.1.
4. After 10 minutes, note the distance the color has moved. Record the final time, length of time in hours, and the distance moved in Table 4.1.
5. Multiply the length of time and the distance moved by 6 to calculate the speed of diffusion:

   _____ mm/hr. Record in Table 4.1.

### Solute Diffusion Through Air

1. Measure the distance from a spot designated by your instructor to your laboratory work area today. Record this distance under Distance Moved in Table 4.1.
2. Record the starting time in Table 4.1 when a perfume or similar substance is released into the air.
3. Note the time when you can smell the perfume. Record this as the final time in Table 4.1. Calculate the length of time since the perfume was released, and record it in Table 4.1.
4. Calculate the speed of diffusion: _____ mm/hr. Record in Table 4.1.

## Table 4.1  Speed of Diffusion

| Medium | Start Time | Final Time | Length of Time (hr) | Distance Moved (mm) | Speed of Diffusion (mm/hr) |
|---|---|---|---|---|---|
| Semisolid | | | | | |
| Liquid | | | | | |
| Air | | | | | |

### Conclusions: Solute Diffusion

- In which experiment was diffusion the fastest? _____

- What accounts for the difference in speed? _____

## Solute Diffusion Across the Plasma Membrane

Some molecules can diffuse across a plasma membrane, and some cannot. In general, small, noncharged molecules can cross a membrane by simple diffusion, but large molecules cannot diffuse across a membrane. The dialysis tube membrane in the Experimental Procedure simulates a plasma membrane.

### Experimental Procedure: Solute Diffusion Across Plasma Membrane

At the start of the experiment,

1. Cut a piece of dialysis tubing approximately 40 cm (approx. 16 in.) long. Soak the tubing in water until it is soft and pliable.
2. Close one end of the dialysis tubing with two knots.
3. Fill the bag halfway with glucose solution.
4. Add four full droppers of starch solution to the bag.
5. Hold the open end while you mix the contents of the dialysis bag. Rinse off the outside of the bag with distilled water.
6. Fill a beaker two-thirds full with distilled water.
7. Add droppers of iodine solution (IKI) to the water in the beaker until an amber (tealike) color is apparent.
8. Record the color of the solution in the beaker in Table 4.2.
9. Place the bag in the beaker with the open end hanging over the edge. Secure the open end of the bag to the beaker with a rubber band as shown (Fig. 4.6). Make sure the contents do not spill into the beaker.

**Figure 4.6  Placement of dialysis bag in water containing iodine.**

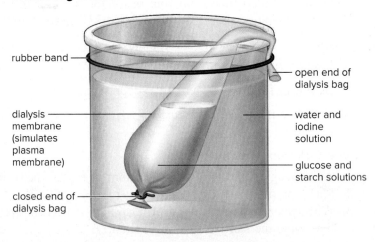

rubber band

open end of dialysis bag

dialysis membrane (simulates plasma membrane)

water and iodine solution

glucose and starch solutions

closed end of dialysis bag

After about 35–45 minutes, at the end of the experiment:

10. You will note a color change. Record the color of the bag contents in Table 4.2.
11. Obtain a small test tube. Using a graduated transfer pipet, draw 1 ml from the bottom of the beaker (near the bag) and place it in the test tube. Using a designated transfer pipet,

> **Benedict's reagent** Exercise care in using this chemical. It is highly corrosive. If any should spill on your skin, wash the area with mild soap and water. Follow your instructor's directions for its disposal.

add 3 ml of Benedict's solution. Heat in a boiling water bath for 5 to 10 minutes, observe any color change, and record your results as positive or negative in Table 4.2. (Optional use of glucose test strip: Dip glucose test strip into beaker. Compare stick with chart provided by instructor.)

12. Remove the dialysis bag from the beaker. Dispose of it and the used Benedict's reagent solution in the manner directed by your instructor.

## Table 4.2   Solute Diffusion Across Plasma Membrane

| At Start of Experiment | At End of Experiment | | |
|---|---|---|---|
| Color of Solution | Color of Solution | Benedict's Test | Conclusion |
| Bag contents: Glucose and Starch | ✕ | | |
| Beaker contents: Water and Iodine | | | |

### Conclusions: Solute Diffusion Across Plasma Membrane

- Based on the color change noted in the bag, conclude what solute diffused across the dialysis membrane from the beaker to the bag, and record your conclusion in Table 4.2.
- From the results of the Benedict's test on the beaker contents, conclude what solute diffused across the dialysis membrane from the bag to the beaker, and record your conclusion in Table 4.2.

- Which solute did not diffuse across the dialysis membrane from the bag to the beaker? _____

  How do you know? _____

  _____

  _____

# 4.4   Osmosis: Diffusion of Water Across Plasma Membrane

### Pre-Lab

6. In terms of osmosis, what would happen if a hospital patient was given pure water as an IV solution rather than a 0.9% sodium chloride (NaCl) solution? _____

   _____

7. In the Tonicity in Potato Strips experiment, do you expect the potato strip placed in 10% NaCl solution to become rigid or limp? Why? _____

**Osmosis** is the diffusion of water across the plasma membrane of a cell. Just like any other molecule, water follows its concentration gradient and moves from the area of higher concentration to the area of lower concentration.

To demonstrate osmosis, a thistle tube is covered with a differentially permeable membrane at its lower opening and partially filled with 50% starch solution. The whole apparatus is placed in a beaker containing distilled water, as described in the legend for Figure 4.7. Therefore, the water concentration in the beaker is 100%. Water molecules can move freely between the thistle tube and the beaker.

**Figure 4.7   Osmosis demonstration.**
**a.** A thistle tube, covered at the broad end by a differentially permeable membrane, contains a starch solution. The beaker contains distilled water. **b.** The solute (starch) is unable to pass through the membrane (red), but the water (arrows) passes through in both directions. There is a net movement of water toward the inside of the thistle tube, where there is a higher solute (and lower water) concentration. **c.** Because of the incoming water molecules, the level of the solution rises in the thistle tube.

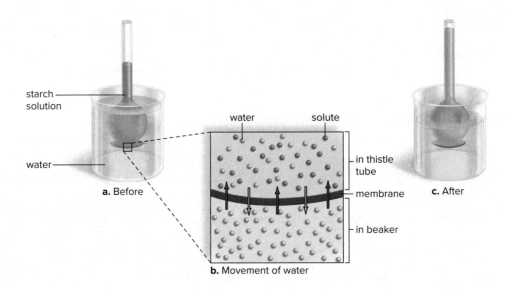

1. Note the level of liquid in the thistle tube, and measure how far it travels up the thistle tube in 10 minutes: _____ mm.

2. Calculate the speed of osmosis under these conditions: _____ mm/hr.

### Conclusions: Osmosis

- In which direction was there a net movement of water? _____

  Explain what is meant by "net movement" after examining the arrows in Figure 4.7b. _____

  _____

- If the starch molecules moved from the thistle tube to the beaker, would there have been a net

  movement of water into the thistle tube? _____ Why wouldn't large starch molecules be able to

  move across the membrane from the thistle tube to the beaker? _____

  _____

- Explain why the water level in the thistle tube rose: In terms of solvent concentration, water moved

  from the area of _____ water concentration to the area of _____ water concentration across a differentially permeable membrane.

# Tonicity in Cells

**Tonicity** is the relative concentration of solute (particles), and therefore also of solvent (water), outside the cell compared with inside the cell.

- An **isotonic solution** has the same concentration of solute (and therefore of water) as the cell. When cells are placed in an isotonic solution, there is no net movement of water.
- A **hypertonic solution** has a higher solute (therefore, lower water) concentration than the cell. When cells are placed in a hypertonic solution, water moves out of the cell into the solution.
- A **hypotonic solution** has a lower solute (therefore, higher water) concentration than the cell. When cells are placed in a hypotonic solution, water moves from the solution into the cell.

## Animal Cells (Red Blood Cells)

A solution of 0.9% NaCl is isotonic to red blood cells. In such a solution, red blood cells maintain their normal appearance (Fig. 4.8*a*). A solution greater than 0.9% NaCl is hypertonic to red blood cells. In such a solution, the cells shrivel up, a process called **crenation** (Fig. 4.8*b*). A solution of less than 0.9% NaCl is hypotonic to red blood cells. In such a solution, the cells swell to bursting, a process called **hemolysis** (Fig. 4.8*c*).

**Figure 4.8  Tonicity and red blood cells.**

(a, c): David M. Phillips/Science Source

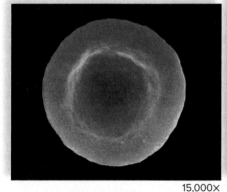
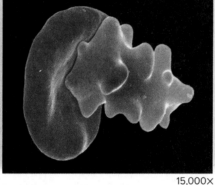

15,000×          15,000×          15,000×

**a. Isotonic solution.**
Red blood cell has normal appearance due to no net gain or loss of water.

**b. Hypertonic solution.**
Red blood cell shrivels due to loss of water.

**c. Hypotonic solution.**
Red blood cell fills to bursting due to gain of water.

---

### Experimental Procedure: Demonstration of Tonicity in Red Blood Cells

Three stoppered test tubes on display have the following contents:
    Tube 1: 0.9% NaCl plus a few drops of whole sheep blood
    Tube 2: 10% NaCl plus a few drops of whole sheep blood
    Tube 3: 0.9% NaCl plus distilled water and a few drops of whole sheep blood

> ⚠️ Do not remove the stoppers of test tubes during this procedure.

1. In the second column of Table 4.3, record the tonicity of each tube in relation to red blood cells.
2. Hold each tube in front of one of the pages of your lab manual. Determine whether you can see the print on the page through the tube. Record your findings in the third column of Table 4.3.
3. In the fourth column of Table 4.3, relate print visibility to the effect of tonicity on cells.

| Table 4.3 | Effect of Tonicity on Red Blood Cells | | |
|-----------|----------|----------------|-------------|
| Tube | Tonicity | Print Visibility | Explanation |
| 1 | | | |
| 2 | | | |
| 3 | | | |

## Plant Cells

When plant cells are in a hypotonic solution, such as fresh water, the large central vacuole gains water and exerts pressure, called **turgor pressure.** The cytoplasm, including the chloroplasts, is pushed up against the cell wall. You observed turgor pressure in Figure 4.4.

When plant cells are in a hypertonic solution, such as 10% NaCl, the central vacuole loses water, and the cytoplasm, including the chloroplasts, pulls away from the cell wall. This is called **plasmolysis.** You will observe plasmolysis in the following Experimental Procedure.

### Experimental Procedure: Tonicity in Elodea Cells

1. If possible, use the *Elodea* slide you prepared earlier in this laboratory. If not, prepare a new wet mount of a small *Elodea* leaf using fresh water. Your slide should look like Figure 4.4.
2. Complete the portion of Table 4.4 that pertains to a hypotonic solution.
3. Prepare a new wet mount of a small *Elodea* leaf using a 10% NaCl solution.
4. After several minutes, focus on the surface of the cells. Note that plasmolysis has occurred, and the cell contents are now in the center in most cells because the cytoplasm has pulled away from the cell wall due to loss of water from the large central vacuole.
5. Complete the portion of Table 4.4 that pertains to a hypertonic solution.

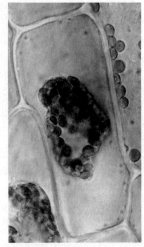

Plasmolysis is visible in most cells.

Ed Reschke/Stone/Getty Images

| Table 4.4 | Effect of Tonicity on *Elodea* Cells | |
|-----------|---------------------|------------------------|
| Tonicity | Appearance of Cells | Due to (Scientific Term) |
| Hypotonic | | |
| Hypertonic | | |

(This Experimental Procedure runs for 1 hour. Prior setup can maximize time efficiency.)

1. Cut two strips of potato, each about 7 cm long and 1.5 cm wide. Make sure you do not have any skin on the potato strips.
2. *Label two test tubes 1 and 2.* Place one potato strip in each tube.
3. Fill tube 1 with water to cover the potato strip.
4. Fill tube 2 with 10% NaCl to cover the potato strip.
5. After 1 hour, remove the potato strips from the test tubes and place them on a paper towel. Observe each strip for limpness (water loss) or stiffness (water gain). Which tube has the limp potato strip? _____ Use tonicity to explain why water diffused out of the potato strip in this tube. _____

_____

Which tube has the stiff potato strip? _____ Use tonicity to explain why water diffused into the potato strip in this tube. _____

_____

### Conclusions: Tonicity

• What occurs when cells are exposed to hypotonic environments?

_____

• What occurs when cells are exposed to hypertonic environments?

_____

## 4.5  pH and Cells

### Pre-Lab

8. What happens to proteins in cells if they are exposed to acidic or basic conditions? _____

_____

9. Is there anywhere in your body where a high or low pH is beneficial? _____

_____

The pH of a solution indicates its hydrogen ion concentration $[H^+]$. The **pH scale** ranges from 0 to 14. A pH of 7 is neutral (Fig. 4.9). A pH lower than 7 indicates that the solution is acidic (has more hydrogen ions than hydroxide ions), whereas a pH greater than 7 indicates that the solution is basic (has more hydroxide ions than hydrogen ions).

The concept of pH is important in biology because organisms are very sensitive to hydrogen ion concentrations. For example, in humans the pH of the blood must be maintained at about 7.4 or we become ill. All organisms need to maintain the hydrogen ion concentration, or pH, at a constant level. A **buffer** is a system of chemicals that takes up excess hydrogen ions or hydroxide ions, as appropriate.

**Figure 4.9   The pH scale.**
The proportion of hydrogen ions ($H^+$) to hydroxide ions ($OH^-$) is indicated by the diagonal line.

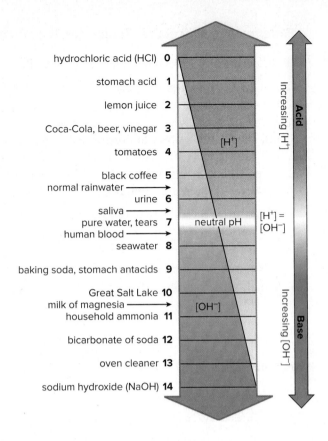

## Experimental Procedure: pH and Cells

1. *Label three test tubes*, and fill them to the halfway mark as follows: tube 1: water; tube 2: buffer (a buffered inorganic solution); and tube 3: simulated cytoplasm (a buffered protein solution).

2. Use pH paper to determine the pH of each tube. Dip the end of a stirring rod into the solution, and then touch the stirring rod to a 5-cm strip of pH paper. Read the current pH by matching the color observed with the color code on the pH paper package. Record your results in the "pH Before Acid" column in Table 4.5.

3. Add 0.1 N hydrochloric acid (HCl) dropwise to each tube until you have added 5 drops—shake or swirl after each drop. Use pH paper as in step 2 to determine the new pH of each solution. Record your results in the "pH After Acid" column in Table 4.5.

> ⚠ **Hydrochloric acid (HCl)** used to produce an acid pH is a strong, caustic acid. Exercise care in using this chemical. If any HCl spills on your skin, rinse immediately with clear water. Follow your instructor's directions for disposal of tubes that contain HCl.

### Table 4.5   pH and Cells

| Tube | Contents | pH Before Acid | pH After Acid | Explanation |
|------|----------|----------------|---------------|-------------|
| 1 | Water | | | |
| 2 | Buffer | | | |
| 3 | Cytoplasm | | | |

## Conclusions: pH and Cells

- Which solution allowed the smallest change in pH after the acid was added?

- Why would you expect cytoplasm to be as effective as the buffer in maintaining pH? _____

_____

## Experimental Procedure: Effectiveness of Antacids

Perform this procedure to test the ability of commercial products such as Alka-Seltzer, Rolaids, Tums, or other antacid tablets to absorb excess $H^+$.

1. Use a mortar and pestle to grind up the amount of antacid that is listed as one dose.
2. For each antacid tested, use 100 ml of phenol red solution diluted to a faint pink to wash the antacid into a 250-ml beaker. Phenol red solution is a pH indicator that turns yellow in an acid and red in a base. Use a stirring rod to get the powder to dissolve.
3. Add and count the number of 0.1 N HCl drops it takes for the solution to turn light yellow.
4. Record your results in Table 4.6.

### Table 4.6 Effectiveness of Antacids

| Antacid | Drops of Acid Needed to Reach End Point | Evaluation |
|---------|------------------------------------------|------------|
| 1 | | |
| 2 | | |
| 3 | | |

## Conclusions: Effectiveness of Antacids

- Did a difference in dosage (convert to mg) have any effect on the results? _____

_____

- Which of the substances on the label could be a buffer? _____

# Laboratory Review 4

_____ 1. What regulates the movement of molecules into and out of the cytoplasm?

_____ 2. Identify the feature(s) that are used to distinguish a prokaryotic cell.

_____ 3. What structures associated with prokaryotic and eukaryotic cells carry out protein synthesis?

_____ 4. What two things are found in the nucleus of a eukaryotic cell?

_____ 5. What plant cell organelle carries out photosynthesis?

_____ 6. What molecule is associated with plant cell walls?

_____ 7. What energy (ATP)-producing organelle is found in animal and plant cells?

_____ 8. What is the movement of molecules from an area of higher to lower concentration called?

_____ 9. What is the movement of water across the plasma membrane called?

_____ 10. Does water move into or out of cells that are placed in a hypotonic solution?

_____ 11. What kind of solution causes crenation to happen to red blood cells: isotonic, hypertonic, or hypotonic?

_____ 12. What kind of solution causes chloroplasts to be pushed outward against the cell wall: isotonic, hypertonic, or hypotonic?

_____ 13. If a solution has more hydrogen ions than hydroxide ions, is the solution acidic or basic?

_____ 14. What is present in human blood that ensures blood pH is maintained at about 7.4?

## Thought Questions

15. What cellular structures are found in both prokaryotic and eukaryotic cells?

16. Ovarian follicle cells produce estrogen, which is a steroid. What organelle will be present in abundance so that the follicle cells might perform this function? Justify your reply.

17. Contact lens solution is described as a sterile, buffered, isotonic aqueous solution. Explain the importance of the adjectives _buffered_ and _isotonic_ to the person buying the solution.

# 5

# How Enzymes Function

## Learning Outcomes

**Introduction**
- Describe how an enzyme functions.

**5.1 Catalase Activity**
- Identify the substrate, enzyme, and product in the reaction studied today.

**5.2 Effect of Temperature on Enzyme Activity**
- Predict the effect of temperature on an enzymatic reaction.

**5.3 Effect of Concentration on Enzyme Activity**
- Predict the effect of enzyme and substrate concentration on an enzymatic reaction.

**5.4 Effect of pH on Enzyme Activity**
- Predict the effect of pH on an enzymatic reaction.

**5.5 Factors That Affect Enzyme Activity**
- Tell how temperature, concentration of substrate or enzyme, and pH can promote or inhibit enzyme activity.

## Introduction

**Pre-Lab**

1. What is the purpose of an enzyme? _____

_____

2. Does catalase participate in a synthesis or degradation reaction? _____

The cell carries out many chemical reactions. The following is a chemical reaction.

$$A + B \longrightarrow C + D$$
$$\text{reactants} \qquad \text{products}$$

In all chemical reactions, the **reactants** are molecules that undergo a change, they can also be called substrates, which are changed during the course of the reaction and are turned into the **products.** The arrow stands for the change that produced the product(s). For example, A + B change and produce C + D. The numbers of reactants and products can vary; in the one you are studying today, a single reactant breaks down to two products. All the reactions that occur in a cell have an enzyme. **Enzymes** are organic catalysts, typically proteins, that speed metabolic reactions. The reactants in an enzymatic chemical reaction are called **substrate(s)** (Fig. 5.1). In today's laboratory, you will be studying the action of the enzyme **catalase,** which breaks hydrogen peroxide $(H_2O_2)$ into water and oxygen.

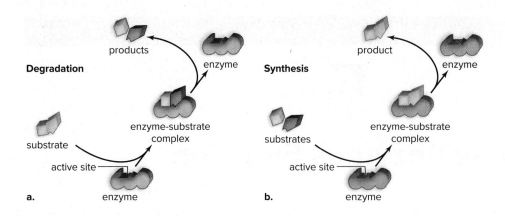

**Figure 5.1 Enzymatic action.**
The reaction occurs on the surface of the enzyme at the active site. The enzyme is reusable. **a.** Degradation: substrate is broken down. **b.** Synthesis: substrates are combined.

Every enzyme has a unique shape to match a particular substrate; therefore, enzymes are specific in their action. Enzymatic reactions can be indicated like this:

$$E + S \longrightarrow ES \longrightarrow E + P$$

In this reaction, E = enzyme, ES = enzyme-substrate complex, and P = product.

Two types of enzymatic reactions in cells are shown in Figure 5.1. During a **degradation reaction,** the substrate is broken down to the product(s), and during a **synthesis reaction,** substrates are joined to form a product. A number of other types of reactions also occur in cells. The location where the enzyme and substrate form an enzyme-substrate complex is called the **active site** because the reaction occurs here. At the end of the reaction, the product is released, and the enzyme can then combine with its substrate again. A cell needs only a small amount of an enzyme because enzymes are used over and over. Some enzymes have turnover rates well in excess of a million product molecules per minute.

> **Planning Ahead** To save time, your instructor may have you start a boiling water bath at the beginning of the laboratory.

## 5.1 Catalase Activity

**Pre-Lab**

3. List the enzyme, substrate, and products of the catalase reaction. _____

_____

4. Why does bubbling occur as the catalase reaction proceeds? _____

_____

5. Which tube (1, 2, or 3) do you expect to have the most activity in the Catalase Activity experiment?

_____

Catalase is involved in a degradation reaction: Catalase speeds the breakdown of hydrogen peroxide ($H_2O_2$) in nearly all organisms, including bacteria, plants, and animals. A cellular organelle called a peroxisome, which contains catalase, is present in every plant and animal organ. This means that we could use any plant or animal organ as our source of catalase today. Commonly, school laboratories use the potato as a source of catalase because potatoes are easily obtained and cut up.

Catalase performs a useful function in organisms because hydrogen peroxide is harmful to cells. Hydrogen peroxide is a powerful oxidizer that can attack and denature cellular molecules like DNA! Knowing its harmful nature, humans use hydrogen peroxide as a commercial antiseptic to kill germs (Fig. 5.2). In reduced concentration, hydrogen peroxide is a whitening agent used to bleach hair and teeth. Skill-ful technicians use it to provide oxygen to aquatic plants and fish, but it is also used industrially to clean most anything from tubs to sewage. It's even put in glow sticks, where it reacts with a dye that then emits light.

When catalase speeds the breakdown of hydrogen peroxide, water and oxygen are released.

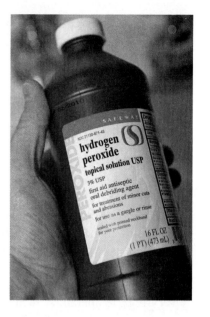

$$2 \text{ H}_2\text{O}_2 \xrightarrow{\text{catalase}} 2 \text{ H}_2\text{O} + \text{O}_2$$
hydrogen peroxide     water   oxygen

**Figure 5.2  Hydrogen peroxide.**
Bubbling occurs when you apply hydrogen peroxide to a cut because oxygen is being released when catalase, an enzyme present in the body's cells, degrades hydrogen peroxide.

Anthony Arena (Chemistry Consultant) and Jill Braaten/McGraw Hill

## Experimental Procedure: Catalase Activity

Number three clean test tubes and use the appropriate graduated transfer pipet to add solutions to the test tubes as follows:

Tube 1
  1. Add 4 ml of hydrogen peroxide to the test tube.
  2. Add 1 ml of catalase. Swirl well to mix, and wait at least 20 seconds for bubbling to develop.
  3. Measure the height of the bubble column (in millimeters), and record your results in Table 5.1.

Tube 2
  1. Add 4 ml of hydrogen peroxide to the test tube.
  2. Add 1 ml of water. Swirl well to mix, and wait at least 20 seconds for bubbling to develop.
  3. Measure the height of the bubble column (in millimeters), and record your results in Table 5.1.

Tube 3
  1. Add 4 ml of sucrose solution to the test tube.
  2. Add 1 ml of catalase. Swirl well to mix, and wait at least 20 seconds for bubbling to develop.
  3. Measure the height of the bubble column (in millimeters), and record your results in Table 5.1.

| Table 5.1 | Catalase Activity | | |
| --- | --- | --- | --- |
| **Tube** | **Contents** | **Bubble Column Height (mm)** | **Explanation of Results** |
| 1 | Catalase Hydrogen peroxide | | |
| 2 | Water Hydrogen peroxide | | |
| 3 | Catalase Sucrose solution | | |

## 5.2 Effect of Temperature on Enzyme Activity

### Pre-Lab

**6.** What are the effects of high temperature on enzymes? _____

_____

**7.** Which tube (1, 2, or 3) do you expect to have the most enzyme activity in the Effect of Temperature

experiment? _____

The active sites on an enzyme allow a specific substrate to bind to it, resulting in a chemical reaction. Therefore, enzymes lower the energy of activation (the temperature needed for a reaction to occur). Still, increasing the temperature is expected to increase the likelihood that active sites will be occupied because molecules move about more rapidly as the temperature rises. In this way, a warm temperature increases enzyme activity.

The shape of an enzyme and its active site must be maintained or else they will no longer be functional. A very high temperature, such as the one that causes water to boil, is likely to cause weak bonds of a protein to break. If this occurs, the enzyme **denatures**—it loses its original shape and the active site will no longer function to bring reactants together. Now enzyme activity plummets.

### Experimental Procedure: Effect of Temperature

Number three clean test tubes and use the appropriate graduated pipet to add solutions to the test tubes as follows:

1. To each tube, add 1 ml of catalase.
2. Place tube 1 in a refrigerator or cold water bath, tube 2 in an incubator or warm water bath, and tube 3 in a boiling water bath. Leave each tube for 15 minutes before proceeding to step 3. Complete the second column in Table 5.2.
3. As soon as you remove the tubes one at a time from the refrigerator, incubator, and boiling water, add 4 ml of hydrogen peroxide.
4. Swirl well to mix, and wait 20 seconds.
5. Measure the height of the bubble column (in millimeters) in each tube, and record your results in Table 5.2. Plot your results in Figure 5.3.

| Table 5.2 Effect of Temperature | | | |
| --- | --- | --- | --- |
| Tube | Temperature (°C) | Bubble Column Height (mm) | Explanation of Results |
| 1  Refrigerator | | | |
| 2  Incubator | | | |
| 3  Boiling water | | | |

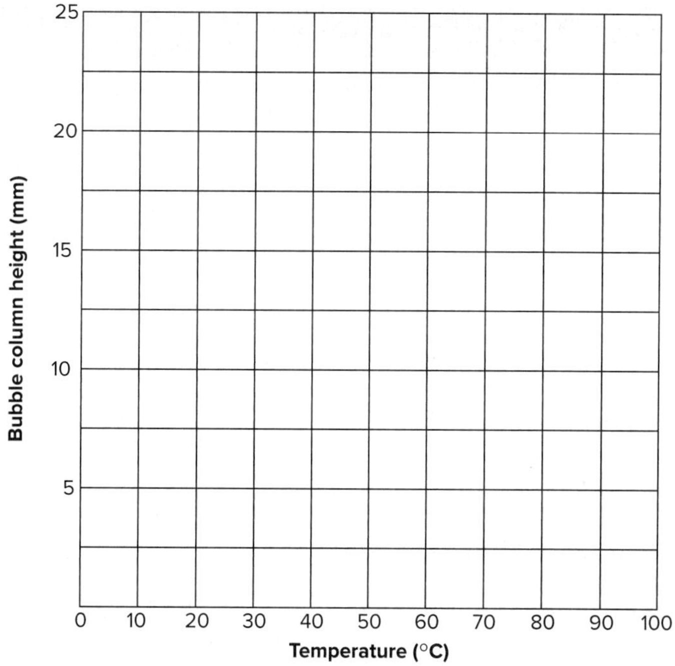

25
20
15
10
5
0
Bubble column height (mm)
10
20
30
40
50
60
70
80
90
100
Temperature (°C)


**Figure 5.3 Effect of temperature on enzyme activity.**

***Conclusion: Effect of Temperature***

- What is your conclusion concerning the effect of temperature on enzyme activity? ____________________

________________________________________________________________________________

## 5.3 Effect of Concentration on Enzyme Activity

**Pre-Lab**

**8.** Which tube (1, 2, or 3) do you expect to have the most enzyme activity in the Effect of Enzyme Concentration experiment? ____________________________________________

Consider that if you increase the number of caretakers per number of children, it is more likely that each child will have quality time with a caretaker. So it is if you increase the amount of enzyme per amount of substrate: It is more likely that substrates will find the active site of an enzyme and a reaction will take place.

***Experimental Procedure: Effect of Enzyme Concentration***

Number three clean test tubes and use the appropriate transfer pipet to add solutions to the test tubes as follows:

1. To each tube, add 4 ml of hydrogen peroxide.
2. To tube 1, add 1 ml of water.
3. To tube 2, add 1 ml of catalase.
4. To tube 3, add 3 ml of catalase.
5. Swirl well to mix before measuring the height of the bubble column.
6. Measure the height of the bubble columns, and record your results in Table 5.3.

## Table 5.3 Effect of Enzyme Concentration

| Tube | Amount of Enzyme | Bubble Column Height (mm) | Explanation of Results |
|------|------------------|---------------------------|------------------------|
| 1 | none | | |
| 2 | 1 ml | | |
| 3 | 3 ml | | |

### Conclusions: Effect of Concentration

- If unlimited time were allotted, would the results be the same in all tubes? _____
  Explain why or why not. _____

- Would you expect similar results if the substrate concentration were varied in the same manner as the enzyme concentration? _____ Why or why not? _____

- What is your conclusion concerning the effect of concentration on enzyme activity? _____

---

## 5.4 Effect of pH on Enzyme Activity

### Pre-Lab

**9.** How do temperature and pH have similar effects on enzymes? _____

**10.** Which tube (1, 2, or 3) do you expect to have the most enzyme activity in the Effect of pH experiment? _____

Each enzyme has an optimal pH in which the speed of the reaction occurs most efficiently. Any higher or lower pH affects hydrogen bonding and the structure of the enzyme, leading to reduced activity or denaturing of the enzyme.

Catalase is an enzyme found in cells where the pH is near 7 (called neutral pH). Other enzymes prefer different pHs. The pancreas secretes a slightly basic (pH 8–8.3) juice into the digestive tract, and the stomach wall releases a very acidic digestive juice, which can be as low as pH 2.

⚠ **Hydrochloric acid (HCl)** used to produce an acid pH is a strong, caustic acid, and sodium hydroxide (NaOH) used to produce a basic pH is a strong, caustic base. Exercise care in using these chemicals, and follow your instructor's directions for disposal of tubes that contain these chemicals. If any acidic or basic solutions spill on your skin, rinse immediately with clear water.

### Experimental Procedure: Effect of pH

Number three clean test tubes and use the appropriate transfer pipet to add solutions to the test tubes as follows:

1. To each tube, add 1 ml of catalase.
2. To tube 1, add 2 ml of water adjusted to pH 3. Swirl to mix.
3. To tube 2, add 2 ml of water adjusted to pH 7. Swirl to mix.
4. To tube 3, add 2 ml of water adjusted to pH 11. Swirl to mix.
5. Wait 1 minute. Now add 4 ml hydrogen peroxide to each tube, and swirl to mix before noting the height of the bubble column.
6. Record your results in Table 5.4, and plot your results in Figure 5.4.

## Table 5.4 Effect of pH

| Tube | pH | Bubble Column Height (mm) | Explanation of Results |
|------|-----|---------------------------|------------------------|
| 1 | 3 | | |
| 2 | 7 | | |
| 3 | 11 | | |

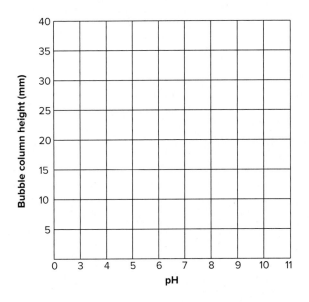

**Figure 5.4 Effect of pH on enzyme activity.**

### Conclusion: Effect of pH

• What is your conclusion concerning the effect of pH on enzyme activity? _____

_____

_____

## 5.5 Factors That Affect Enzyme Activity

In Table 5.5, summarize what you have learned about factors that affect the speed of an enzymatic reaction. For example, in general, what type of temperature promotes enzyme activity, and what type inhibits enzyme activity? Answer similarly for enzyme or substrate concentration and pH.

### Table 5.5 Factors That Affect Enzyme Activity

| Factors | Promote Enzyme Activity | Inhibit Enzyme Activity |
|---------|-------------------------|-------------------------|
| Enzyme specificity | | |
| Temperature | | |
| Enzyme or substrate concentration | | |
| pH | | |

## Conclusions: Factors That Affect Enzyme Activity

- Why does enzyme specificity promote enzyme activity? _____

  _____

- Why does a warm temperature promote enzyme activity? _____

  _____

- Why does increasing enzyme concentration promote enzyme activity? _____

  _____

- Why does optimum pH promote enzyme activity? _____

  _____

_____ 1. Which symbol in a chemical reaction represents the change from reactants to products?

_____ 2. Are the substrates in an enzymatic reaction reactants or products?

_____ 3. What is the name of the location on an enzyme where its substrate or substrates bind?

_____ 4. What cellular organelle contains catalase and is present in all plant and animal organs?

_____ 5. What did measuring the bubble column heights in all the Experimental Procedures indicate?

_____ 6. What purpose did the tube containing water and hydrogen peroxide serve in the catalase activity Experimental Procedure?

_____ 7. What word describes the loss of an enzyme's original shape and ability of its active site to function?

_____ 8. What happens to enzyme activity if the temperature increases a few degrees?

_____ 9. What did the control tube lack when the effect of enzyme concentration was studied?

_____ 10. If enzyme concentration increases, will products from the enzyme reaction increase, decrease, or stay the same?

_____ 11. Enzymes work best at their _____ pH.

_____ 12. How does a significant change in pH affect the structure of an enzyme?

_____ 13. In the Experimental Procedure on the effect of pH on enzyme activity, which tube contained water with an acidic pH?

_____ 14. _____ is the maintenance of internal balance in the body for conditions such as temperature and pH.

## Thought Questions

15. People with Crohn's disease may develop lactose intolerance caused by damage to their small intestine. These people may take a commercial product called Lactaid to prevent digestive upset when consuming dairy products. What is in Lactaid that helps these people avoid those symptoms? Explain how this relates to what you learned about the specificity of enzymes for substrates in this lab.

16. High-grade fevers (in humans, 103°F or higher) are of concern, and medical care should be obtained. Explain why, based on your understanding of enzyme function.

# 6

# Photosynthesis

## Learning Outcomes

**6.1 Photosynthetic Pigments**
- Describe the separation of plant pigments by paper chromatography.

**6.2 Solar Energy**
- Calculate the rate of photosynthesis using white and green light.

**6.3 Carbon Dioxide Uptake**
- Measure carbon dioxide uptake during photosynthesis.

## Introduction

### Pre-Lab

**1.** List the reactants and products of photosynthesis. _____

_____

**Photosynthesis** involves the use of solar energy to produce a carbohydrate:

$$CO_2 + H_2O \xrightarrow{\text{solar energy}} (CH_2O) + O_2$$

In this equation, $(CH_2O)$ represents any general carbohydrate.

Photosynthesis takes place in chloroplasts (Fig. 6.1). Here membranous thylakoids are stacked in grana surrounded by the stroma. During the light reactions, pigments within the thylakoid membranes absorb solar energy, water is split, and oxygen is released. The Calvin cycle reactions occur within the stroma. During these reactions, carbon dioxide ($CO_2$) is reduced and solar energy is now stored in a carbohydrate ($CH_2O$).

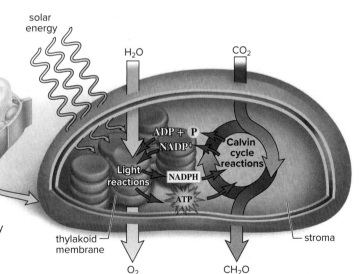

**Figure 6.1 Overview of photosynthesis.**
Photosynthesis includes the light reactions when energy is collected and $O_2$ is released and the Calvin cycle reactions when $CO_2$ is reduced and an energy-rich carbohydrate ($CH_2O$) is formed.

## 6.1 Photosynthetic Pigments

### Pre-Lab

2. What color pigments do you expect to see in this experiment? _____

3. What determines how far a pigment will move up the chromatography paper? _____
_____

Visible light is composed of different colors of light. Plants contain various pigments to absorb the solar energy of these different colors. We will use a technique called chromatography to separate the pigments found in plant leaves. Chromatography separates molecules from each other on the basis of their solubility in particular solvents. The solvents used in the following Experimental Procedure are petroleum ether and acetone, which have no charged groups and are therefore nonpolar. As a nonpolar solvent moves up the chromatography paper, the pigment moves along with it. The more nonpolar a pigment, the more soluble it is in a nonpolar solvent and the faster and farther it proceeds up the chromatography paper.

### Experimental Procedure: Photosynthetic Pigments

1. Assemble a **chromatography apparatus** (Fig. 6.2):
   • Obtain a large, dry test tube and a cork with a hook.
   • Attach a strip of precut **chromatography paper** (hold from the top) to the hook and test for fit. The paper should hang straight down and barely touch the bottom of the test tube; trim if necessary.
   • Measure 2 cm from the bottom of the paper and place a small dot with a pencil (not a pen).
   • With the stopper in place, mark the test tube with a wax pencil 1 cm below where the dot is on the paper.
   • Set the chromatography apparatus in a test tube rack.

2. Prepare the chromatography paper:
   • Remove the chromatography paper from the test tube; place on a paper towel and apply **plant pigments** to the dot on the paper as directed by your instructor.
   • In the **fume hood,** add **chromatography solution** up to the wax pencil mark you made on the test tube. Place the chromatography paper attached to the hook back in the test tube. The pigment spot should remain above the chromatography solution. Close the chromatography apparatus tightly.

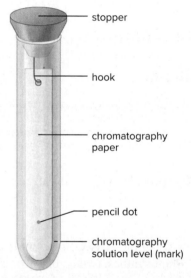

**Figure 6.2  Chromatography apparatus.**
For a chromatography apparatus, the paper must be cut to size and arranged to hang down without touching the sides of a dry tube.

3. Develop the chromatogram:
   • Place the reassembled chromatography apparatus in the test tube rack and allow 10 minutes for the chromatogram to develop, but check frequently so that the solution does not reach the top of the paper.
   • When the solvent has moved to within 1 cm of the upper edge of the paper, remove the paper. Close the empty apparatus tightly. With a pencil, lightly mark the location of the solvent front (where the solvent stopped on the paper) and allow the chromatogram to dry in the fume hood.

> ⚠ The chromatography solution is toxic and extremely flammable. Do not breathe the fumes, and do not place the chromatography solution near any source of heat. A fume (ventilation) hood is recommended.

4. Read the chromatogram:
   • Observe the different pigments separated on the paper. Measure the distance in millimeters (mm) from the dark green pigment spot to the top of each individual pigment band, and record these values in Table 6.1.
   • Measure the distance the solvent moved from the dark green pigment spot to the solvent front and add this value to Table 6.1.
   • Use this formula to calculate the $R_f$ (ratio-factor) values for each pigment, and record these values in Table 6.1:

$$R_f = \frac{\text{distance moved by pigment}}{\text{distance moved by solvent}}$$

### Table 6.1  $R_f$ (Ratio-Factor) Values for Each Pigment

| Pigments | Distance Moved (mm) | $R_f$ Values |
|---|---|---|
| Carotenes (yellow) | | |
| Xanthophylls (orange) | | |
| Chlorophyll *a* (bluish-green) | | |
| Chlorophyll *b* (light green) | | |
| Solvent | | ——— |

## Conclusion: Photosynthetic Pigments

• Why do plants contain various colors of pigments?

_____

_____

## 6.2  Solar Energy

### Pre-Lab

4. What can you measure to determine if photosynthesis is taking place? _____

5. With the help of the equation for photosynthesis, list everything necessary for photosynthesis to occur in the volumeter. _____

_____

6. Which substance do you hypothesize provides the $CO_2$? _____

During light reactions of photosynthesis, solar energy is absorbed by the photosynthetic pigments and is transformed into the chemical energy of a carbohydrate ($CH_2O$). Without solar energy, photosynthesis would be impossible.

Release of oxygen from a plant indicates that the light reactions of photosynthesis are occurring. The oxygen released during photosynthesis is taken up by a plant when cellular respiration occurs. This must be taken into account when the rate of photosynthesis is calculated.

## Role of White Light

White (sun) light contains different wavelengths of light, as is demonstrated when white light passes through a prism (Fig. 6.3). White light is the best for photosynthesis because it contains all the wavelengths of visible light.

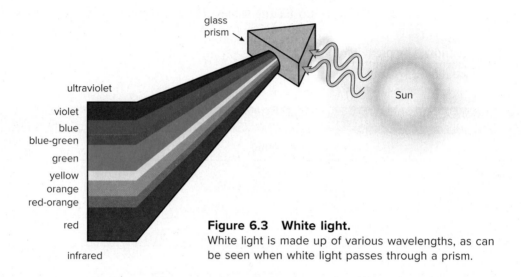

**Figure 6.3   White light.**
White light is made up of various wavelengths, as can be seen when white light passes through a prism.

### Experimental Procedure: White Light

1. Place a generous quantity of *Elodea* with the cut end up (make sure the cuts are fresh) in a test tube with a rubber stopper containing a piece of glass tubing, as illustrated in Figure 6.4. When assembled, this is your volumeter for studying the need for light in photosynthesis. (Do not hold the volumeter in your hand, as body heat will also drive the reaction forward.) Your instructor will show you how to fix the volumeter in an upright position.

2. Before stoppering the test tube, add sufficient 3% sodium bicarbonate ($NaHCO_3$) solution so that when the rubber stopper is inserted into the tube, the solution comes to rest at about ¼ the length of the upright glass tubing. Mark this location on the glass tubing with a wax pencil.

3. Place a beaker of plain water next to the *Elodea* tube to serve as a heat absorber. Place a lamp (150 watt) next to the beaker. The tube, beaker, and lamp should be as close to one another as possible.

4. Turn on the lamp. As soon as the edge of the solution in the tubing begins to move, time the reaction for 10 minutes. Be careful not to bump the tubing or to readjust the stopper, or your readings will be altered. After 10 minutes, mark the edge of the solution, and measure in millimeters the distance the edge moved upward: _____ mm/10 min. This is **net photosynthesis,** a measurement that does not take into account the oxygen that was used up for cellular respiration. Record your results in Table 6.2.

    Why did the edge move upward? _____

**Figure 6.4  Volumeter.**
A volumeter apparatus is used to study the role of light in photosynthesis.

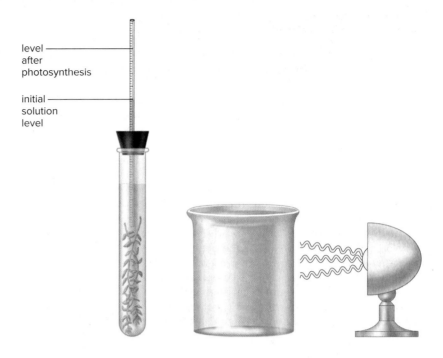

level
after
photosynthesis

initial
solution
level

5. Carefully wrap the tube containing *Elodea* in aluminum foil, and record here the length of time it takes for the edge of the solution in the tubing to move downward 1 mm: _____. Convert your measurement to _____ mm/10 min, and record this value for **cellular respiration** in Table 6.2. (Do not use a minus sign, even though the edge moved downward.) Why does cellular respiration, which occurs in a plant whether it is light or dark, cause the edge to move downward? _____

_____

6. If the *Elodea* had *not* been respiring in step 4, how far would the edge have moved upward? _____ mm/10 min. This is **gross photosynthesis** (net photosynthesis + cellular respiration). Record this number in Table 6.2.

7. Calculate the **rate of photosynthesis** (mm/hr) by multiplying gross photosynthesis (mm/10 min) by 6 (i.e., 10 min × 6 = 60 min = 1 hr): _____ mm/hr. Record this value in Table 6.2.

| Table 6.2  Rate of Photosynthesis (White Light) | | |
|---|---|---|
| | Movement of Edge (mm/10 min) | Rate of Photosynthesis (mm/hr) |
| Net photosynthesis (white light) | | _____ |
| Cellular respiration (no light) | | _____ |
| **Gross photosynthesis** (net + cellular respiration) | | |

## Role of Green Light

Green light is only one part of white light (see Fig. 6.3). The photosynthetic pigments absorb certain colors of light better than other colors (Fig. 6.5). According to Figure 6.5, what color light do the chlorophylls absorb

best? _____ Least? _____

What color light do the carotenoids (carotenes and xanthophylls) absorb best? _____

Least? _____

**Figure 6.5   Action spectrum for photosynthesis.**
The action spectrum for photosynthesis is the sum of the absorption spectrums for the pigments chlorophyll *a*, chlorophyll *b*, and carotenoids. The peaks in this diagram represent wavelengths of sunlight absorbed by photosynthetic pigments. The chlorophylls absorb predominantly violet-blue and orange-red light and reflect green light. The carotenoids (carotenes and xanthophylls) absorb mostly blue-green light and reflect yellow-red light.

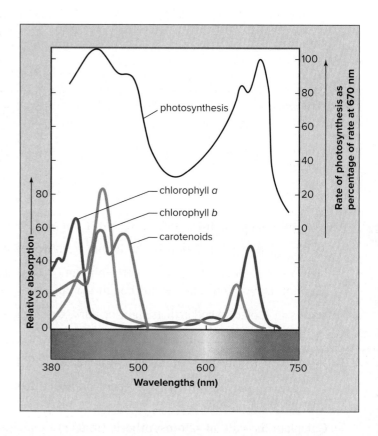

---

## *Experimental Procedure: Green Light*

1. Add three drops of green dye (or use a green cellophane wrapper) to the beaker of water used in the previous Experimental Procedure until there is a distinctive green color. Remove all previous wax pencil marks from the glass tubing.

2. Record in Table 6.3 your data for gross photosynthesis (mm/10 min) and for rate of photosynthesis for white light (mm/hr) from Table 6.2.

3. Turn on the lamp. Mark the location of the edge of the solution on the glass tubing. As soon as the edge begins to move, time the reaction for 10 minutes. After 10 minutes, mark the edge of the solution, and

   measure in millimeters the distance the edge moved. Net photosynthesis for green light = _____ mm/10 min. As before, net photosynthesis does not take into account that oxygen was used for cellular respiration.

4. Carefully wrap the tube containing *Elodea* in aluminum foil, and record here the length of time it takes for the edge of the solution in the tubing to recede 1 mm: _____. Convert your measurement to _____ mm/10 min. As before, this reading shows how much oxygen was used for cellular respiration.

5. Calculate gross photosynthesis for green light (mm/10 min) as you did for white light, and record your data in Table 6.3. (As before, add your data for net photosynthesis to those for cellular respiration.)

6. Calculate rate of photosynthesis for green light (mm/hr) as you did for white light, and record your data in Table 6.3. (As before, convert your data for mm/10 min to mm/hr.)

7. Collect the white and green light rates of photosynthesis (mm/hr) from each group in your lab; then average all the rates including your own. In the last column of Table 6.3, record the average for white and green light rates of photosynthesis (mm/hr).

### Table 6.3   Rate of Photosynthesis (White and Green Light)

| Gross Photosynthesis | Your Data | Class Data |
|---|---|---|
| White (from Table 6.2) | mm/10 min | _____ |
| Green | mm/10 min | _____ |
| **Rate of photosynthesis** | | |
| White (from Table 6.2) | mm/hr | mm/hr |
| Green | mm/hr | mm/hr |

8. Calculate the rate of photosynthesis (green light) as a percentage of the rate of photosynthesis (white light) using this equation:

$$\text{Percentage} = \frac{\text{rate of photosynthesis (green light)}}{\text{rate of photosynthesis (white light)}} \times 100$$

Percentage based on your data recorded in Table 6.3 = _____.

Percentage based on class data recorded in Table 6.3 = _____.

### Conclusion: Rate of Photosynthesis

- Do your results support the hypothesis that green light is minimally used by a land plant for photosynthesis? _____ Explain why. _____

_____

_____

# 6.3  Carbon Dioxide Uptake

## Pre-Lab

**7.** Why will blowing on a solution of phenol red turn the solution yellow? _____

_____

**8.** How is carbon dioxide supplied to plants in the environment? _____

_____

During the Calvin cycle reactions of photosynthesis, the plant takes up carbon dioxide ($CO_2$) and reduces it to a carbohydrate, such as glucose ($C_6H_{12}O_6$). Therefore, the carbon dioxide in the solution surrounding *Elodea* should disappear as photosynthesis takes place.

## Experimental Procedure: Carbon Dioxide Uptake

1. Temporarily remove the *Elodea* from the test tube. Empty the sodium bicarbonate ($NaHCO_3$) solution from the test tube, rinse the test tube thoroughly, and fill with a phenol red solution diluted to a faint pink. (Add more water if the solution is too dark.) Phenol red is a pH indicator that turns yellow in an acid and red in a base.

   > ⚠ **Phenol red**  Avoid ingestion, inhalation, and contact with skin, eyes, and mucous membranes. Follow your instructor's directions for disposal of this chemical. Use protective eyewear when performing this experiment.

2. Blow *lightly* on the surface of the solution. Stop blowing as soon as the surface color changes to yellow. Then shake the test tube until the rest of the solution turns yellow.

   Blowing onto the solution adds what gas to the test tube? _____ When carbon dioxide combines with water, it forms carbonic acid; therefore, the solution appears yellow.

3. Thoroughly rinse the *Elodea* with distilled water, return it to the test tube, which now contains a yellow solution, and assemble your volumeter as before.

4. The water in the beaker used to absorb heat should be clear.

5. Turn on the lamp, and wait until the edge of the solution just begins to move upward. Note the time.

   Observe until you note a change in color. How long did it take for the color to change? _____

6. Why did the solution eventually turn red? _____

_____

**7.** The carbon cycle includes all the many ways that organisms exchange carbon dioxide with the atmosphere. Figure 6.6 notes the relationship between cellular respiration and photosynthesis. Animals produce carbon dioxide used by plants to carry out photosynthesis. Plants produce the food (and oxygen) that they and animals require to carry out cellular respiration. Therefore, the same carbon atoms pass between animals and plants and between plants and animals (Fig. 6.6).

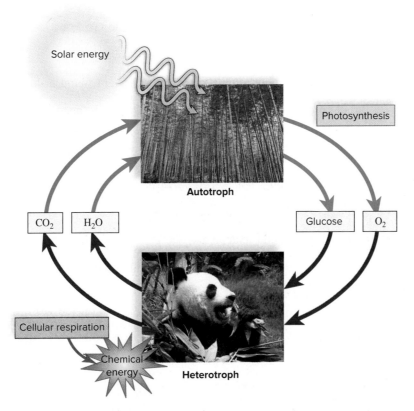

**Figure 6.6  Photosynthesis and cellular respiration.**
Animals are dependent on plants for a supply of oxygen, and plants are dependent on animals for a supply of carbon dioxide.

(bamboo trees) Zhao jian kang/Shutterstock; (panda) Melinda Chan/Moment/Getty Images

_____  **1.** List the reactants necessary for photosynthesis.

_____  **2.** What organelle, present in eukaryotic photosynthetic organisms, is the location of photosynthesis?

_____  **3.** During photosynthesis, where is water split and oxygen released?

_____  **4.** Where is $CO_2$ reduced during the Calvin cycle reactions?

_____  **5.** What technique was used to separate the leaf pigments?

_____  **6.** List the various photosynthetic pigments that are found in leaves.

_____  **7.** Based on distance traveled up the chromatography paper, which pigment is the most nonpolar pigment?

_____  **8.** What color of light is *not* absorbed by green plants?

_____  **9.** What color of light is absorbed by carotenes?

_____  **10.** What gas is released during the light reactions of photosynthesis?

_____  **11.** How will plants and animals use the gas released from photosynthesis?

_____  **12.** If net photosynthesis is 15 mm/10 min and cellular respiration is 1.5 mm/10 min, how much is gross photosynthesis?

_____  **13.** When $CO_2$ combines with $H_2O$, what causes the phenol red to turn yellow?

_____  **14.** $CO_2$ is broken apart during photsynthesis. Where will the carbon molecules end up?

## Thought Questions

**15.** Global deforestation has decreased the amount of plants on earth available to run photosynthesis. What impact is this having on the pH of the oceans? Explain your reply.

**16.** Cyanobacteria are photosynthetic bacteria and appear green in color. As prokaryotic organisms, they lack chloroplasts but have photosynthetic pigments. Predict where their photosynthetic pigments are located and what photosynthetic pigment gives them their color.

**17.** What is the importance of the light reactions of photosynthesis since they do not directly produce energy-rich carbohydrates?

# 7

# Mitosis: Cellular Reproduction

## Learning Outcomes

**7.1 The Cell Cycle**
- Name and describe the stages of the cell cycle.
- Explain how DNA replication allows the chromosome number to stay constant during mitosis.

**7.2 Animal Cell Mitosis and Cytokinesis**
- Explain the importance of the mitotic spindle to mitosis.
- Identify the phases of mitosis and cytokinesis in animal cell models and in a prepared animal mitosis slide.

**7.3 Plant Cell Mitosis and Cytokinesis**
- Identify the phases of mitosis and cytokinesis in plant cell models and in a prepared plant mitosis slide.
- Use the frequency of mitotic phases in a root tip slide to determine the time span for each phase of mitosis.

## Introduction

### Pre-Lab

**1.** Why do cells need to divide? _____

Dividing cells experience nuclear division, cytoplasmic division, and a period of time between divisions called interphase. During **interphase,** the nucleus appears normal, and the cell is performing its usual cellular functions. Also, the cell is increasing all of its components, including such organelles as the mitochondria, ribosomes, and centrioles (if present). DNA replication (making an exact copy of the DNA) occurs toward the end of interphase. Then, during nuclear division, called **mitosis,** the new nuclei receive the same number of chromosomes as the parental nucleus. When the cytoplasm divides, a process called **cytokinesis,** two daughter cells are produced (Fig. 7.1).

In multicellular organisms, mitosis permits growth and repair of tissues. In eukaryotic, unicellular organisms, mitosis is a form of asexual reproduction. Sexually reproducing organisms utilize another form of nuclear division, called **meiosis.** In animals, meiosis is a part of gametogenesis, the production of gametes (sex cells). The gametes are sperm in male animals and eggs in female animals. Meiosis is the topic for Laboratory 8.

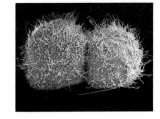

**Figure 7.1 Cytokinesis.**
Following chromosome distribution during mitosis, two daughter cells are produced by cytokinesis, a division of the cytoplasm.

National Institutes of Health (NIH)/USHHS

# 7.1 The Cell Cycle

As stated in the Introduction, the period of time between cell divisions is known as interphase. Because early investigators noted little visible activity between cell divisions, they dismissed this period of time as a resting state. But when later investigators discovered that DNA replication and chromosome duplication occur during interphase, the **cell cycle** concept was proposed. The cell cycle is divided into the four stages noted in Figure 7.2.

**Figure 7.2  The cell cycle.**
Immature cells go through a cycle that consists of four stages: $G_1$, S (for synthesis), $G_2$, and M. The cell divides during the M stage, which consists of mitosis and cytokinesis.

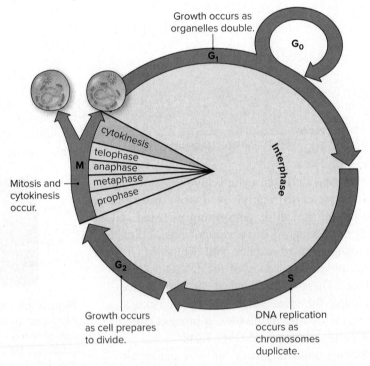

## The S Stage of the Cell Cycle

While each stage of the cell cycle is critical to successful mitosis (division of the nucleus), the S stage is particularly important because it is the stage during which DNA replicates. At the completion of replication, there are two double helices and each chromosome is duplicated.

When mitosis is about to occur, we can see each duplicated chromosome because chromatin has condensed and compacted to form two sister chromatids held together at a centromere. Consult Table 7.1, and *label the sister chromatids and centromere in Figure 7.3.*

What happens to the sister chromatids during mitosis? During mitosis, the sister chromatids separate, and then they are called daughter chromosomes. This is the manner in which DNA replication causes each body (somatic) cell of an organism to contain the same number of chromosomes. In humans, each cell in the body contains 46 chromosomes.

Mitosis creates identical cells, but it is customary to call the cell that divides the **parent cell** and the resulting two cells, the **daughter cells.**

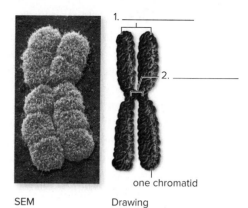

SEM  Drawing

**Figure 7.3  Duplicated chromosomes.**
DNA replication results in duplicated chromosomes that consist of two sister chromatids held together at a centromere.

Andrew Syred/Science Source

### The M Stage of the Cell Cycle

Study the terms in Table 7.1. These terms all pertain to mitosis. During mitosis, sister chromatids (now called daughter chromosomes) move into the daughter nuclei.

| Table 7.1 | Structures Associated with Mitosis |
|---|---|
| **Structure** | **Description** |
| Nucleus | A large organelle containing the chromosomes and acting as a control center for the cells |
| Nucleolus | An organelle found inside the nucleus that produces the subunits of ribosomes |
| Chromosome | Rod-shaped body in the nucleus that is seen during mitosis and meiosis and that contains DNA, and therefore the hereditary units, or genes |
| Chromatids | The two identical parts of a chromosome following DNA replication |
| Centromere | A constriction where duplicates (sister chromatids) of a chromosome are held together |
| Kinetochore | A body that appears during cell division on either side of a centromere; attaches a chromatid to a spindle fiber |
| Spindle | Microtubule structure that brings about chromosome movement during cell division |
| Metaphase plate | The center of the fully formed spindle |
| Centrosome | The central microtubule-organizing center of cells; in animal cells, contains two centrioles |
| Centrioles* | Short, cylindrical organelles at the spindle poles in animal cells |
| Aster* | Short, radiating fibers surrounding the centrioles in dividing cells |

*Animal cells only

## 7.2  Animal Cell Mitosis and Cytokinesis

### Pre-Lab

**6.** Why is it important for the spindle to carefully organize chromosomes during mitosis?

_____

_____

**7.** What is the difference in purpose between mitosis and cytokinesis?

_____

_____

When an animal cell divides, it begins with mitosis and then is completed once cytokinesis, or division of the cytoplasm, has occurred. Mitosis is called duplication division because the daughter cells have the same chromosome makeup as the parent cell. As we now know, a spindle, sometimes called the mitotic spindle, occurs during mitosis. A **spindle** is composed of spindle fibers (microtubules) formed by the centrosomes, which are located at the **poles** of the spindle (Fig. 7.4). In animal cells, the poles contain centrioles surrounded by an **aster,** which is an array of fibers. The kinetochores of the duplicated chromosomes are attached to spindle fibers at the metaphase plate of the spindle. When spindle fibers shorten, the chromatids (now daughter chromosomes) move toward the poles.

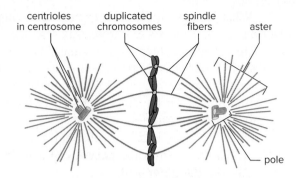

**Figure 7.4 The mitotic spindle.**
The kinetochores of a duplicated chromosome are attached to spindle fibers at the metaphase plate of the spindle. When spindle fibers shorten, the sister chromatids, now called daughter chromosomes, move toward opposite spindle poles.

## Animal Cell Mitosis

You will now have the opportunity to observe the stages of mitosis in animal cell models and a prepared slide. Mitosis is the type of nuclear division that occurs when embryos develop, organisms grow larger, and injuries heal. Without mitosis, none of these important events could occur.

### Observation: Animal Cell Mitosis

#### Animal Mitosis Models

1. Each species has its own chromosome number. Counting the number of centromeres tells you the number of chromosomes in models or slides. What is the number of chromosomes in the parent cell and in the daughter cells in this model series? _____

2. Examine the phases of mitosis as depicted by the models and arrange them in the correct order.

3. In the space below, *draw and label each stage*. Be sure to include labels for the prominent structures visible during mitosis.

#### Whitefish Blastula Slide

1. Examine a prepared slide of whitefish blastula cells undergoing mitosis. The blastula is an early embryonic stage in the development of animals.

2. Try to find a cell in each phase of mitosis using Figure 7.5 as a guide. Have a partner or your instructor check your ability to identify these cells.

3. Match these statements to the correct phase of animal cell mitosis in Figure 7.5, and *write the correct statements on the lines provided.*

   **Statements:**
   - Daughter chromosomes are moving to the poles of the spindle.
   - Duplicated chromosomes have no particular arrangement in the cell.
   - Two daughter cells are now forming.
   - Duplicated chromosomes are aligned at the metaphase plate of the spindle.

4. The prophase cell in Figure 7.5 has the same number of chromosomes as the telophase nuclei. Explain the different appearance of the chromosomes. _____

_____

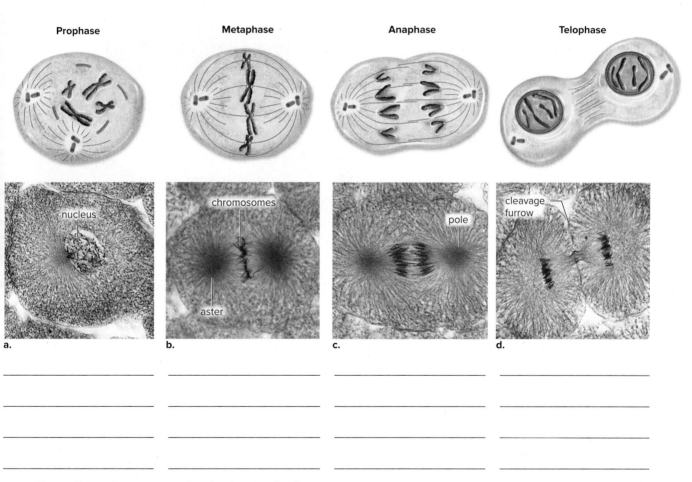

**Figure 7.5    The phases of animal cell mitosis.**

Mitosis always has these four main phases. Others, designated by the terms *early* or *late*, as in *early prophase* or *late anaphase*, can also be cited.

(a–d) Ed Reschke

## Cytokinesis in Animal Cells

**Cytokinesis,** division of the cytoplasm, usually accompanies mitosis. During cytokinesis, each daughter cell receives a share of the organelles that duplicated during interphase. Cytokinesis begins in anaphase, continues in telophase, and reaches completion by the start of the next interphase.

In animal cells, a **cleavage furrow,** an indentation of the membrane between the daughter nuclei, begins as anaphase draws to a close (Fig. 7.6). The cleavage furrow deepens as a band of actin filaments called the contractile ring slowly constricts the cell, forming two daughter cells. Note in Figure 7.5 that "cleavage furrow" is labeled in telophase. Are any of the cells in your whitefish blastula slide undergoing cytokinesis? _____

In the space below, *draw and label the cell.*

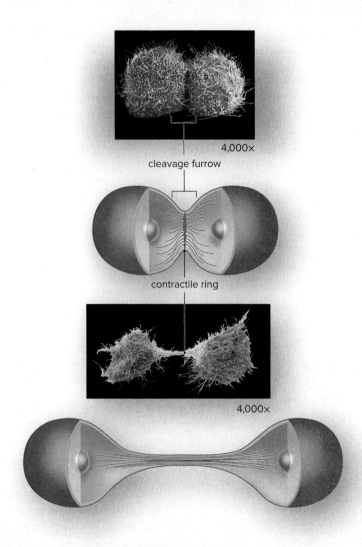

**Figure 7.6  Cytokinesis in animal cells.**
A single cell becomes two cells by a furrowing process. A contractile ring composed of actin filaments gradually gets smaller, and the cleavage furrow pinches the cell into two cells.

(a) National Institutes of Health (NIH)/USHHS; (b) Steve Gschmeissner/Brand X Pictures/Science Photo Library/Getty Images

## 7.3  Plant Cell Mitosis and Cytokinesis

### Pre-Lab

**8.** How does division of plant cells differ from animal cells? _____

_____

**9.** At any given time, would you expect most cells in a plant to be in interphase, prophase, metaphase, anaphase, or telophase? Why? _____

_____

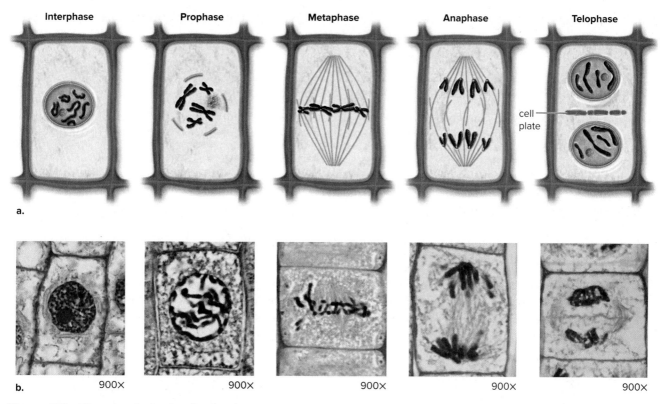

**Figure 7.7 Phases of plant cell mitosis.**
Mitosis has these phases in all eukaryotic cells. Plant cells are recognizable by the presence of a cell wall and the absence of centrioles and asters. **a.** Drawings of the phases to be identified. **b.** Micrographs of these phases.

b (left) Ed Reschke; b (center left–right) Kent Wood/Science Source

Division of a plant cell resembles that of an animal cell. The plant cell first undergoes mitosis, then it undergoes cytokinesis. However, plant cells, as you know, are surrounded by a plant cell wall. This feature has no effect on mitosis, but it does affect cytokinesis as we shall observe.

## Plant Cell Mitosis

Although plant cells do utilize a spindle to divide duplicated chromosomes, they do not have well-defined spindle poles because they lack centrioles and asters.

### Observation: Plant Cell Mitosis

**Note:** Interphase is not a phase *of* mitosis, but a preliminary stage *to* mitosis.

*Plant Mitosis Models*

1. Identify interphase and the phases of plant cell mitosis using models of plant cell mitosis and Figure 7.7 as a guide.
2. As mentioned previously, plant cells do not have centrioles and asters, which are short radiating fibers produced by centrioles.
3. What is the number of chromosomes in each of the cells in your model series? _____

*Onion Root Tip Slide*

1. Examine a prepared slide of onion root tip cells (*Allium*) undergoing mitosis. In plants, the root tip contains tissue that is continually dividing and producing new cells.
2. Focus in low power and then switch to high power. Practice identifying phases that correspond to those shown in Figure 7.7.

3. Using high power, focus up and down on a cell in telophase. You may be able to just make out the cell plate, the region where a plasma membrane is forming between the two prospective daughter cells. Later, cell walls will appear in this region.

## Time Span for Phases of the Cell Cycle in the Onion Root Tip

Knowing that the cell cycle consists of interphase plus four phases of mitosis and that the cell cycle typically lasts about 24 hours (1,440 minutes), hypothesize how many minutes the cell spends during each of these phases of the cell cycle.

Interphase _____ Prophase _____ Metaphase _____ Anaphase _____ Telophase _____

### Experimental Procedure: Time Span for Phases of the Cell Cycle in the Onion Root Tip

1. Select an area of the onion root tip slide that contains cells in all phases of mitosis and also interphase. Concentrate on examining a confined area of about 20 to 30 cells.

2. Don't stray beyond a confined region, and as you identify phases, put a slash mark on the lines provided in this chart. Preferably work with a partner who will enter the slash marks as you call out observed phases in a confined region of the root tip. Convert your 20 to 30 slashes to Arabic numbers on the lines provided under the title Total. Also record these numbers in the second column of Table 7.2.

| Phase | Slash Marks | Total |
|-------|-------------|-------|
| Interphase | _____ | _____ |
| Prophase | _____ | _____ |
| Metaphase | _____ | _____ |
| Anaphase | _____ | _____ |
| Telophase | _____ | _____ |

3. Calculate the percentage of total number of cells that are in each of the phases, and record the percentages in the third column of Table 7.2. (To do this, divide the number of cells in each phase by the total number of cells observed and multiply by 100.)

4. Assuming that the cell cycle lasts 24 hours (1,440 minutes), use these percentages to calculate the time span for each phase of the cell cycle. Enter the time span for each phase in the fourth column of Table 7.2.

| Table 7.2 Time Span for Phases of the Cell Cycle in the Onion Root Tip | | | |
|-------|-------------|----------|-----------------|
| Phase | Number Seen | % of Total | Time Span (min) |
| Interphase | | | |
| Prophase | | | |
| Metaphase | | | |
| Anaphase | | | |
| Telophase | | | |
| Total | | | |

## Conclusions: Time Span for Phases of the Cell Cycle in the Onion Root Tip

- Were your hypotheses supported or not supported by your observation of onion root tip cells undergoing the cell cycle? _____ Describe any specific discrepancies.

  _____

  _____

- Suggest a possible explanation for the length of time a cell spends on different phases of the cell cycle. _____

  _____

## Cytokinesis in Plant Cells

After mitosis, the cytoplasm divides by cytokinesis. In plant cells, membrane vesicles derived from the Golgi apparatus migrate to the center of the cell and form a **cell plate** (Fig. 7.8), which is the location of a new plasma membrane for each daughter cell. Later, individual cell walls appear in this area.

Were any of the cells of the onion root tip slide undergoing cytokinesis? _____

*Draw and label the cell in the space below.*

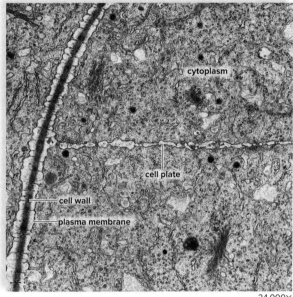

24,000×

**Figure 7.8 Electron micrograph showing cytokinesis in plant cells.**
During plant cell cytokinesis, vesicles fuse to form a cell plate that separates the daughter nuclei. Later, the cell plate gives rise to a new cell wall.

Biophoto Associates/Science Source

_____  1. What is the name for the period of time when a cell is performing its usual functions and replicating DNA?

_____  2. When does nuclear division occur?

_____  3. What is the purpose of mitosis in a eukaryotic organism?

_____  4. List the four stages of the cell cycle.

_____  5. What occurs during the S stage of interphase?

_____  6. What is the term used to describe duplicated chromosomes?

_____  7. If asters are observed in cells undergoing mitosis, are the cells animal or plant cells?

_____  8. Name the structure that helps chromosomes move and is formed during prophase.

_____  9. When do chromosomes align at the equator of the spindle?

_____  10. What happens to sister chromatids during anaphase of mitosis?

_____  11. During which phase of mitosis does the nuclear envelope reform and the nucleolus reappear?

_____  12. When are organelles divided up between daughter cells?

_____  13. In which phase of the cell cycle do cells spend most of their time?

_____  14. What structure gives rise to new plant cell walls?

## Thought Questions

15. Why is it necessary for chromosomes to duplicate before mitosis?

16. What are the potential consequences to an individual if they have cells in which interphase is shortened and the cell is continuously undergoing the cell cycle?

17. List several organs in your body where you would expect to find actively dividing cells and several organs where you would find mostly resting cells.

## Cell Cycle Diagrams

**18.** Draw an animal cell in Interphase and each of the four key phases of Mitosis. Where appropriate, label the following structures: cell membrane, nuclear envelope, chromatin, chromosomes, contractile ring, cleavage furrow, centrioles, and aster.

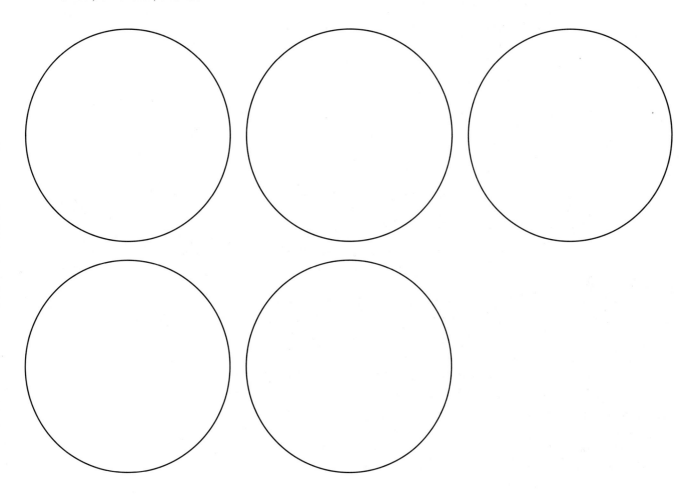

# 8

# Meiosis: Sexual Reproduction

## Learning Outcomes

**8.1 Meiosis: Reduction Division**
- Name and describe the phases of meiosis I and meiosis II.
- Recognize the various phases of meiosis when examining slides of meiosis.

**8.2 Production of Variation During Meiosis**
- Using pop bead chromosomes, demonstrate two ways meiosis introduces variations among the daughter cells of meiosis and therefore the gametes.

**8.3 Human Life Cycle**
- State the function of mitosis and meiosis in the human life cycle.
- Contrast the process and the events of mitosis and meiosis.

## Introduction

### Pre-Lab

1. What are the differences in purpose between meiosis and mitosis? _____

_____

In sexually reproducing organisms, meiosis results in the production of gametes (sex cells). The gametes are sperm (the smaller gamete) and egg (the larger gamete). Fusion of sperm and egg results in a zygote that develops into a new genetically unique individual (Fig. 8.1).

**Meiosis** is nuclear division that reduces the chromosome number, so each gamete contains half the species number of chromosomes. At the start of meiosis, the parent cells have the full number of chromosomes and each is duplicated. Following two divisions called **meiosis I** and **meiosis II,** the four daughter cells have only one copy of each type of chromosome, and these chromosomes consist of only one chromatid.

Not only does meiosis reduce chromosome number, it also introduces variation, and this laboratory will show you exactly how the chromosomes are shuffled during meiosis and how the genetic material is recombined during the process of meiosis.

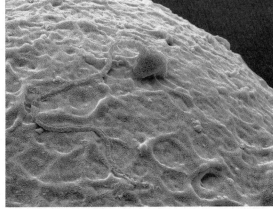

4,200×

**Figure 8.1  Zygote formation.**
During sexual reproduction, union of the sperm and egg produces a zygote that becomes a new individual.
Thierry Berrod/Mona Lisa Production/Science Source

# 8.1 Meiosis: Reduction Division

### Pre-Lab

**2.** Why is the chromosome number reduced by half during meiosis? _____

_____

**3.** How many daughter cells are produced during meiosis? _____

Meiosis is a form of nuclear division that reduces the chromosome number by half. Therefore, when a sperm fertilizes an egg, the zygote will receive the normal number of chromosomes for that species.

tetrad

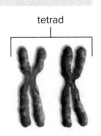

**Figure 8.2   Homologues.**
A pair of homologues have the same shape and size. They form a tetrad because each homologue has two chromatids.

Before meiosis begins, the parent cell is **diploid (2n)**; it contains pairs of chromosomes called **homologues.** Consider the pair of homologues shown in Figure 8.2. A pair of homologues is called a **tetrad** because it contains two pairs of sister chromatids or four chromatids altogether. Following meiosis, the daughter cells are **haploid (n);** they contain only one from each pair of homologues.

### Observation: Meiosis in Lily Anther

Almost all eukaryotes practice meiosis during sexual reproduction. The anther is the male part of the flower where meiosis, consisting of **meiosis I** (Fig. 8.3) and **meiosis II** (Fig. 8.4), occurs preparatory to producing sperm.

#### Phases of Meiosis I

**1.** Examine a prepared slide of a lily anther where cells are undergoing meiosis I. Homologues pair up (called **synapsis**) during prophase I. During metaphase I, tetrads are at the metaphase plate and during anaphase I, homologues separate. Separation of homologues makes the daughter cells following meiosis I haploid because each daughter cell receives only one chromosome from each pair of homologues.

**Figure 8.3   Meiosis I.**
Phases of meiosis I in plant cell micrographs and drawings that depict the movement of the chromosomes. (The blue chromosomes were inherited from one parent and the red chromosomes were inherited from the other parent.)

(all) Ed Reschke

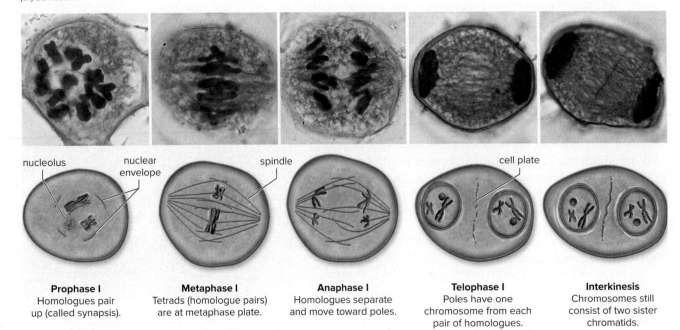

| **Prophase I** | **Metaphase I** | **Anaphase I** | **Telophase I** | **Interkinesis** |
|---|---|---|---|---|
| Homologues pair up (called synapsis). | Tetrads (homologue pairs) are at metaphase plate. | Homologues separate and move toward poles. | Poles have one chromosome from each pair of homologues. | Chromosomes still consist of two sister chromatids. |

2. Using Figure 8.3 as a guide, try to find a cell in each phase of meiosis I.

3. *In Figure 8.3, place a 2n or n beside each drawing.*

### Phases of Meiosis II

1. Examine a prepared slide of a lily anther where cells are undergoing meiosis II. A brief period of time called **interkinesis** occurs between meiosis I and meiosis II. What happens during meiosis II? Separation of the sister chromatids, of course, just as in mitosis (Fig. 8.4).

2. Using Figure 8.4 as a guide, try to find a cell in each phase of meiosis II.

3. *In Figure 8.4, place a 2n or n beside each drawing.* Why is it important for both sperm and egg to have the haploid number of chromosomes? _____

_____

**Figure 8.4   Meiosis II.**
Phases of meiosis II in plant cell micrographs and drawings that depict the movement of the chromosomes.

(all) Ed Reschke

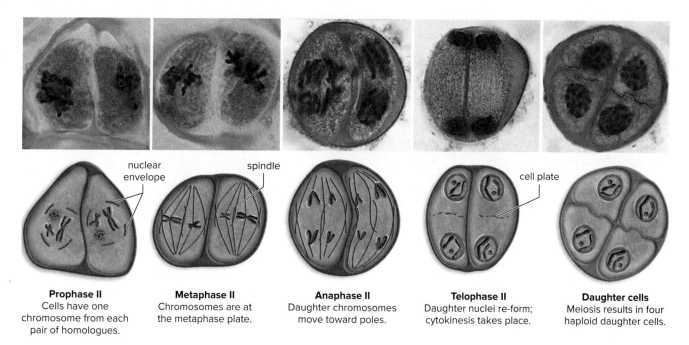

| **Prophase II** | **Metaphase II** | **Anaphase II** | **Telophase II** | **Daughter cells** |
| Cells have one chromosome from each pair of homologues. | Chromosomes are at the metaphase plate. | Daughter chromosomes move toward poles. | Daughter nuclei re-form; cytokinesis takes place. | Meiosis results in four haploid daughter cells. |

## 8.2   Production of Variation During Meiosis

### Pre-Lab

4. What materials are used to simulate meiosis for this exercise? _____

5. During which part of the cell cycle are chromosomes duplicated prior to meiosis? _____

_____

6. Which two events lead to genetic variation produced during meiosis? _____

_____

The following Experimental Procedure is designed to show that during meiosis, crossing-over and independent separation of homologues lead to diversity of genetic material in the gametes.

## Experimental Procedure: Production of Variation During Meiosis

First, you will build four chromosomes: two pairs of homologues, as in Figure 8.5. In other words, the parent cell is 2n = 4.

### Building Chromosomes

1. Obtain the following materials: 48 pop beads of one color (e.g., red) and 48 pop beads of another color (e.g., blue) for a total of 96 beads; eight magnetic centromeres.
2. Build a homologue pair of duplicated chromosomes using Figure 8.5a as a guide. Each chromatid will have 16 beads. Be sure to bring the centromeres of the same color together so that they form one duplicated chromosome.
3. Build another homologue pair of duplicated chromosomes using Figure 8.5b as a guide. Each chromatid will have eight beads.
4. Note that your chromosomes look the same as those in Figure 8.5.

**Figure 8.5 Two pairs of homologues.**
The red chromosomes were inherited from one parent, and the blue chromosomes were inherited from the other parent. Color does not signify homologues; size and shape signify homologues.

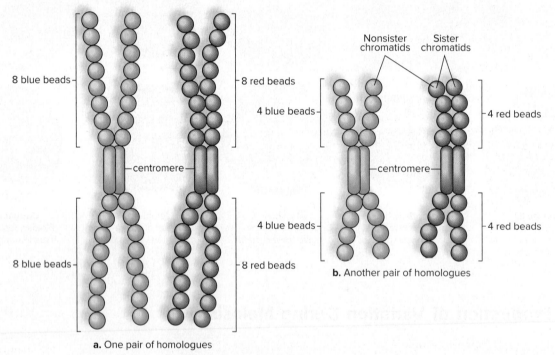

### Meiosis I

You will now have your chromosomes undergo meiosis I.

### Prophase I

5. Put all four of your chromosomes in the center of your work area: This area represents the nucleus. Synapsis, a very important event, occurs during meiosis. To simulate synapsis, place the long blue chromosome next to the long red chromosome and the short blue chromosome next to the short red chromosome to show that the homologues pair up as in Figure 8.3 (prophase I). Now **crossing-over** occurs. During crossing-over, an exchange of genetic material occurs between nonsister chromatids. Perform crossing-over by switching some blue beads for red beads between the inside chromatids of the homologues. The genetic material contains genes and therefore crossing-over recombines genes.

> **Genetic Variation Due to Meiosis**  As a result of crossing-over, the genetic material on a chromosome in a gamete can be different from that in the parent cell.

### Metaphase I

6. Keep the homologues together and align them at the metaphase plate. Add centrioles to represent the poles of a spindle apparatus. These are animal cells.

### Anaphase I and Telophase I

7. Separate the homologues, so that each pole of the spindle receives one chromosome from each pair of homologues. This separation causes each pole to receive the haploid number of chromosomes.

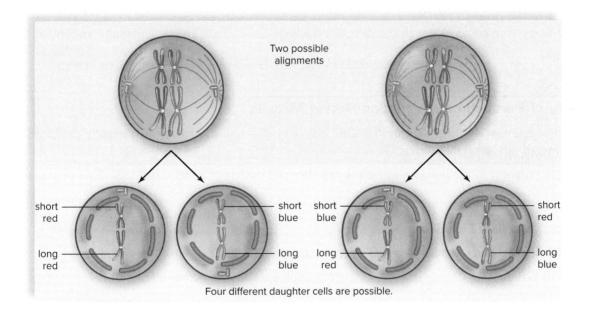

Four different daughter cells are possible.

Only two daughter cells result from meiosis I, but two possible alignments of chromosomes can occur at the metaphase plate during metaphase I. Therefore, four possible combinations of haploid chromosomes are possible in the daughter cells following meiosis I in a 2n = 4 parent cell. The possible combinations increase as the chromosome number increases.

> **Genetic Variation Due to Meiosis**  As a result of independent homologue separation, all possible combinations of the haploid number of chromosomes can occur among the gametes.

### Meiosis II

Follow these directions to simulate meiosis II.

#### Prophase II

**8.** Choose one daughter nucleus to be the parent nucleus undergoing meiosis II.

#### Metaphase II

**9.** Move the duplicated chromosomes to the metaphase II metaphase plate, as shown in the art to the right.

#### Anaphase II

**10.** Pull the two magnets of each duplicated chromosome apart. What does this action represent? _____

_____

#### Telophase II

**11.** Put the chromosomes—each having one chromatid—at the poles near the centrioles. At the end of telophase, the daughter nuclei reform.

- You chose only one daughter nucleus from meiosis I to be the nucleus that divides. In reality, both daughter nuclei go on to divide again. Therefore, how many nuclei are usually present when meiosis II is complete? _____

- In this exercise, how many chromosomes were in the parent cell nucleus undergoing meiosis II? _____

- How many chromosomes are in the daughter nuclei? _____ Explain how this is possible. _____

### Summary of Production of Variation During Meiosis

Sexual reproduction results in offspring that can look very different as a result of variation produced during meiosis through the following events:

1. During prophase I, the homologues come together and exchange genetic material. This process is called _____.

2. During metaphase I, the homologues align _____ and therefore differently.

3. During fertilization variant sperm fertilize variant eggs, further helping to ensure that the new individual inherits different _____ of homologues than a parent had.

## 8.3  Human Life Cycle

### Pre-Lab

**7.** How many chromosomes does each parent contribute during fertilization? _____

**8.** Which type of cell division reduces the chromosome number to produce gametes? _____

_____

**9.** Which type of cell division creates identical cells for growth and repair? _____

**Figure 8.6  Life cycle of humans.**

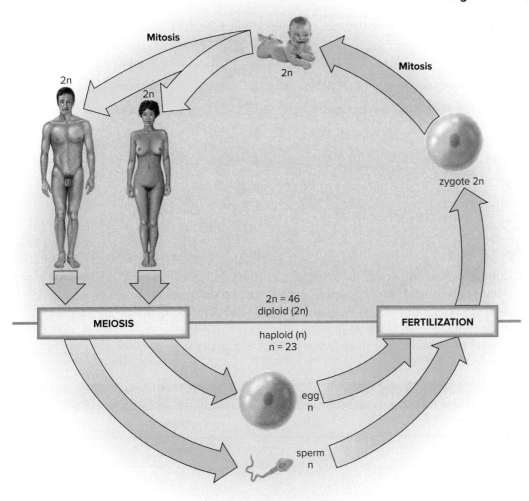

The term **life cycle** in sexually reproducing organisms refers to all the reproductive events that occur from one generation to the next. The human life cycle involves both mitosis and meiosis (Fig. 8.6).

During development and throughout our entire life mitosis is involved in the growth, maintenance and repair of tissues. As a result of mitosis, each somatic (body) cell has the diploid number of chromosomes (2n), which is 46 chromosomes.

During gamete formation, meiosis reduces the chromosome number from the diploid to the haploid number (n) in such a way that the gametes (sperm and egg) have one chromosome derived from each pair of homologues. In males, meiosis is a part of **spermatogenesis,** which occurs in the testes and produces sperm. In females, meiosis is a part of **oogenesis,** which occurs in the ovaries and produces eggs. After the sperm and egg join during fertilization, the resulting zygote is 2n. The zygote then undergoes mitosis with differentiation of cells to become a fetus, and eventually a new human being.

Meiosis keeps the number of chromosomes constant between the generations, and it also, as we have seen, causes the gametes to be different from one another. Therefore, because of sexual reproduction, there are more variations among individuals.

## Summary of Human Life Cycle

Fill in the blanks to ensure your understanding of the role of meiosis and mitosis in humans.

1. Name of organ that produces gametes      in males _____      in females _____

2. Name of process that produces gametes      in males _____      in females _____

3. Type of cell division involved in process      in males _____      in females _____

4. Name of gamete      in males _____      in females _____

5. Number of chromosomes in gamete      in males _____      in females _____

6. Results of fertilization      _____

7. Number of chromosomes provided      _____

## Mitosis Versus Meiosis

When comparing mitosis to meiosis it is important to note that meiosis requires two nuclear divisions but mitosis requires only one nuclear division. Therefore, mitosis produces two daughter cells and meiosis produces four daughter cells. Following mitosis, the daughter cells are still diploid but following meiosis, the daughter cells are haploid. Figure 8.7 explains why.

Fill in Table 8.1 to indicate general differences between mitosis and meiosis.

### Table 8.1 Differences Between Mitosis and Meiosis

|  | Mitosis | Meiosis |
|---|---|---|
| 1. Number of divisions |  |  |
| 2. Chromosome number in daughter cells |  |  |
| 3. Number of daughter cells |  |  |

Complete Table 8.2 to indicate specific differences between mitosis and meiosis I.

### Table 8.2 Mitosis Compared with Meiosis I

| Mitosis | Meiosis I |
|---|---|
| Prophase: No pairing of chromosomes | Prophase I: _____ |
| Metaphase: Duplicated chromosomes at metaphase plate | Metaphase I: _____ |
| Anaphase: Sister chromatids separate | Anaphase I: _____ |
| Telophase: Chromosomes have one chromatid | Telophase I: _____ |

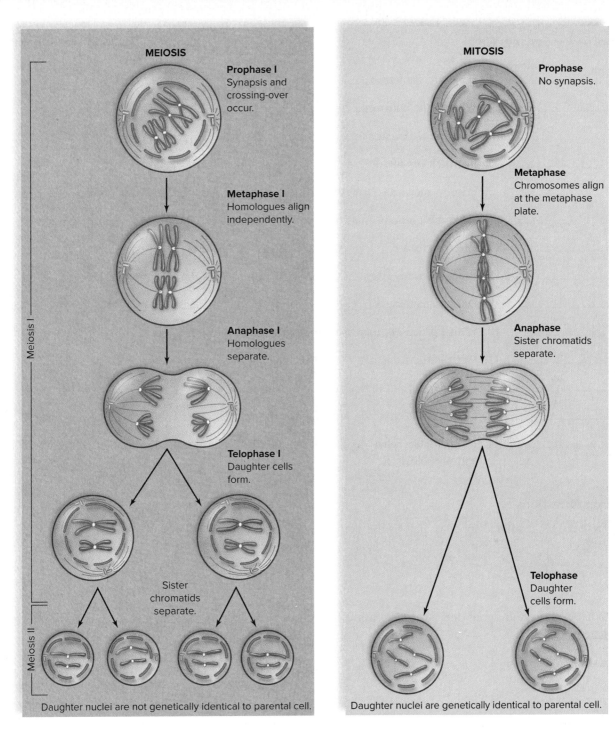

**Figure 8.7  Comparison of mitosis and meiosis.**
The final products of meiosis and mitosis differ in the number of chromosomes, genetic variation, and number of daughter cells produced.

_____  **1.** What are gametes?

_____  **2.** What is the end result of meiosis?

_____  **3.** How many divisions occur during meiosis?

_____  **4.** In which organs does meiosis take place in humans?

_____  **5.** Prior to meiosis, during which stage of the cell cycle does DNA replication occur?

_____  **6.** What are chromosomes that look alike and carry genes for the same traits called?

_____  **7.** Parent cells are _____ (2n) and daughter cells are _____ (n).

_____  **8.** What term describes the pairing of chromosomes during prophase I?

_____  **9.** During anaphase I _____ separate; during anaphase II _____ _____ separate.

_____  **10.** Which processes increase variation during meiosis?

_____  **11.** What is the exchange of genetic material between nonsister chromatids called?

_____  **12.** During what phase does crossing-over occur?

_____  **13.** During what phase do homologues separate?

_____  **14.** If a parent cell has 48 chromosomes, how many does each daughter cell have at the end of meiosis II?

**Thought Questions**

**15.** Describe what would happen if meiosis did not reduce the chromosome number of gametes.

**16.** What events occur during meiosis and fertilization that result in offspring having a mix of their parents' traits?

**17.** What are the similarities and differences between mitosis and meiosis?

# 9

# Patterns of Inheritance

## Learning Outcomes

**9.1 One-Trait Crosses**
- State Mendel's law of segregation, and relate this law to laboratory exercises.
- Describe and predict the results of a one-trait cross in both tobacco seedlings and corn.

**9.2 Two-Trait Crosses**
- State Mendel's law of independent assortment, and relate this law to laboratory exercises.
- Explain and predict the results of a two-trait cross in corn plants and *Drosophila*.

**9.3 X-Linked Crosses**
- Explain and predict the results of an X-linked cross in *Drosophila*.

## Introduction

### Pre-Lab

1. Uppercase letters indicate the _____ genotype and lowercase letters indicate the _____ genotype.

Gregor Mendel, sometimes called the "founder of genetics," formulated the basic laws of genetics examined in this laboratory. He determined that individuals have two alternate forms of a gene (now called **alleles**) for each trait in their body cells. Today, we know that alleles are located on the chromosomes. An individual can be **homozygous dominant** (two dominant alleles, *GG*), **homozygous recessive** (two recessive alleles, *gg*), or **heterozygous** (one dominant and one recessive allele, *Gg*). **Genotype** refers to an individual's genes, while **phenotype** refers to an individual's physical appearance (Fig. 9.1). In normal Mendelian genetic situations, homozygous dominant and heterozygous individuals express the dominant phenotype; homozygous recessive individuals show the recessive phenotype.

**Allele Key**
*T* = tall plant
*t* = short plant

| Phenotype | tall | tall | short |
|---|---|---|---|
| Genotype | *TT* | *Tt* | *tt* |

**Allele Key**
*L* = long wings
*l* = short wings

| Phenotype | long wings | long wings | short wings |
|---|---|---|---|
| Genotype | *LL* | *Ll* | *ll* |

**Figure 9.1 Genotype versus phenotype.**
Only with homozygous recessive do you immediately know the genotype.

# 9.1 One-Trait Crosses

**Pre-Lab**

2. What process separates the parents, two alleles to create gametes (sperm and egg) with a single allele?

_____

3. A _____ _____ is useful for determining the potential traits of offspring.

A single pair of alleles is involved in one-trait crosses. Mendel found that reproduction between two hetero-zygous individuals (*Aa*), called a **monohybrid cross,** results in both dominant and recessive phenotypes among the offspring. In Figure 9.2, the expected phenotypic ratio among the offspring is 3:1. Three offspring have the dominant phenotype for every one that has the recessive phenotype.

Mendel realized that these results are obtainable only if the alleles of each parent segregate (separate from each other) during meiosis. Therefore, Mendel formulated his first law of inheritance:

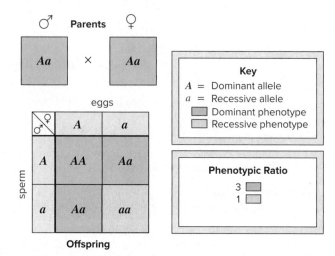

**Figure 9.2 What are the expected results of a cross?**
A Punnett square allows you to determine the expected phenotypic ratio for a cross.

## Law of Segregation

Each organism contains two alleles for each trait, and the alleles segregate during the formation of gametes. Each gamete (egg or sperm) then contains only one allele for each trait. When fertilization occurs, the new organism has two alleles for each trait, one from each parent.

## Punnett Squares

**Punnett squares,** named after the man who first used them, allow you to easily determine the results of a cross between individuals whose genotypes are known. Consider that when fertilization occurs, two gametes, such as a sperm and an egg, join together. Whereas individuals have two alleles for every trait, gametes have only one allele because alleles are on the chromosomes and homologues separate during meiosis. Heterozygous parents with the genotype *Aa* produce two types of gametes: 50% of the gametes contain an *A* and 50% contain an *a*. A Punnett square allows you to vertically line up all possible types of sperm and to horizontally line up all possible types of eggs. Every possible combination of gametes occurs within the squares, and these combinations indicate the possible genotypes of the offspring. In Figure 9.2, one possible pairing is *AA* = homozygous dominant, two are *Aa* = heterozygous, and one is *aa* = homozygous recessive. Therefore, there is a 3/4 chance of inheriting the dominant phenotype and a 1/4 chance of inheriting the recessive phenotype. This is said to be a **phenotypic ratio** of 3:1.

A Punnett square can be used for any cross regardless of the trait(s) and the genotypes of the parents. All you need to do is use the correct letters for the particular trait(s), and make sure you have given the parents the correct genotypes and correct proportion of each type of gamete. Then you can determine the possible genotypes of the offspring and the expected phenotypic ratio among the offspring.

Inheritance is a game of chance. Just as there is a 50% probability of heads or tails when tossing a coin, there is a 50% probability that a sperm or egg will have an *A* or an *a* when the parent is *Aa*. The chance of an equal number of heads or tails improves as the number of tosses increases. In the same way, the chance of an equal number of gametes with *A* and *a* improves as the number of gametes increases. Therefore, the 3:1 ratio among offspring is more likely to occur when you have a large sample size.

## Color of Tobacco Seedlings

In tobacco plants, a dominant allele (*C*) for chlorophyll gives the plants a green color, and a recessive allele (*c*) for chlorophyll causes a plant to appear white. If a tobacco plant is homozygous for the recessive allele (*c*), it cannot manufacture chlorophyll and thus appears white (Fig. 9.3).

**Figure 9.3 Monohybrid cross.**
These tobacco seedlings are growing on an agar plate. The white plants cannot manufacture chlorophyll.

---

### Experimental Procedure: Color of Tobacco Seedlings

1. Obtain a numbered agar plate on which tobacco seedlings are growing. They are the offspring of a cross between heterozygous parents: the cross *Cc* × *Cc*. Complete the Punnett square to determine the expected phenotypic ratio.

   What is the expected phenotypic ratio? _____

2. Record the plate number and using a stereomicroscope, view the seedlings. Count the number that are green and the number that are white. Record your results in Table 9.1.

3. Repeat this procedure for two additional plates. Total the number that are green and the number that are white.

4. Complete Table 9.1 by recording the class data. Total the number that are green and the number that are white per class.

**Key:**
C – green
c = white

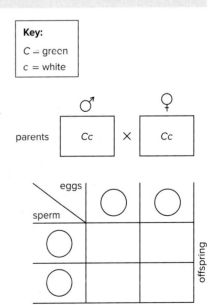

## Table 9.1  Color of Tobacco Seedlings

| | Number of Offspring | | |
| | Green Color | White Color | Phenotypic Ratio |
|---|---|---|---|
| Plate # _____ | | | |
| Plate # _____ | | | |
| Plate # _____ | | | |
| Totals | | | |
| Class data | | | |

### Conclusions: Color of Tobacco Seedlings

- In the last column of Table 9.1, record the actual phenotypic ratio per observed plate, per total number of green versus white plants you counted, and per the entire class. To determine the actual phenotypic ratio, divide the number of green color seedlings in a plate by the number of white color seedlings in a plate.

  Do your results differ from the expected phenotypic ratio? If so, explain your answer. _____
  _____
  _____

- Mendel found that the more plants he counted, the closer he came to the expected phenotypic ratio.

  Were your class data closer to the expected phenotypic ratio than your individual data? _____

## Color of Corn Kernels

In corn plants, the allele for purple kernel (*P*) is dominant over the allele for yellow kernel (*p*) (Fig. 9.4).

**Figure 9.4  One-trait cross.**
Two types of kernels are seen on an ear of corn following a cross between *Pp* and *pp*.

Evelyn Jo Johnson

## Experimental Procedure: Color of Corn Kernels

1. Obtain an ear of corn from the supply table. You will be examining the results of the cross $Pp \times pp$. Complete the Punnett square to determine the expected phenotypic ratio. Note that when one parent has only one possible type of gamete, only one column is needed in the Punnett square.

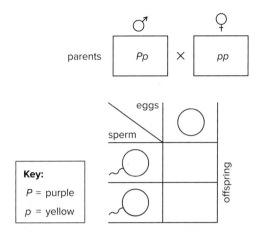

What is the expected phenotypic ratio among the offspring? _____

2. Record the sample number and then count the number of kernels that are purple and the number that are yellow. As before, use two more samples, and record your results in Table 9.2.

### Table 9.2  Color of Corn Kernels

| | Number of Kernels | | |
| --- | --- | --- | --- |
| | **Purple** | **Yellow** | **Phenotypic Ratio** |
| Sample # _____ | | | |
| Sample # _____ | | | |
| Sample # _____ | | | |
| Totals | | | |
| Class data | | | |

## Conclusions: Color of Corn Kernels

- In the last column of Table 9.2, record the actual phenotypic ratio per observed sample, per total number of purple versus yellow kernels you counted, and per the entire class. Do your results differ from the expected phenotypic ratio? If so, explain your answer. _____

  _____

- Were your class data closer to the expected phenotypic ratio than your individual data? _____
  If so, explain your answer. _____

  _____

**Laboratory 9**  Patterns of Inheritance     **91**

1. In pea plants, purple flowers (*P*) is dominant and white flowers (*p*) is recessive. What is the genotype of pure-breeding white plants? Pure-breeding means that they produce plants with only one phenotype. \_\_\_\_ If pure-breeding purple plants are crossed with these white plants, what phenotype is expected? _____

2. In pea plants, tall (*T*) is dominant and short (*t*) is recessive. A heterozygous tall plant is crossed with a short plant. What is the expected phenotypic ratio? _____

3. Unexpectedly to the farmer, two tall plants have some short offspring. What is the genotype of the parent plants and the short offspring? parent _____ offspring _____

4. In horses, two trotters are mated to each other and produce only trotters; two pacers are mated to each other and produce only pacers. When one of these trotters is mated to one of the pacers, all the horses are trotters. Create a key and show the cross. key _____ cross _____

5. A brown dog is crossed with two different black dogs. The first cross produces only black dogs, and the second cross produces equal numbers of black and brown dogs. What is the genotype of the brown dog? _____ The first black dog? _____ The second black dog? _____

6. In pea plants, green pods (*G*) is dominant and yellow pods (*g*) is recessive. When two pea plants with green pods are crossed, 25% of the offspring have yellow pods. What is the genotype of all plants involved? plants with green pods _____ plants with yellow pods _____

7. A breeder wants to know if a dog is homozygous black or heterozygous black. If the dog is heterozygous, which cross is more likely to produce a brown dog, *Bb* × *bb* or *Bb* × *Bb*? Explain your answer. _____

_____

_____

8. If the cross in problem 6 produces 220 plants, how many offspring have green pods and how many have yellow pods? _____ If the cross in problem 2 produces 220 plants, how many offspring are tall and how many are short? _____

## 9.2 Two-Trait Crosses

### Pre-Lab

4. Which two traits of corn kernels will be examined in this activity? _____

5. What is the phenotype for a corn kernel with genotype *ppss*? _____

Two-trait crosses involve two pairs of alleles. Mendel found that during a **dihybrid cross,** when two dihybrid individuals (*AaBb*) reproduce, the phenotypic ratio among the offspring is 9:3:3:1, representing four possible phenotypes. He realized that these results could be obtained only if the alleles of the parents segregated independently of one another when the gametes were formed. From this, Mendel formulated his second law of inheritance:

# Law of Independent Assortment

Members of an allelic pair segregate (assort) independently of members of another allelic pair. Therefore, all possible combinations of alleles can occur in the gametes.

The FOIL method is a way to determine the gametes. FOIL stands for *F*irst two alleles from each trait; *O*uter two alleles from each trait; *I*nner two alleles from each trait; *L*ast two alleles from each trait. Here is how the FOIL method can help you determine the gametes for the genotype *PpSs:*

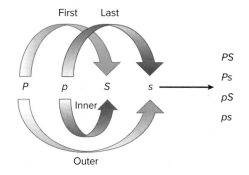

## Color and Texture of Corn

In corn plants, the allele for purple kernel (*P*) is dominant over the allele for yellow kernel (*p*), and the allele for smooth kernel (*S*) is dominant over the allele for rough kernel (*s*) (Fig. 9.5).

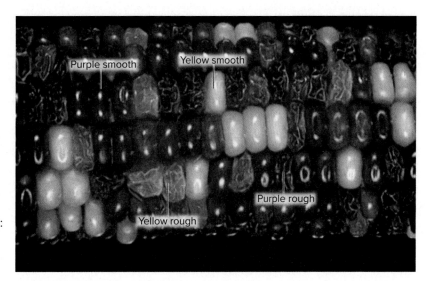

**Figure 9.5  Dihybrid cross.**
Four types of kernels are seen on an ear of corn following a dihybrid cross (*PpSs* × *PpSs*): purple smooth, purple rough, yellow smooth, and yellow rough.

Dan Guravich/Science Source

### Experimental Procedure: Color and Texture of Corn

1. Obtain an ear of corn from the supply table. You will be examining the results of the cross *PpSs* × *PpSs*.

2. Complete the Punnett square that follows this question. Use the results to count the number of kernels of each possible phenotype listed in Table 9.3. Record the sample number and your results in Table 9.3. Use three samples, and total your results for all samples. Also record the class data (i.e., the number of kernels that are the four phenotypes per class).

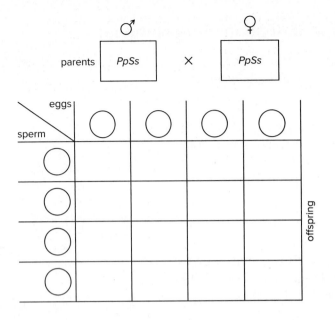

## Table 9.3  Color and Texture of Corn

| | Number of Kernels | | | | |
| --- | --- | --- | --- | --- | --- |
| | Purple Smooth | Purple Rough | Yellow Smooth | Yellow Rough | Phenotypic Ratio |
| Sample # _____ | | | | | |
| Sample # _____ | | | | | |
| Sample # _____ | | | | | |
| Totals | | | | | |
| Class data | | | | | |

### Conclusion: Color and Texture of Corn

- Calculate the actual phenotypic ratios based on the data and record them in Table 9.3. Do the results differ from the expected ratio per individual data? _____ Per class data? _____
  If so, explain your answer. _____
  _____
  _____

## Wing Length and Body Color in *Drosophila*

*Drosophila* are the tiny flies you often see flying around ripe fruit; therefore, they are called fruit flies. If a culture bottle of fruit flies is on display, take a look at it. Because so many flies can be grown in a small culture bottle, fruit flies have contributed substantially to our knowledge of genetics. If you were to examine *Drosophila* flies under the stereomicroscope, they would appear like this:

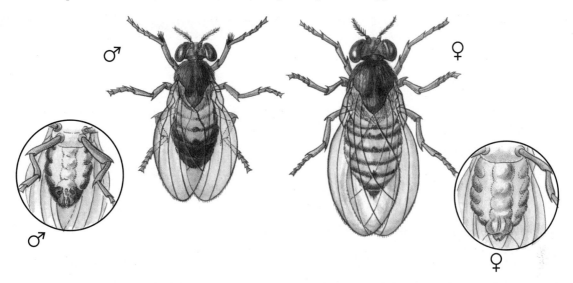

In *Drosophila*, long wings (*L*) are dominant over short (vestigial) wings (*l*), and gray body (*G*) is dominant over black (ebony) body (*g*). Consider the cross *LlGg* × *llgg* and complete this Punnett square:

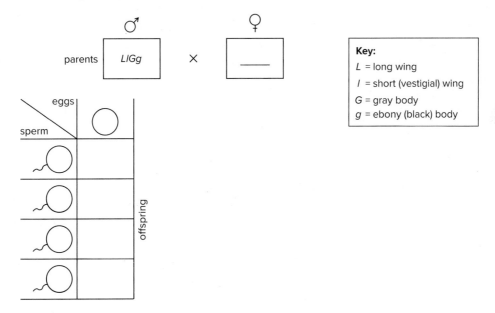

What is the expected phenotypic ratio for this cross? _____

## Experimental Procedure: Wing Length and Body Color in Drosophila

If your instructor has frozen flies available, cross out the numbers in Table 9.4 and use the stereomicroscope or a hand lens to count the flies of each type given in Table 9.4. Otherwise simply use the data supplied for you in Table 9.4.

| Table 9.4   Wing Length and Body Color in *Drosophila** | | | | | |
| --- | --- | --- | --- | --- | --- |
| | **Phenotypes** | | | | |
| | Long Gray | Long Ebony | Short Gray | Short Ebony | Phenotypic Ratio |
| Number of offspring | 28 | 32 | 28 | 30 | |
| Class data | 128 | 120 | 120 | 120 | |

*Wings and body are understood in this table.

### Conclusion: Wing Length and Body Color in Drosophila

- Calculate the actual phenotypic ratio based on the data and record in Table 9.4. Do the results differ from the expected ratio per individual data? _____ Per class data? _____ Explain your answer. _____

_____

### Two-Trait Genetics Problems

1. In tomatoes, tall is dominant and short is recessive. Red fruit is dominant and yellow fruit is recessive. Choose a key for height _____ for color of fruit _____ What is the genotype of a plant heterozygous for both traits? _____ What are the possible gametes for this plant? _____

_____

2. Using words, what are the likely parental genotypes if the results of a two-trait problem are 1:1:1:1 among the offspring? _____ × _____

3. In horses, black (*B*) and a trotting gait (*T*) are dominant, while brown (*b*) and a pacing gait (*t*) are recessive. If a black trotter (homozygous for both traits) is mated to a brown pacer, what phenotypic ratio is expected among the offspring? _____

4. Two black trotters have a brown pacer offspring. What is the genotype of all horses involved? black trotter parents _____ brown pacer offspring _____

5. The phenotypic ratio among the offspring for two corn plants producing purple and smooth kernels is 9:3:3:1. (See page 94 of this lab for the key.) What is the genotype of these plants? parental plants _____ the 9 offspring _____ 3 of the offspring _____ the other 3 _____ and the 1 offspring _____

**6.** Which matings could produce at least some fruit flies heterozygous in both traits? *Write* yes *or* no *beside each*. (You do not need a key.)

*ggLl × Ggll* _____      *GGLl × ggLl* _____      *GGLL × ggll* _____

Explain your answer. _____

_____

**7.** Create two new crosses that would not be able to produce offspring that are heterozygous for both of their traits.

_____ × _____        _____ × _____

**8.** Chimpanzees are not deaf if they inherit both an allele *E* and an allele *G*. A cross between two deaf chimpanzees produces only chimpanzees that can hear. What are the genotypes of all chimpanzees involved? parents _____ × _____ offspring _____

## 9.3   X-Linked Crosses

**Pre-Lab**

  **6.** Which letters are used to represent sex chromosomes? _____

  **7.** Where are X-linked traits found? _____

In animals such as fruit flies, chromosomes differ between the sexes. All but one pair of chromosomes in males and females are the same; these are called **autosomes** because they do not actively determine sex. The pair that is different is called the **sex chromosomes.** In fruit flies and humans, the sex chromosomes in females are XX and those in males are XY.

Some alleles on the X chromosome have nothing to do with gender, and these genes are said to be X-linked. The Y chromosome does not carry these genes and indeed carries very few genes. Males with a normal chromosome inheritance are never heterozygous for X-linked alleles, and if they inherit a recessive X-linked allele it will be expressed.

### Red/White Eye Color in *Drosophila*

In fruit flies, the alleles for eye color are located on the X chromosome. Red eyes ($X^R$) are dominant over white eyes ($X^r$). You will be examining the results of the cross $X^R Y \times X^R X^r$. Complete this Punnett square and state the expected phenotypic ratio for this cross.

females _____      males _____

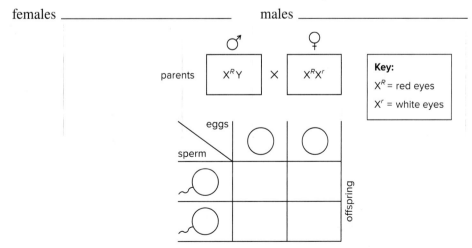

## Experimental Procedure: Red/White Eye Color in *Drosophila*

If your instructor has frozen flies available, cross out the numbers in Table 9.5 and use the stereomicroscope or a hand lens to count the flies of each type given in Table 9.5. Use the art on page 95 to tell males from females, and record male and female data separately. If frozen flies are not available, simply use the data supplied for you in Table 9.5.

| Table 9.5 Red/White Eye Color in *Drosophila* | | | |
|---|---|---|---|
| | **Number of Offspring** | | |
| **Your Data:** | **Red Eyes** | **White Eyes** | **Phenotypic Ratio** |
| Males | 16 | 17 | |
| Females | 63 | 0 | |
| **Class Data:** | | | |
| Males | 45 | 48 | |
| Females | 215 | 0 | |

### Conclusions: Red/White Eye Color in *Drosophila*

- Calculate the phenotypic ratios based on the data for males and females separately and record them in Table 9.5. Do the results differ from the expected ratio per individual data? _____ Per class data? _____ If so, explain your answer. _____

_____

_____

- Using the Punnett square provided, calculate the expected phenotypic results for the cross $X^R Y \times X^r X^r$. What is the expected phenotypic ratio among the offspring? males _____ females _____

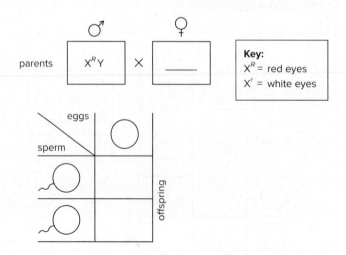

## X-Linked Genetics Problems

1. State the genotypes and gametes for each of these fruit flies:

   |  | genotype | gamete(s) |
   |---|---|---|
   | white-eyed male | _____ | _____ |
   | white-eyed female | _____ | _____ |
   | red-eyed male | _____ | _____ |
   | homozygous red-eyed female | _____ | _____ |
   | heterozygous red-eyed female | _____ | _____ |

2. What are the phenotypic ratios if a white-eyed female is crossed with a red-eyed male?

   males _____ females _____

3. Regardless of any type of cross, do white-eyed males inherit the allele for white eyes from the male

   parent or the female parent? _____ Explain your answer. _____

4. In sheep, horns are sex linked; $H$ = horns and $h$ = no horns. Using symbols, what cross do you recommend

   if a farmer wants to produce hornless males? _____ × _____

5. In *Drosophila,* bar eye is sex linked; $B$ = bar eye and $b$ = no bar eye. What are the genotypes of the parents and the phenotypic ratios for these crosses?

   bar-eyed male × non-bar-eyed female _____ , _____

   bar-eyed male × heterozygous female _____ , _____

   non-bar-eyed male × heterozygous female _____ , _____

6. A female fruit fly has white eyes. What is the genotype of the male parent? _____ What could

   be the genotype of the female parent? _____ or _____

7. In a cross between fruit flies, all of the male offspring have white eyes and all of the female offspring have

   red eyes. What is the genotype of the parents? female parent _____ male parent _____

8. In a cross between fruit flies, a white-eyed male and a red-eyed female produce no offspring that have

   white eyes. What is the genotype of the parents? male parent _____ female parent _____

9. Make up a sex-linked genetic cross using words and parental genotypes. Create a key and show a Punnett square and the phenotypic ratios.

_____ 1. What are the alternate forms of a gene called?

_____ 2. What word is used to describe an individual who has one dominant allele and one recessive allele?

_____ 3. What word is used to refer to the physical appearance of an individual?

_____ 4. Which of Mendel's laws says alleles separate during gamete formation so that each gamete contains only one allele for each trait?

_____ 5. What is used to determine the results of a cross between individuals with known genotypes?

_____ 6. If the phenotypic ratio among the offspring is 3:1, what are the genotypes of the parents?

_____ 7. What is the phenotype of a tobacco seeding with the genotype _cc_ for chlorophyll?

_____ 8. If the allele for a purple corn kernel (_P_) is dominant over the allele for a yellow corn kernel (_p_), what is the genotype of a yellow kernel?

_____ 9. What is the cross between two individuals with _GgRr_ genotypes called?

_____ 10. How many of the offspring of the cross between two _GgRr_ individuals will show the recessive phenotype for both traits?

_____ 11. Which of Mendel's laws says the separation of one pair of alleles into gametes does not impact the separation of a second pair of alleles into gametes?

_____ 12. How many different kinds of gametes can be produced by a parent with a _Ggrr_ genotype?

_____ 13. What is the name given to the chromosomes that do not actively determine the gender of an organism?

_____ 14. Can a female fruit fly that is homozygous dominant for red eyes have any offspring (male or female) with white eyes?

## Thought Questions

15. If an organism displays the dominant trait phenotype (e.g., green color in tobacco plants), it is uncertain whether the organism's genotype is homozygous dominant (_CC_) or heterozygous (_Cc_). What kind of cross (a known genotype) could be done to determine the genotype of the unknown organism? Explain your reply by describing the expected offsprings' phenotypes.

16. If you count 52 purple rough corn kernels and 50 yellow smooth corn kernels, how many purple smooth corn kernels should you expect to count? How many yellow rough corn kernels should you expect to count? Explain your reply.

17. If a trait is an X-linked recessive trait, why is it unlikely that many females with the recessive trait (e.g., white eyes in fruit flies) phenotype will exist?

# 10
# DNA Biology and Technology

## Learning Outcomes

**10.1 DNA Structure and Replication**
- Explain how the structure of DNA facilitates replication.
- Explain how DNA replication is semiconservative.

**10.2 RNA Structure**
- List the ways in which RNA structure differs from DNA structure.

**10.3 DNA and Protein Synthesis**
- Describe how DNA is able to store information.
- State the function of transcription and translation during protein synthesis.

**10.4 Isolation of DNA and Biotechnology**
- Describe the procedure for isolating DNA.
- Describe the process of DNA gel electrophoresis.

**10.5 Detecting Genetic Disorders**
- Understand the relationship between an abnormal DNA base sequence and a genetic disorder.
- Suggest two ways to detect a genetic disorder.

## Introduction

### Pre-Lab

1. During which phase of the cell cycle is DNA replicated? _____

2. Where is DNA located in a cell? _____

This laboratory pertains to molecular genetics and biotechnology. Molecular genetics is the study of the structure and function of **DNA (deoxyribonucleic acid),** the genetic material. Biotechnology is the manipulation of DNA for the benefit of humans and other organisms.

First, we will study the structure of DNA and see how that structure facilitates DNA replication in the nucleus of cells. DNA replicates during the S phase of the cell cycle to ensure each daughter cell has a complete copy of the genetic material. DNA replication also precedes meiosis.

The structure of DNA also facilitates the process of protein synthesis. The sequence of nucleotides in DNA is used to code for the sequence of amino acids in a protein. **RNA (ribonucleic acid)** plays a large role in protein synthesis. Three types of RNA are involved, including messenger RNA (mRNA), ribosomal RNA (rRNA), and transfer RNA (tRNA).

We now understand that a mutated gene has an altered DNA base sequence, and altered sequences can cause genetic disorders. You will have an opportunity to carry out a laboratory procedure that detects whether an individual is normal, has sickle-cell disease, or is a carrier.

# 10.1 DNA Structure and Replication

The structure of DNA lends itself to **replication,** the process that makes a copy of a DNA molecule. DNA replication is a necessary part of chromosome duplication, which precedes mitosis and meiosis. Therefore, DNA replication is needed for growth and repair as well as for sexual reproduction.

## DNA Structure

DNA is a polymer of nucleotide monomers (Fig. 10.1). Each nucleotide is composed of three molecules: deoxyribose (a 5-carbon sugar), a phosphate, and a nitrogen-containing base.

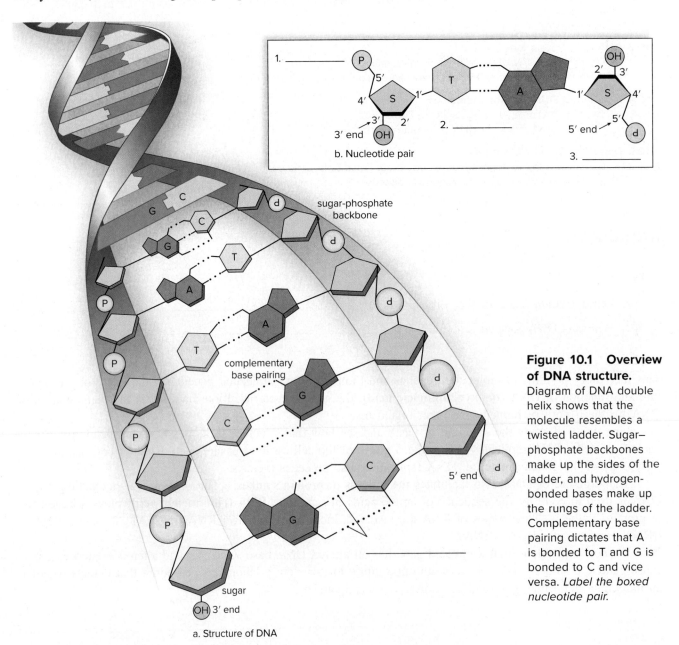

**Figure 10.1 Overview of DNA structure.** Diagram of DNA double helix shows that the molecule resembles a twisted ladder. Sugar–phosphate backbones make up the sides of the ladder, and hydrogen-bonded bases make up the rungs of the ladder. Complementary base pairing dictates that A is bonded to T and G is bonded to C and vice versa. *Label the boxed nucleotide pair.*

1. A boxed nucleotide pair is shown in Figure 10.1. If you are working with a kit, draw a representation of one of your nucleotides here. *Label the phosphate, the base pair, and the deoxyribose in your drawing.*

2. Notice the four types of bases: cytosine (C), thymine (T), adenine (A), and guanine (G). What is the color of each of the four types of bases in Figure 10.1? In your kit? Complete Table 10.1 by writing in the colors of the bases.

### Table 10.1 Base Colors

|  | In Figure 10.1 | In Your Kit |
| --- | --- | --- |
| Cytosine |  |  |
| Thymine |  |  |
| Adenine |  |  |
| Guanine |  |  |

3. Using Figure 10.1 as a guide, join several nucleotides together. Observe the entire DNA molecule.

   What types of molecules make up the backbone (uprights of ladder) of DNA (Fig. 10.1*a*)? _____

   and _____ In the backbone, the phosphate of one nucleotide is bonded to a sugar of the next nucleotide.

4. Using Figure 10.1 as a guide, join the bases together with hydrogen bonds. *Label a hydrogen bond in Figure 10.1.* Dashes are used to represent hydrogen bonds in Figure 10.1*a* because hydrogen bonds are

   (strong or weak) _____.

5. Notice in Figure 10.1*a* and in your model that the base A is always paired with the

   base _____, and the base C is always paired with the base _____. This is called complementary base pairing.

6. In Figure 10.1*a*, what molecules make up the rungs of the ladder? _____

7. Each half of the DNA molecule is a DNA strand. Why is DNA also called a double helix (Fig. 10.1*a*)?

   _____

## DNA Replication

During replication, the DNA molecule is duplicated so that there are two identical DNA molecules. We will see that complementary base pairing makes replication possible.

## Observation: DNA Replication

1. Before replication begins, DNA is unzipped. Using Figure 10.2*a* as a guide, break apart your two DNA strands. What bonds are broken in order to unzip the DNA strands? _____

2. Using Figure 10.2*b* as a guide, attach new complementary nucleotides to each strand using complementary base pairing.

3. Show that you understand complementary base pairing by completing Table 10.2.
   You now have two DNA molecules (Fig. 10.2*c*). Are your molecules identical?

   _____
   _____

4. Because of complementary base pairing, each new double helix is composed of

   an _____ strand and a _____

   strand. *Write* old *or* new *in 1–10, Figure 10.2*a, b, *and* c. *Conservative* means to save something from the past. Why is DNA replication called semiconservative?

   _____
   _____
   _____

**Figure 10.2   DNA replication.**
Use of the ladder configuration better illustrates how replication takes place. **a.** The parental DNA molecule. **b.** The "old" strands of the parental DNA molecule have separated. New complementary nucleotides available in the cell are pairing with those of each old strand. **c.** Replication is complete.

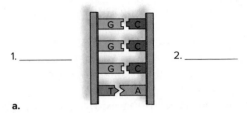

a.

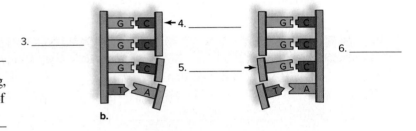

b.

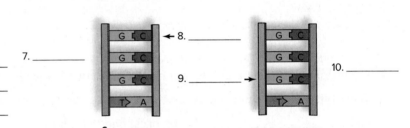

c.

5. Genetic material has to be inherited from cell to cell and organism to organism. Consider that because of DNA replication, a chromosome is composed of two chromatids and each chromatid is a DNA double helix. The chromatids separate during cell division so that each daughter cell receives a copy of each chromosome. In your own words, how does replication provide a means for

   passing DNA from cell to cell and organism to organism? _____
   _____

| Table 10.2   DNA Replication | | | | | | | | | | | | | | | | | | | | | | | | | | | |
|---|---|---|---|---|---|---|---|---|---|---|---|---|---|---|---|---|---|---|---|---|---|---|---|---|---|---|---|
| Old strand | G | G | G | T | T | C | C | A | T | T | A | A | A | T | T | C | C | A | G | A | A | A | T | C | A | T | A |
| New strand | | | | | | | | | | | | | | | | | | | | | | | | | | | |

# 10.2 RNA Structure

## Pre-Lab

**5.** List the differences between DNA and RNA. _____

_____

**6.** *Label Figure 10.3 using the terms* base, ribose, *and* phosphate.

Like DNA, RNA is a polymer of nucleotides (Fig. 10.3). In an RNA nucleotide, the sugar ribose is attached to a phosphate molecule and to a nitrogen-containing base, C, U, A, or G. In RNA, the base uracil replaces thymine as one of the pyrimidine bases. RNA is single-stranded, whereas DNA is double-stranded.

**Figure 10.3   Overview of RNA structure.**
RNA is a single strand of nucleotides. *Label the boxed nucleotide.*

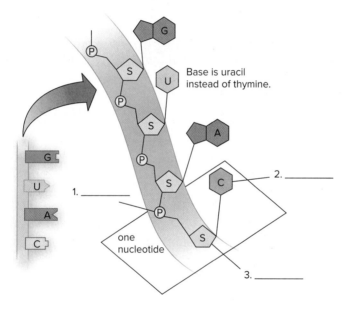

**1.** Describe the backbone of an RNA molecule. _____

**2.** Where are the bases located in an RNA molecule? _____

## Observation: RNA Structure

**1.** If you are using a kit, construct a nucleotide for RNA. *Label the ribose (the sugar in RNA), the phosphate, and the base in 1–3, Figure 10.3.*

**2.** Complete Table 10.3 by writing in the colors of the bases for Figure 10.3 and for your kit.

## Table 10.3　Base Colors

| | In Figure 10.3 | In Your Kit |
|---|---|---|
| Cytosine | | |
| Uracil | | |
| Adenine | | |
| Guanine | | |

## Complementary Base Pairing

Complementary base pairing occurs between DNA and RNA. The RNA base uracil pairs with the DNA base adenine; the other bases pair as shown previously. Complete Table 10.4 to show the complementary DNA bases for the RNA bases.

## Table 10.4　DNA and RNA Bases

| RNA bases | C | U | A | G |
|---|---|---|---|---|
| DNA bases | | | | |

## 10.3　DNA and Protein Synthesis

### Pre-Lab

7. Where in the cell do transcription and translation occur? _____

_____

8. How many bases are needed to code for an amino acid during translation? _____

Protein synthesis involves the processes of transcription and translation. During **transcription,** which takes place in the nucleus, an RNA molecule called **messenger RNA (mRNA)** is made complementary to one of the DNA strands. This mRNA leaves the nucleus and goes to the ribosomes in the cytoplasm. Ribosomes are composed of **ribosomal RNA (rRNA)** and proteins in two subunits. They provide a location for protein synthesis to occur.

During **translation,** RNA molecules called **transfer RNA (tRNA)** bring amino acids to the ribosome, and the amino acids join in the order prescribed by mRNA. In this way, the sequence of amino acids in a new polypeptide was originally specified by DNA. This is the information that DNA, the genetic material, stores.

## Transcription

During transcription, the complementary RNA strand is made from a DNA template (Fig. 10.4). A portion of DNA unwinds and unzips at the point of attachment of the enzyme **RNA polymerase.** A strand of mRNA is produced when complementary nucleotides join in the order dictated by the sequence of bases in DNA. Transcription occurs in the nucleus, and the mRNA passes out of the nucleus to enter the cytoplasm.

### Observation: Transcription

1. If you are using a kit, unzip your DNA model so that only one strand remains. This strand is called the **template strand** because it will be used as a template to construct an mRNA molecule. The **gene strand** is preferred terminology for the complementary DNA strand you discarded because as you can verify in Figure 10.4, it has the same sequence of bases as the mRNA molecule, except that uracil substitutes for thymine.

2. Using Figure 10.4 as a guide, construct a messenger RNA (mRNA) molecule by first lining up RNA nucleotides complementary to the template strand of your DNA molecule. Join the nucleotides together to form mRNA.

3. A portion of DNA has the sequence of bases shown in Table 10.5. *Complete Table 10.5 to show the sequence of bases in mRNA.*

4. If you are using a kit, unzip mRNA transcript from the DNA. Locate the end of the strand that will move to the _____ in the cytoplasm.

**Figure 10.4   Messenger RNA (mRNA).**
Messenger RNA, which is complementary to a section of DNA, forms during transcription.

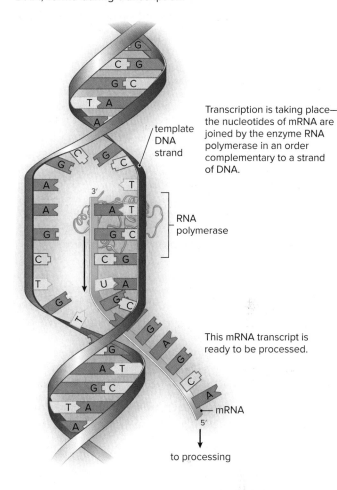

Transcription is taking place—the nucleotides of mRNA are joined by the enzyme RNA polymerase in an order complementary to a strand of DNA.

template DNA strand

RNA polymerase

This mRNA transcript is ready to be processed.

mRNA

to processing

| Table 10.5 | Transcription |
|---|---|
| DNA | T A C A C G A G C A A C T |
| mRNA | |

# Translation

During translation, a polypeptide is made. DNA specifies the sequence of amino acids in a polypeptide because every three bases code for an amino acid. Therefore, DNA is said to have a **triplet code.** The bases in mRNA are complementary to the bases in DNA. Every three bases in mRNA are called a **codon.** One codon of mRNA represents one amino acid. Thus, the sequence of DNA bases serves as the blueprint for the sequence of amino acids assembled to make a protein. The correct sequence of amino acids in a polypeptide is the message that mRNA carries.

Messenger RNA leaves the nucleus and proceeds to the ribosomes, where protein synthesis occurs. As previously mentioned, transfer RNA (tRNA) molecules transfer amino acids to the ribosomes. Each RNA has one particular tRNA amino acid at one end and a specific **anticodon** at the other end (Fig. 10.5). *Label Figure 10.5,* where the amino acid is represented as a colored ball, the tRNA is green, and the anticodon is the sequence of three bases. (The anticodon is complementary to the mRNA codon.)

**Figure 10.5  Transfer RNA (tRNA).**
Each type of transfer RNA with a specific anticodon carries a particular amino acid to the ribosomes.

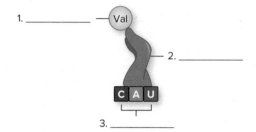

## Observation: Translation

1. Figure 10.6 shows seven tRNA–amino acid complexes. Every amino acid has a name; in the figure, only the first three letters of the name are inside the ball.

2. If you are using a kit, arrange your tRNA–amino acid complexes in the order consistent with Table 10.6. Complete Table 10.6. Why are the codons and anticodons in groups of three? _____

**Figure 10.6  Transfer RNA diversity.**
Each type of tRNA carries only one particular amino acid, designated here by the first three letters of its name.

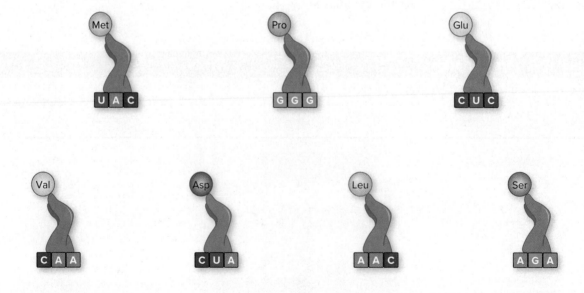

| Table 10.6 | Translation | | | | | | | |
|---|---|---|---|---|---|---|---|---|
| mRNA codons | | AUG | CCC | GAG | GUU | GAU | UUG | UCU |
| tRNA anticodons | | | | | | | | |
| Amino acid* | | | | | | | | |

*Use three letters only. See Table 10.7 for the full names of these amino acids.

| Table 10.7 | Names of Amino Acids |
|---|---|
| **Abbreviation** | **Name** |
| Met | methionine |
| Pro | proline |
| Asp | aspartate |
| Val | valine |
| Glu | glutamic acid |
| Leu | leucine |
| Ser | serine |

3. Figure 10.7 shows the manner in which the polypeptide grows. A ribosome has three binding sites. From left to right, they are the A (amino acid) site, the P (peptide) site, and the E (exit) site. A tRNA leaves from the E site after it has passed its amino acid or peptide to the newly arrived tRNA–amino acid complex. Then the ribosome moves forward, making room for the next tRNA–amino acid. This sequence of events occurs over and over until the entire polypeptide is borne by the last tRNA to come to the ribosome. Then a release factor releases the polypeptide chain from the ribosome. *In Figure 10.7, label the ribosome, the mRNA, and the peptide.*

**Figure 10.7 Protein synthesis.**
**1.** A ribosome has a site for two tRNA–amino acid complexes. **2.** Before a tRNA leaves, an RNA passes its attached peptide to a newly arrived tRNA–amino acid complex. **3.** The ribosome moves forward, and the next tRNA–amino acid complex arrives.

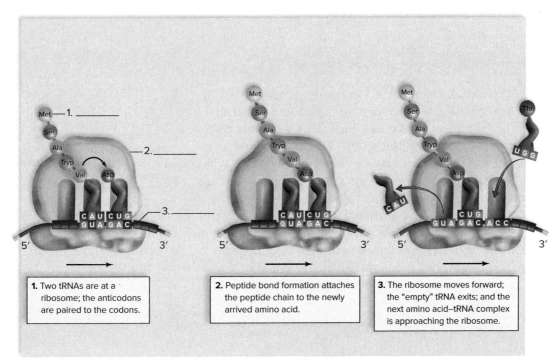

# 10.4  Isolation of DNA and Biotechnology

## Pre-Lab

**9.** What process can be used to separate fragments of DNA? _____

In the following Experimental Procedure, you will isolate DNA from the cells of a fruit or vegetable. It will only be necessary to expose the cells to an agent (dishwasher detergent) that emulsifies membrane in order to "free" the DNA from its enclosures (plasma membrane and nuclear envelope). When transferred to a tube, the presence of NaCl allows DNA to precipitate as a sodium salt. The precipitate forms at the interface between ethanol and the salt solution, and then it floats to the top of the tube where it may be collected.

### Experimental Procedure: Isolation of DNA

1. You will need a slice of fruit or vegetable (i.e., tomato, onion, or apple), a mortar and pestle, and a large clean glass test tube on ice.
2. Crush and grind a slice of fruit or vegetable in a mortar and pestle. Remove the pestle and set aside. Add enough 0.9% NaCl solution to achieve a "soupy" consistency.
3. Add two drops of dishwasher detergent (such as Blue Dawn) to the mixture in the mortar. Swirl until the color of the detergent disappears. Wait 3 to 5 minutes. The solution becomes clear and viscous as the DNA escapes its enclosures.
4. Check the viscosity of the solution in the mortar, and if it is too viscous for you to pipet, add a few more milliliters of NaCl solution. Use a transfer pipet to move 2 to 3 ml of the DNA from the mortar to the glass tube. Try to pick up clear zones of the solution WITHOUT CELL DEBRIS.
5. Slowly add 5 to 7 ml of ice-cold ethanol down the side of the tube on ice. As you do so, DNA will precipitate *between* the ethanol and the NaCl solutions. Then it will float to the top, where it can be picked up by a fresh transfer pipet. Any "white dust" you see consists of RNA molecules.
6. Use a transfer pipet to place the DNA in a small, clean test tube. Pipet away any extra water/ethanol and air-dry the DNA for a few minutes. Dissolve the DNA in 3 to 4 ml distilled $H_2O$.
7. Add five drops of phenol red, a pH indicator. The resulting dark pink color confirms the presence of nucleic acid (i.e., the DNA).

### Experimental Procedure: Gel Electrophoresis

During gel electrophoresis, charged DNA molecules migrate across a span of gel (gelatinous slab) because they are placed in a powerful electrical field. In the present experiment, each DNA sample is placed in a small depression in the gel called a well. The gel is placed in a powerful electrical field. The electricity causes DNA fragments, which are negatively charged, to move through the gel according to their size.

> ⚠ For Experimental Procedures, wear safety goggles, gloves, and protective clothing. If any chemical spills on your skin, wash immediately with mild soap and water; flood eyes with water only. Report any spills immediately to your instructor.

   Almost all DNA gel electrophoresis is carried out using horizontal gel slabs (Fig. 10.8). First, the gel is poured onto a plastic plate, and the wells are formed. After the samples are added to the wells, the gel and the plastic plate are put into an electrophoresis chamber, and buffer is added. The DNA samples begin to migrate after the electrical current is turned on. With staining, the DNA fragments appear as a series of bands spread from one end of the gel to the other according to their size because smaller fragments move faster than larger fragments.

**Figure 10.8  Equipment and procedure for gel electrophoresis.**

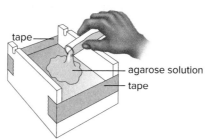

tape

agarose solution

tape

**a.** Agarose solution poured into casting tray

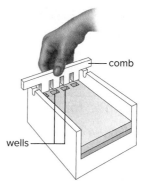

comb

wells

**b.** Comb that forms wells for samples

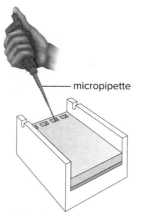

micropipette

**c.** Wells that can be loaded with samples

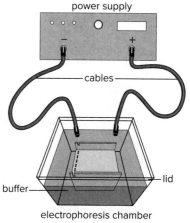

power supply

cables

buffer

lid

electrophoresis chamber

**d.** Electrophoresis chamber and power supply

## 10.5  Detecting Genetic Disorders

### Pre-Lab

**10.** What methods can be used to detect genetic disorders? _____

_____

The base sequence of DNA in all the chromosomes is an organism's genome. The Human Genome Project was a monumental effort to determine the normal order of all the 3.6 billion nucleotide bases in the human genome. Someday it will be possible to sequence anyone's genome within a relatively short time; this will enable us to determine what particular base sequence alterations signify that he or she has a disorder or will have one in the future. In this laboratory, you will study the alteration in base sequence that causes a person to have sickle-cell disease.

In persons with sickle-cell disease, the red blood cells are sickle-shaped cells instead of the biconcave disk shape found in normal red blood cells. Sickle-shaped cells can't pass along narrow capillary passageways. They clog the vessels and break down, causing the person to suffer from poor circulation, anemia, and poor resistance to infection. Internal hemorrhaging leads to further complications, such as jaundice, episodic pain in the abdomen and joints, and damage to internal organs.

Sickle-shaped red blood cells are caused by an abnormal hemoglobin ($Hb^S$). Individuals with the $Hb^A Hb^A$ genotype are normal; those with the $Hb^S Hb^S$ genotype have sickle-cell disease, and those with the $Hb^A Hb^S$ have sickle-cell trait. Persons with sickle-cell trait do not usually have sickle-shaped cells unless they experience dehydration or mild oxygen deprivation.

## Genetic Sequence for Sickle-Cell Disease

Examine Figure 10.9a and 10.9b, which show the DNA base sequence, the mRNA codons, and the amino acid sequence for a portion of the gene for $Hb^A$ and the same portion for $Hb^S$.

1. In what one base does $Hb^A$ differ from $Hb^S$?    $Hb^A$ _____    $Hb^S$ _____

2. What are the codons that contain this base?    $Hb^A$ _____    $Hb^S$ _____

3. What is the amino acid difference?    $Hb^A$ _____    $Hb^S$ _____

This amino acid difference causes the polypeptide chain in sickle-cell hemoglobin to pile up as firm rods that push against the plasma membrane and deform the red blood cell into a sickle shape:

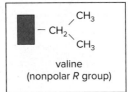

glutamic acid
(polar R group)

valine
(nonpolar R group)

**Figure 10.9   Sickle-cell disease.**
**a.** When red blood cells are normal, the base sequence (in one location) for $Hb^A$ alleles is CTC. **b.** In sickle-cell disease at these locations, it is CAC.

(a–b) Bill Longcore/Science Source

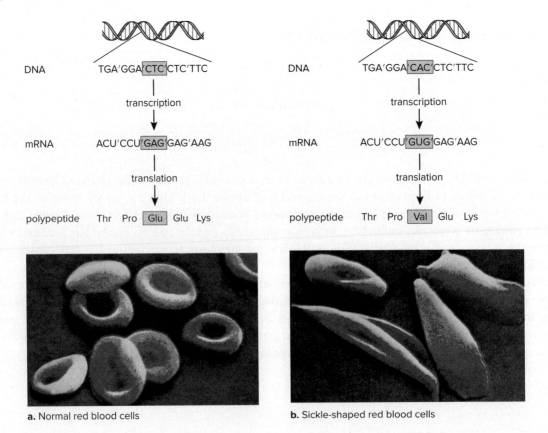

a. Normal red blood cells

b. Sickle-shaped red blood cells

## Detection of Sickle-Cell Disease by Gel Electrophoresis

Three samples of hemoglobin have been subjected to protein gel electrophoresis. Protein gel electrophoresis (Fig. 10.10) is carried out in the same manner as DNA gel electrophoresis (see Fig. 10.8) except the gel has a different composition.

1. Sickle-cell hemoglobin ($Hb^S$) migrates more slowly toward the positive pole than normal hemoglobin ($Hb^A$) because the amino acid valine has no polar $R$ groups, whereas the amino acid glutamic acid does have a polar $R$ group.

2. In Figure 10.10, which lane contains only $Hb^S$, signifying that the individual is $Hb^S Hb^S$?

   ⎯⎯⎯⎯

3. Which lane contains only $Hb^A$, signifying that the individual is $Hb^A Hb^A$? ⎯⎯⎯⎯

4. Which lane contains both $Hb^S$ and $Hb^A$, signifying that the individual is $Hb^A Hb^S$? ⎯⎯⎯⎯

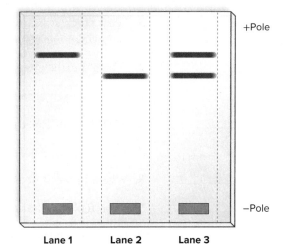

**Figure 10.10  Gel electrophoresis of hemoglobins.**

+Pole

−Pole

Lane 1    Lane 2    Lane 3

### Detection by Genomic Sequencing

You are a genetic counselor. A young couple seeks your advice because sickle-cell disease occurs within both of their families. You order DNA base sequencing to be done. The results come back that at one of the loci for normal hemoglobin, each individual has the abnormal sequence CAC instead of CTC. The other locus is normal. What are the chances that this couple will have a child with sickle-cell disease? ⎯⎯⎯⎯⎯⎯⎯⎯⎯⎯⎯⎯⎯⎯⎯⎯⎯⎯⎯⎯⎯⎯⎯

_____ 1. What biomolecule is composed of deoxyribose sugars, phosphates, and the nitrogen-containing bases adenine, guanine, cytosine, and thymine?

_____ 2. What holds one nitrogen base to another in the DNA strand?

_____ 3. What process makes a copy of DNA so that daughter cells that result from mitosis and cytokinesis each have a copy of DNA?

_____ 4. Which 5-carbon sugar is characteristic of RNA?

_____ 5. What nitrogen base is associated with RNA but not with DNA?

_____ 6. What kind of RNA is made when transcription occurs?

_____ 7. Which strand of DNA acts as the template when making the mRNA?

_____ 8. Where in a cell does protein synthesis occur?

_____ 9. What molecule transports amino acids to the ribosomes?

_____ 10. How many nitrogen bases make up a codon, and what does each codon represent?

_____ 11. What connects one amino acid to another in the polypeptide chain?

_____ 12. Which size DNA fragments, smaller or larger, move more quickly through a gel during electrophoresis?

_____ 13. How are many genetic disorders now identified?

_____ 14. What protein is affected when an individual has sickle-cell disease?

## Thought Questions

15. Students sometimes get the impression that DNA must be replicated prior to protein synthesis. Explain why DNA is only replicated prior to the start of mitosis and meiosis and what the result would be if DNA was replicated each time a new protein needed to be made.

16. If the portion of DNA to be transcribed has 66 bases, how many amino acids will the resulting protein have? Explain your reply.

17. a. What concerns should high school and college athletic coaches have about the potential for their players to have sickle-cell trait?

    b. How could players be identified as having sickle-cell trait or not?

# 11

# Human Genetics

## Introduction

### Pre-Lab

1. If two parents are homozygous recessive for earlobes, will they have children with attached or unattached earlobes? _____

In this laboratory, you will discover that many of the principles of genetics apply to humans as they do to plants and fruit flies. A gene has two alternate forms, called **alleles,** for a variety of different traits, such as hairline, finger length, and so on. One possible allele, designated by an uppercase letter, is **dominant** over the **recessive** allele, designated by a lowercase letter. An individual can be **homozygous dominant** (two dominant alleles, *EE*), **homozygous recessive** (two recessive alleles, *ee*), or **heterozygous** (one dominant and one recessive allele, *Ee*). **Genotype** refers to an individual's alleles, and **phenotype** refers to an individual's appearance (Fig. 11.1). Homozygous dominant and also heterozygous individuals show the dominant phenotype; homozygous recessive individuals show the recessive phenotype.

**Figure 11.1  Genotype versus phenotype.**
Unattached earlobes (*E*) are dominant over attached earlobes (*e*). **a.** Homozygous dominant individuals have unattached earlobes. **b.** Homozygous recessive individuals have attached earlobes. **c.** Heterozygous individuals have unattached earlobes.

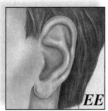
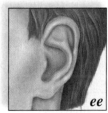
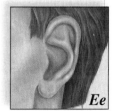

a. Unattached earlobe     b. Attached earlobe     c. Unattached earlobe

# 11.1   Determining the Genotype

Humans inherit 46 chromosomes that occur in 23 pairs. Twenty-two of these pairs are called autosomes and one pair is the sex chromosomes. Autosomal traits are determined by alleles on the autosomal chromosomes.

## Autosomal Dominant and Recessive Traits

Figure 11.2 shows a few human traits.

1. What is the homozygous dominant genotype for type of hairline? _____ What is the phenotype? _____

2. What is the homozygous recessive genotype for finger length? _____ What is the phenotype? _____

3. Why does the heterozygous individual *Ff* have freckles? _____

**Figure 11.2   Commonly inherited traits in humans.**
The alleles indicate which traits are dominant and which are recessive.   (a) SuperStock; (b) John Foxx/Dynamic Graphics/Getty Images; (c–f) Bob Coyle/McGraw Hill; (g) Hero/Corbis/Fancy/Glow Images; (h) Wayhome Studio/Shutterstock

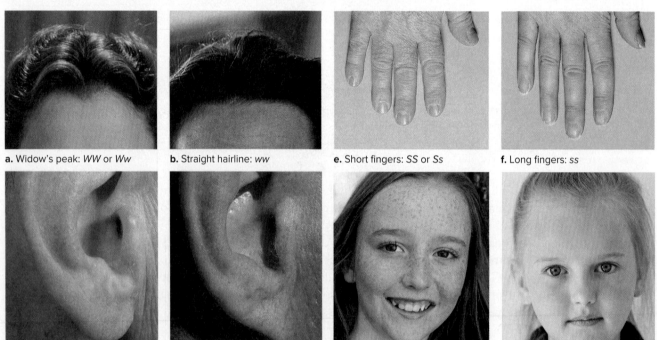

**a.** Widow's peak: *WW* or *Ww*      **b.** Straight hairline: *ww*      **e.** Short fingers: *SS* or *Ss*      **f.** Long fingers: *ss*

**c.** Unattached earlobes: *EE* or *Ee*      **d.** Attached earlobes: *ee*      **g.** Freckles: *FF* or *Ff*      **h.** No freckles: *ff*

These genetic problems use the alleles from Figure 11.2 and Table 11.1.

4. A woman and the members of her immediate family have attached earlobes. What is the woman's genotype? _____ Her maternal grandfather has unattached earlobes. Determine the genotype of her maternal grandfather. _____ Explain your answer. _____

**5.** A male does not have short fingers, but his parents do. Determine the genotype of his parents

_____ , _____ and of the male. _____ Explain your answer. _____

_____

**6.** A male child is adopted. He has a widow's peak hairline. Could both of his parents have had widow's

peak hairlines? _____ Could both of his parents have had straight hairlines? _____

Explain your answer. _____

_____

## Experimental Procedure: Human Traits

**1.** For this Experimental Procedure, you will need a lab partner to help you determine your phenotype for the traits listed in the first column of Table 11.1.

**2.** Determine your probable genotype. If you have the recessive phenotype, you know your genotype. If you have the dominant phenotype, you may be able to decide whether you are homozygous dominant or

### Table 11.1  Autosomal Human Traits

| Trait: d = Dominant<br>r = Recessive | Probable<br>Genotypes | Number in<br>Class | Percentage of<br>Class with Trait |
|---|---|---|---|
| Hairline: | | | |
| Widow's peak (d) | WW or Ww | _____ | _____ |
| Straight hairline (r) | ww | _____ | _____ |
| Earlobes: | | | |
| Unattached (d) | UU or Uu | _____ | _____ |
| Attached (r) | uu | _____ | _____ |
| Skin pigmentation: | | | |
| Freckles (d) | FF or Ff | _____ | _____ |
| No freckles (r) | ff | _____ | _____ |
| Hair on back of hand: | | | |
| Present (d) | HH or Hh | _____ | _____ |
| Absent (r) | hh | _____ | _____ |
| Thumb hyperextension—"hitchhiker's thumb": | | | |
| Last segment cannot be bent backward (d). | TT or Tt | _____ | _____ |
| Last segment can be bent back to 60° (r). | tt | _____ | _____ |
| Bent little finger: | | | |
| Little finger bends toward ring finger (d). | LL or Ll | _____ | _____ |
| Straight little finger (r) | ll | _____ | _____ |
| Interlacing of fingers: | | | |
| Left thumb over right (d) | II or Ii | _____ | _____ |
| Right thumb over left (r) | ii | _____ | _____ |

heterozygous by recalling the phenotype of your parents, siblings, or children. Circle your probable genotype in the second column of Table 11.1.

3. Your instructor will tally the class's phenotypes for each trait so that you can complete the third column of Table 11.1.

4. Complete Table 11.1 by calculating the percentage of the class with each trait. Are dominant phenotypes always the most common in a population? _____ Explain your answer. _____

---

## 11.2 Determining Inheritance

### Pre-Lab

3. If the parents shown in Figure 11.3 have two children, is it possible both children have a widow's peak? _____

4. What is a carrier? _____

Recall that a Punnett square is a means to determine the genetic inheritance of offspring if the genotypes of both parents are known. In a **Punnett square,** all possible types of sperm are lined up with all possible types of eggs so that every possible combination of gametes occurs within the square. Figure 11.3 shows how to construct a Punnett square when autosomal alleles are involved.

**Figure 11.3  Punnett square.**
In a Punnett square, all possible sperm are displayed vertically and all possible eggs are displayed horizontally, or vice versa. The genotypes of the offspring (in this case, also the phenotypes) are in the squares.

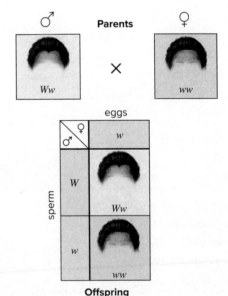

**Results of Cross**
Phenotypic ratio 1:1
☐ Chance of widow's peak ½ = 50%
☐ Chance of straight hairline ½ = 50%

---

### Inheritance of Genetic Disorders

Figure 11.4 can be used to learn the chances that a particular phenotype will occur.

In Figure 11.4*a*,

$\frac{1}{4}$ of the offspring have the recessive phenotype = _____ % chance

$\frac{3}{4}$ of the offspring have the dominant phenotype = _____ % chance

**Figure 11.4  Two common patterns of autosomal inheritance in humans.**
**a.** Both parents are heterozygous. **b.** One parent is heterozygous and the other is homozygous recessive. The letter *A* stands for any trait that is dominant and the letter *a* stands for any trait that is recessive. Substitute the correct alleles for the problem you are working on. For example, *C* = normal; *c* = cystic fibrosis.

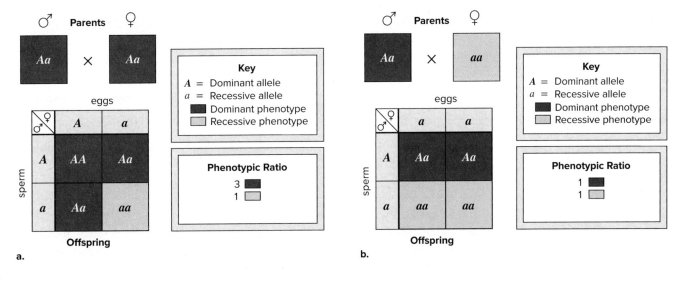

In Figure 11.4*b*,

$\frac{1}{2}$ of the offspring have the recessive or the dominant phenotype = _____ % chance

*In all the following genetic problems, use letters to fill in the parentheses with the genotype of the parents.*

1. **a.** With reference to Figure 11.4*a*, if a genetic disorder is recessive and both parents are heterozygous (_____), what are the chances that an offspring will have the disorder? _____

   **b.** With reference to Figure 11.4*a*, if a genetic disorder is dominant and both parents are heterozygous (_____), what are the chances that an offspring will have the disorder? _____

2. **a.** With reference to Figure 11.4*b*, if the parents are heterozygous (_____) by homozygous recessive (_____), and the genetic disorder is recessive, what are the chances that the offspring will have the disorder? _____

   **b.** With reference to Figure 11.4*b*, if the parents are heterozygous (_____) by homozygous recessive (_____), and the genetic disorder is dominant, what are the chances that an offspring will have the disorder? _____

## Autosomal Disorders

1. **Neurofibromatosis (NF),** sometimes called von Recklinghausen disease, is one of the most common genetic disorders. It affects roughly 1 in 3,000 people. At birth or later, the affected individual may have six or more large tan spots on the skin. Such spots may increase in size and number and become darker. Small benign tumors (lumps) called neurofibromas may occur under the skin or in the muscles.

   Neurofibromatosis is a dominant disorder. If a heterozygous woman reproduces with a homozygous man who does not express neurofibromatosis, what are the chances a child will have neurofibromatosis?

   _____

2. **Cystic fibrosis** is due to abnormal mucus-secreting tissues. At first, the infant may have difficulty regaining the birth weight despite good appetite and vigor. Symptoms include a cough with rapid respiratory rate and large, frequent, and foul-smelling stools due to abnormal pancreatic secretions. Whereas children previously died in infancy due to infections, they now often survive because of antibiotic therapy.

Cystic fibrosis is a recessive disorder. A **carrier** is an individual who appears to be normal but carries a recessive allele for a genetic disorder. A male and female parent are both carriers (_____) for cystic fibrosis. What are the chances they will produce a child that will have cystic fibrosis? _____

3. **Huntington disease** does not appear until the 30s or early 40s. There is a progressive deterioration of the individual's nervous system that eventually leads to involuntary movement, mental illness, and death. Studies suggest that Huntington disease is pleiotropic. This is when a single gene has multiple effects upon the individual's phenotype.

Huntington disease is a dominant disorder. Drina is 25 years old and as yet has no signs of Huntington disease. Her mother is heterozygous for Huntington disease (_____), but her father does not express the characteristics (_____) of the disorder. What are the chances that Drina will develop Huntington disease? _____

_____

4. **Phenylketonuria (PKU)** is characterized by severe intellectual impairment due to an abnormal accumulation of the common amino acid phenylalanine within cells, including neurons. The disorder takes its name from the presence of a breakdown product, phenylketone, in the urine and blood. Newborn babies are routinely tested at the hospital and, if necessary, are placed on a diet low in phenylalanine.

PKU is a recessive disorder. Mr. and Mrs. Martinez appear to be normal, but they have a child with PKU. What are the genotypes of Mr. and Mrs. Martinez? _____

5. **Tay–Sachs disease** is caused by the inability of the lysosome to break down a certain type of fat molecule that accumulates around nerve cells leading to their destruction. Newborns with Tay–Sachs do not exhibit symptoms at birth. At first, they may learn to sit up and stand, but later they develop intellectual disability, blindness, and paralysis. Death usually occurs between ages 3 and 4.

Tay–Sachs is an autosomal recessive disorder. Is it possible for two individuals who do not have Tay–Sachs to have a child with the disorder? Explain your answer. _____

_____

### X-Linked Disorders

The sex chromosomes designated X and Y carry genes just like the autosomal chromosomes. Some genes, particularly on the X chromosome, have nothing to do with biological sex inheritance and are said to be X-linked. **X-linked recessive disorders** are due to recessive genes carried on the X chromosomes. Males are more likely to have an X-linked recessive disorder than females because the Y chromosome is blank for this trait. Does a color-blind male give his male offspring a recessive-bearing X or a Y that is blank for the recessive allele? _____

The possible genotypes and phenotypes for an X-linked recessive disorder are as follows:

| **Females** | **Males** |
|---|---|
| $X^B X^B$ = normal vision | $X^B Y$ = normal vision |
| $X^B X^b$ = normal vision (carrier) | $X^b Y$ = color blindness |
| $X^b X^b$ = color blindness | |

An X-linked recessive disorder in a male is always inherited from their female parent. Most likely, the male's female parent is heterozygous and therefore does not show the disorder. The female parent is designated a carrier for the disorder. Figure 11.5 shows how females can become carriers. *Use the key to genotypes and phenotypes for X-linked recessive disorders to complete the parentheses of questions 2 and 3 with genotypes of the parents.*

**1. a.** What is the genotype for a color-blind female? _____ How many recessive alleles does a female need to inherit to be color blind? _____

**b.** What is the genotype for a color-blind male? _____ How many recessive alleles does a male need to inherit to be color blind? _____

**2. a.** With reference to Figure 11.5a, if the female parent has normal vision (_____) and the male parent is color blind (_____), what are the chances that a female offspring will be color blind? _____

**b.** A female offspring will be a carrier? _____ **c.** A male offspring will be color blind? _____

**3. a.** With reference to Figure 11.5b, if the female parent is a carrier (_____) and the male parent has normal vision (_____), what are the chances that a female offspring will be color blind? _____

**b.** A female offspring will be a carrier? _____ **c.** A male offspring will be color blind? _____

**Figure 11.5  Two common patterns of X-linked inheritance in humans.**
**a.** A color blind male parent has carrier female offspring. **b.** The male offspring of a carrier female parent have a 50% chance of being color blind.

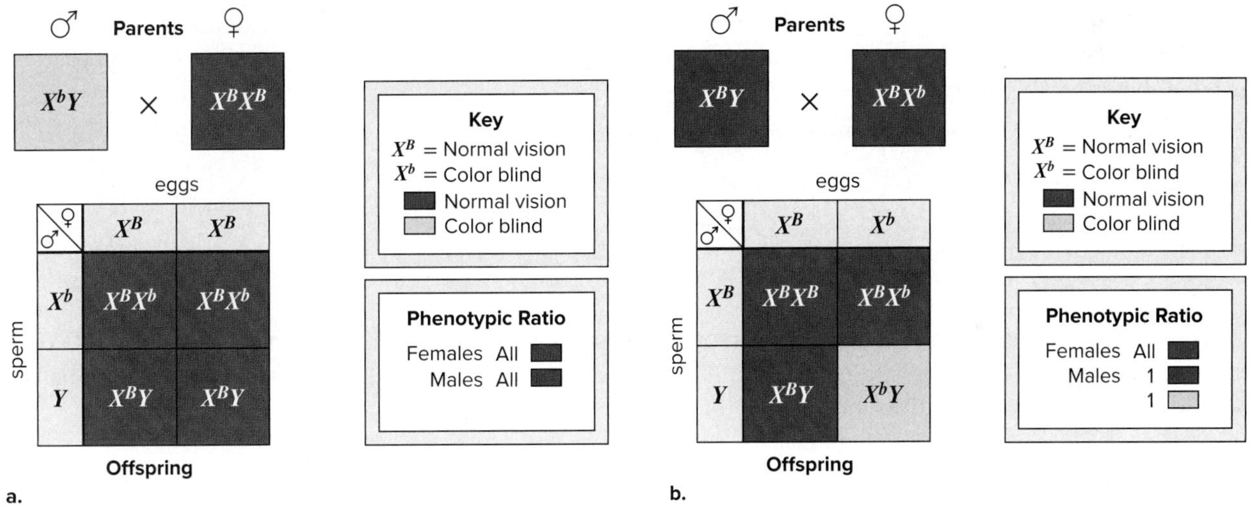

## X-Linked Genetics Problems

For **color blindness,** there are two possible X-linked alleles involved. One affects the green-sensitive cones, whereas the other affects the red-sensitive cones. About 6% of males in the United States are color blind due to a mutation involving green perception, and about 2% are color blind due to a mutation involving red perception.

**1.** A biological female with normal color vision (_____), whose male parent was color blind (_____), marries a biological male with normal color vision (_____). What genotypes could occur among their offspring? _____

What genotypes could occur if the woman's male parent was color blind? _____

**2.** Antonio's male parent is color blind (_____) but his female parent is not color blind (_____).

Is Antonio necessarily color blind? _____

Could he be color blind? _____

**Laboratory 11** Human Genetics   **121**

**Hemophilia** is a disorder in which a person's blood is unable to clot. Although individuals with hemophilia do bleed externally after an injury, they also suffer from internal bleeding, particularly around joints. Hemorrhages can be checked with transfusions of fresh blood (or plasma) or concentrates of the clotting protein. The most common type of hemophilia is hemophilia A, due to the absence or minimal presence of a particular clotting factor called factor VIII.

3. Make up a cross involving hemophilia that could be answered by a Punnett square, as in Figure 11.5a or b.

_____

_____

What is the answer to your genetics problem? _____

_____

## Multiple Alleles

When a trait is controlled by **multiple alleles,** the gene has more than two possible alleles. But each person has only two of the possible alleles. For example, ABO blood type is determined by multiple alleles: $I^A$, $I^B$, $i$. Red blood cells have surface molecules called antigens that indicate they belong to the person. The $I^A$ allele causes red blood cells to carry an A antigen, the $I^B$ allele causes red blood cells to carry a B antigen, and the $i$ allele causes the red blood cells to have neither of these antigens. $I^A$ and $I^B$ are dominant to $i$. Remembering that each person can have any two of the possible alleles, these are possible genotypes and phenotypes for blood types.

> ⚠ **Protective clothing** Wear protective laboratory clothing, latex gloves, and goggles. If the chemicals touch the skin, eyes, or mouth, wash immediately. If inhaled, seek fresh air.

| Genotypes | Antigens on Red Cells | Blood Types |
|---|---|---|
| $I^A I^A$, $I^A i$ | A | A |
| $I^B I^B$, $I^B i$ | B | B |
| $I^A I^B$ | A and B | AB |
| $ii$ | none | O |

Blood type also indicates whether the person is Rh positive or Rh negative. If the genotype is *DD* or *Dd,* the person is Rh positive, and if the genotype is *dd,* the person is Rh negative. It is customary to simply attach a + or − superscript to the ABO blood type, as in A⁻.

### Blood Typing Problems

1. A male with type A blood reproduces with a female who has type B blood. Their child has blood type O. Using $I^A$, $I^B$, and $i$, give the genotype of all persons involved. male parent _____ female parent _____ child _____

2. If a child has type AB blood and the male parent has type B blood, what could the genotype of the female parent be? _____ or _____

3. If both the female and male parents have type AB blood, they cannot be the parents of a child who has what blood type? _____

4. What blood types are possible among the children if the parents are $I^A i \times I^B i$? (*Hint:* Do a Punnett square using the possible gametes for each parent.)

In this Experimental Procedure, a mother, Wanda, is seeking the identity of the male parent of her child, Sophia. We will use blood typing to decide which of three men could possibly be the male parent.

1. Obtain three testing plates, each of which contains three depressions; vials of blood from possible male parents 1, 2, and 3, respectively; vials of anti-A serum, anti-B serum, and anti-Rh serum. (All of these are synthetic.)
2. Using a wax pencil, number the plates so you know which plate is for each possible male parent #1, #2, or #3. Look carefully at a plate and notice the wells are designated as A, B, or Rh.
3. Being sure to close the cap to each vial in turn, do the following using plate #1:

   Add a drop of male parent #1s blood to all three wells—close the cap.
   Add a drop of anti-A (blue) to the well designated A—close the cap.
   Add a drop of anti-B (yellow) to the well designated B—close the cap.
   Add a drop of anti-Rh (clear) to the well designated Rh—close the cap.

4. Stir the contents of each well with a mixing stick of the correct color. After a few minutes, examine the wells for agglutination, that is, granular appearances that indicate the blood type. (It may take a few minutes for the reaction to occur.) If a person had AB$^+$ blood, which wells would show agglutination? _____

   _____

5. Repeat steps 3 and 4 for plates #2 and #3.
6. Record the blood type results for each of the men in Table 11.2.

### Table 11.2   Blood Types of Involved Persons

| | Mother* | Child* | | Father? | |
| --- | --- | --- | --- | --- | --- |
| | Wanda | Sophia | #1 | #2 | #3 |
| Blood type | | | | | |

*Record the blood types of the Mother and Child that your instructor supplies.

### Conclusion

Which man is Sophia's male parent based on the results of your blood typing experiment? _____

_____

_____

_____

# 11.3   Genetic Counseling

## Pre-Lab

5. What tool can be used to determine chromosomal abnormalities? _____

6. What tool can be used to determine inheritance patterns of a genetic disorder? _____

Potential parents are becoming aware that many illnesses are caused by abnormal chromosomal inheritance or by gene mutations. Therefore, they are seeking genetic counseling, which is available in many major hospitals. The counselor helps the couple understand the mode of inheritance for a condition of concern as well as the potential risk of inheritance for a given disorder. This provides the parents with information so they can make an informed decision about how to proceed.

## Determining Chromosomal Inheritance

If a genetic counselor suspects that a condition is due to a chromosome anomaly, the counselor may suggest that the parents' chromosomes be examined. It is possible to view the chromosomes of an individual because cells can be microscopically examined and photographed just before cell division occurs. A computer is then used to arrange the chromosomes by pairs. The resulting pattern of chromosomes is called a **karyotype.**

A trisomy occurs when the individual has three chromosomes instead of two chromosomes at one karyotype location. **Trisomy 21** (Down syndrome) is the most common autosomal trisomy in humans. Survival to adulthood is common. Characteristic facial features include an eyelid fold, a flat face, and a large, fissured tongue. Some degree of intellectual impairment is common as is early-onset Alzheimer disease. Sterility due to sexual underdevelopment may also be present.

### Observation: Sex Chromosome Anomalies

A female with **Turner syndrome** (XO) has only one sex chromosome, an X chromosome; the O signifies the absence of the second sex chromosome. Because their ovaries never become functional, females with Turner syndrome do not undergo puberty or menstruation, and their breasts do not develop. Generally, females with Turner syndrome have a short build, folds of skin on the back of the neck, and difficulty recognizing various spatial patterns.

When an egg having two X chromosomes is fertilized by an X-bearing sperm, an individual with **poly-X syndrome** results. The body cells have three X chromosomes and therefore 47 chromosomes. Although they tend to have learning disabilities, females with poly-X syndrome have no apparent physical anomalies, and many are fertile and have children with a normal chromosome count.

When an egg having two X chromosomes is fertilized by a Y-bearing sperm, a male with **Klinefelter syndrome** results. This syndrome causes underdeveloped testes and can result in enlarged breasts, longer than average limbs, impaired muscular development, sparse body hair, and learning disabilities.

**Jacob syndrome** (XYY) occurs in males and is associated with a taller stature, persistent acne, and speech and reading problems.

*Label each karyotype in Figure 11.6 as one of the syndromes just discussed.*

**Figure 11.6   Sex chromosome anomalies.**

(1–4) CNRI/Science Source

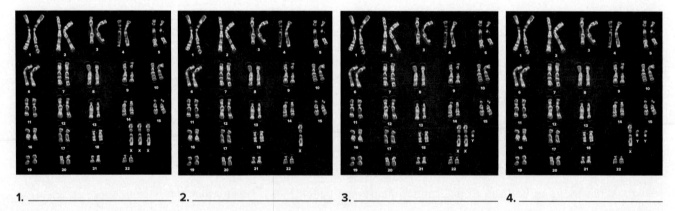

1. _____   2. _____   3. _____   4. _____

## Determining the Pedigree

A pedigree shows the inheritance of a genetic disorder within a family and can help determine the inheritance pattern and whether any particular individual has an allele for that disorder. Then a Punnett square can be done to determine the chances of a couple producing an affected child.

The symbols used to indicate unaffected and affected males and females, reproductive partners, and siblings in a pedigree are shown in Figure 11.7.

For example, suppose you wanted to determine the inheritance pattern for straight hairline and you knew which members of a generational family had the trait (Fig. 11.8a).

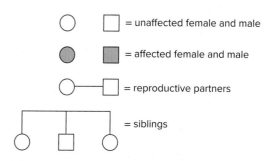

**Figure 11.7  Pedigree symbols.**

**Figure 11.8  Autosomal pedigrees.**
**a.** Child with recessive phenotype can have parents without the recessive phenotype. **b.** Child with the dominant phenotype has parent(s) with the dominant phenotype; heterozygous parents can also have a child without the dominant phenotype.

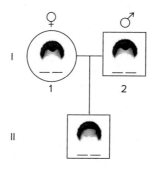

a. Straight hairline is recessive.

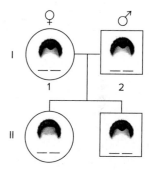

b. Widow's peak is dominant.

A pedigree allows you to determine that straight hairline is autosomal recessive because two parents without this phenotype have a child with the phenotype. This can happen only if the parents are heterozygous and straight hairline is recessive. Similarly, a pedigree allows you to determine that widow's peak is autosomal dominant (Fig. 11.8b): A child with this phenotype has at least one parent with the dominant phenotype, but again, heterozygous parents can produce a child without widow's peak. *Give each person in Figure 11.8*a *and 11.8*b *a genotype.*

Not shown is an X-linked recessive pedigree. An X-linked recessive phenotype occurs mainly in males, and it skips a generation because a female who inherits a recessive allele for the condition from their male parent may have a male offspring with the condition.

### *Pedigree Analysis*

For each of the following pedigrees, decide whether a trait is inherited as an autosomal dominant, autosomal recessive, or X-linked recessive. Then decide the genotype of particular individuals in the pedigree.

**1.** Study the following pedigree:

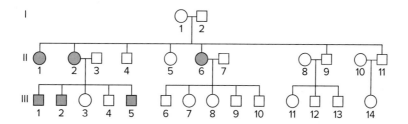

**a.** Notice that neither of the original parents is affected, but several children are affected. This could happen only if the trait were _____.

**b.** What is the genotype of the following individuals? Use *A* for the dominant allele and *a* for the recessive allele and explain your answer.

Generation I, individual 1: _____

Generation II, individual 1: _____

Generation III, individual 8: _____

**2.** Study the following pedigree:

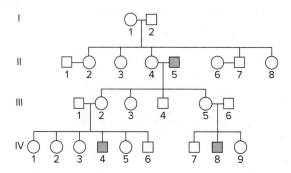

**a.** Notice that only males are affected. This could happen only if the trait were _____.

**b.** What is the genotype of the following individuals? Explain your answer.

Generation I, individual 1: _____

Generation II, individual 4: _____

Generation III, individual 1: _____

## Construction of a Pedigree

You are a genetic counselor who has been given the following information from which you will construct a pedigree.

1. Your data: <u>Henry</u> has a double row of eyelashes, which is a dominant trait. Both his <u>maternal grandfather</u> and his <u>mother</u> have double eyelashes. <u>Their spouses</u> have normal eyelashes. Henry is married to <u>Isabella</u> and their <u>first child, Polly,</u> has normal eyelashes. The couple wants to know the chances of any child having a double row of eyelashes.

2. What is your key for this trait?

   Key: _____ normal eyelashes       _____ double row of eyelashes

3. *Construct two identical pedigrees with symbols (circles and squares) only for the underlined persons in #1. The pedigrees start with the maternal grandfather and grandmother and end with Polly.*

   **Pedigree 1**                                                              **Pedigree 2**

4. Pedigree 1: Try out a pattern of autosomal dominant inheritance by assigning appropriate genotypes for an autosomal dominant pattern of inheritance to each person in this pedigree. Pedigree 2: Try out a pattern of X-linked dominant inheritance by assigning appropriate genotypes for this pattern of inheritance to each person in your pedigree. Which pattern is correct? _____

5. Use correct genotypes to show a cross between Henry and Isabella and from experience with crosses, state the expected phenotypic ratio among the offspring:

   **Cross**          Henry          Isabella          **Phenotypic ratio:**

   _____ × _____                    _____

6. What are the percentage chances that Henry and Isabella will have a child with double eyelashes? _____
   "Chance has no memory," and each child has the same chance for double eyelashes. Explain why.

   _____

   _____

_____  1. What term refers to an individual with two dominant alleles?

_____  2. What term refers to the alleles possessed by an individual?

_____  3. How many autosomes do humans have?

_____  4. An individual will have a widow's peak if the genotype is *WW* or *Ww*. Is widow's peak an autosomal dominant or autosomal recessive trait?

_____  5. What technique can be used to determine the genetic inheritance of offspring when the parents have known genotypes?

_____  6. Cystic fibrosis is an autosomal recessive disease. What word refers to an individual who has one recessive allele for cystic fibrosis?

_____  7. What is the genotype of individuals who will not develop Huntington disease (autosomal dominant)?

_____  8. From whom does a male child inherit color blindness—the male or female parent?

_____  9. What is the genotype of a carrier for hemophilia?

_____  10. How many dominant alleles does someone with AB blood type have?

_____  11. Explain how a child with AB blood type can have a parent with O blood?

_____  12. What is the presence of three chromosomes (instead of two) called, and what human chromosome is most commonly affected in this fashion?

_____  13. What is created when chromosomes are paired by size and shape?

_____  14. What technique is used to determine the inheritance pattern of specific traits within a family?

## Thought Questions

15. The blood types of parents and their two children are determined. The male parent has type A blood. The female parent has type O blood. One child has type A blood and the other has type O blood. What are the genotypes of all people involved? Explain your reply.

16. What are the chances that a male with color-blindness will have a male grandchild with color-blindness (the child of the male's normal-vision female offspring and a normal-vision male)? What are the genotypes of all people involved?

17. Recall what you learned about the fate of homologous chromosomes during meiosis I and the sister chromatids during meiosis II, and explain how trisomies like trisomy 21 or poly-X syndrome could occur.

# 12

# Evidences of Evolution

## Learning Outcomes

**12.1 Evidence from the Fossil Record**
- Use the geologic timescale to trace the evolution of life in broad outline.
- Explain how fossils help establish the sequence in the evolution of life.

**12.2 Evidence from Comparative Anatomy**
- Explain how comparative anatomy provides evidence that humans are related to other groups of vertebrates.
- Compare skeletons and skulls of humans, chimpanzees, and hominids.

**12.3 Molecular Evidence**
- Explain how molecular evidence aids the study of how humans are related to other groups of organisms on Earth.

## Introduction

### Pre-Lab

1. List three pieces of evidence that support evolution. _____

_____

**Evolution** is the process by which organisms are related by **common descent:** All organisms can trace their ancestry to the first cells. The process of evolution is amazingly simple. A group of organisms changes over time because the members of the group most suited to the natural environment have more offspring than the others in the group. So, for example, among bacteria, those which can withstand an antibiotic leave more offspring, and with time the entire group of bacteria becomes resistant to the antibiotic. Reproduction and therefore evolution have been going on since the first cells appeared on Earth. By studying (1) the fossil record, (2) anatomical and embryological comparative anatomy, and (3) molecular evidence, science is able to show that all organisms are related to one another.

Students digging for fossils

Mel Evans/AP Images

## 12.1  Evidence from the Fossil Record

**Pre-Lab**

2. How are dates on the geologic timescale determined? _____

3. How many mass extinctions have occurred in the history of the Earth? _____

The geologic timescale, which was developed by both geologists and paleontologists, depicts the history of life based on the fossil record (Table 12.1). A **fossil** is any evidence of the existence of an organism in ancient times as opposed to modern times. Paleontologists specialize in removing fossils from the Earth's crust. In this section, we will study the geologic timescale and then examine different fossils.

## Geologic Timescale

### Divisions of the Timescale

Notice that the timescale divides the history of Earth into eras, then periods, and then epochs. The four eras span the greatest amounts of time, and the epochs are the shortest time frames. Notice that only the periods of the Cenozoic era are divided into epochs, meaning that more attention is given to the evolution of primates and flowering plants than to the earlier evolving organisms. List the four eras in the geological timescale starting with Precambrian time: _____

1. Using the geologic timescale, you can trace the history of life by beginning with Precambrian time at the bottom of the timescale. The prokaryotes evolved before any other group. According to the timescale, how long ago did they arise? _____

2. The Precambrian time was very long, lasting from the time the Earth first formed until 542 MYA. The fossil record during the Precambrian time is meager, but the fossil record from the Cambrian period onward is rich (for reasons still being determined). This helps explain why the timescale usually does not show any periods until the Cambrian period of the Paleozoic era.

   During the _____ era and the _____ period, the first flowering plants appear. How many million years ago was this? _____

3. On the timescale, note the Carboniferous period. During this period great swamp forests covered the land. These are also called coal-forming forests because, with time, they became the coal we burn today. Were the plants in these forests flowering? _____

   Which group of animals were diversifying at this time? _____

4. You should associate the evolution of humans with the Cenozoic era. Among mammals, humans are primates. During what period and epoch did primates appear? _____

   Among primates, humans are hominids. During what period and epoch did hominids appear? _____

   _____ The scientific name for humans is *Homo sapiens*.

   What period and epoch is the age of *Homo sapiens*? _____

### Dating Within the Timescale

Dates on the timescale are obtained by measuring the amount of a radioactive isotope in the rocks surrounding the fossils. Typically, fossils are found in sedimentary rock that was slowly created as fine sediment settled and hardened at the bottoms of lakes, rivers, and oceans. How could you tell the relative age of two fossils found at different depths in sedimentary rock? _____

## Table 12.1 The Geological Timescale: Major Divisions of Geological Time and Some of the Major Evolutionary Events of Each Time Period

| Era | Period | Epoch | Million Years Ago (MYA) | Plant Life | Animal Life |
|---|---|---|---|---|---|
| **Cenozoic** | Quaternary | Holocene | current | Humans influence plant life. | Age of *Homo sapiens* |
| | | | **Significant Extinction Event Underway** | | |
| | Quaternary | Pleistocene | 0.01 | Herbaceous plants spread and diversify. | Presence of Ice-Age mammals. Modern humans appear. |
| | Neogene | Pliocene | 2.6 | Herbaceous angiosperms flourish. | First hominids appear. |
| | Neogene | Miocene | 5.3 | Grasslands spread as forests contract. | Apelike mammals and insects grazing mammals flourish; flourish. |
| | Neogene | Oligocene | 23.0 | Many modern families of flowering plants evolve; appearance of grasses. | Browsing mammals and monkeylike primates appear. |
| | Paleogene | Eocene | 33.9 | Subtropical forests with heavy rainfall thrive. | All modern orders of mammals are represented. |
| | Paleogene | Paleocene | 55.8 | Flowering plants continue to diversify. | Ancestral primates, herbivores, carnivores, and insectivores appear. |
| | | | **Mass Extinction: 50% of all species, including dinosaurs and most reptiles** | | |
| **Mesozoic** | Cretaceous | | 65.5 | Flowering plants spread; conifers persist. | Placental mammals appear; modern insect groups appear. |
| | Jurassic | | 145.5 | Flowering plants appear. | Dinosaurs flourish; birds appear. |
| | | | **Mass Extinction: 48% of all species, including corals and ferns** | | |
| | Triassic | | 199.6 | Forests of conifers and cycads dominate. | First mammals appear; first dinosaurs appear; corals and molluscs dominate seas. |
| | | | **Mass Extinction ("The Great Dying"): 83% of all species on land and sea** | | |
| **Paleozoic** | Permian | | 251.0 | Gymnosperms diversify. | Reptiles diversify; amphibians decline. |
| | Carboniferous | | 299.0 | Age of great coal-forming forests: ferns, club mosses, and horsetails flourish. | Amphibians diversify; first reptiles appear; first great radiation of insects. |
| | | | **Mass Extinction: Over 50% of coastal marine species, corals** | | |
| | Devonian | | 359.2 | First seed plants appear. Seedless vascular plants diversify. | First insects and first amphibians appear on land. |
| | Silurian | | 416.0 | Seedless vascular plants appear. | Jawed fishes diversify and dominate the seas. |
| | | | **Mass Extinction: Over 57% of marine species** | | |
| | Ordovician | | 443.7 | Nonvascular land plants appear. | Invertebrates spread and diversify; first jawless and then jawed fishes appear. |
| | Cambrian | | 488.3 | Marine algae flourish. | All invertebrate phyla present; first chordates appear. |
| | | | 630 | | First soft-bodied invertebrates evolve. |
| | | | 1,000 | Protists diversify. | |
| | | | 2,100 | First eukaryotic cells evolve. | |
| | | | 2,700 | O$_2$ accumulates in atmosphere. | |
| | | | 3,500 | First prokaryotic cells evolve. | |
| | | | 4,570 | Earth forms. | |

## Limitations of the Timescale

Because the timescale tells when various groups evolved and flourished, it might seem that evolution has been a series of events leading only from the first cells to humans. For example, prokaryotes (bacteria and archaea) never declined and are still the most abundant and successful organisms on Earth. Even today, they constitute up to 90% of the total weight of organisms.

Then, too, the timescale lists mass extinctions, but it doesn't tell when specific groups became extinct. **Extinction** is the total disappearance of a species or a higher group; **mass extinction** occurs when a large number of species disappear in a few million years or less. For lack of space, the geologic timescale can't depict in detail what happened to the members of every group mentioned. Figure 12.1 does show how mass extinction affected certain groups of animals. Which group of animals shown in Figure 12.1 suffered the most during the **P-T extinction** (Permian-Triassic extinction)? _____

The **K-T extinction** occurred between the Cretaceous and the Tertiary periods. Which group of animals shown in Figure 12.1 became extinct during the K-T extinction? _____

Figure 12.1 shows only periods and no eras. *Fill in the eras on the lines provided in the figure.*

**Figure 12.1  Mass extinctions.**
Five significant mass extinctions and their effects on the abundance of certain forms of marine and terrestrial life. The width of the horizontal bars indicates the varying number of each life-form considered.

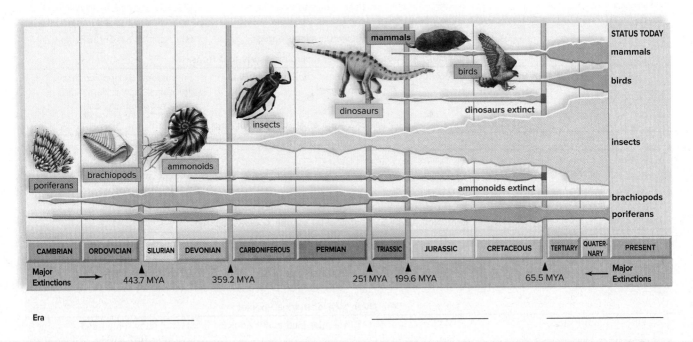

Era  _____          _____     _____

## Fossils

The fossil record uses anatomical data to show the evolutionary changes that have occurred as one group gives rise to another group.

### Invertebrate Fossils

The **invertebrates** are animals without a backbone (Fig. 12.2) while the **vertebrates** are animals with a backbone.

In order to familiarize yourself with possible invertebrate fossils, a few of the most common fossilized invertebrate groups are depicted in Figure 12.2.

## Figure 12.2 Invertebrate fossils.

(ammonite) Carl Pendle/Photographer's Choice/Getty Images; (echinoderms) DEA/G. Nimatallah/De Agostini/Getty Images; (mayfly) Quan Long/iStockphoto/Getty Images; (horseshoe crab): Koi88/Alamy Stock Photo; (snails) Ed Reschke/Stone/Getty Images; (trilobites): Danita Delimont/Gallo Images/Getty Images

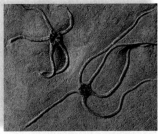

Fossilized brittle stars, echinoderms

Fossilized horseshoe crab, an arthropod

Fossilized snails, mollusks

Fossilized mayfly, an arthropod

Fossilized ammonite, a mollusc

Fossilized trilobites, arthropods

### Observation: Invertebrate Fossils

1. Obtain a box of selected fossils. If the fossils are embedded in rocks, examine the rock until you have found the fossil. Fossils are embedded in rocks because the sediment that originally surrounded them hardened over time. Most fossils consist of hard parts such as shells, bones, or teeth because these parts are not consumed or decomposed over time. One possible reason the Cambrian might be rich in fossils is

   that organisms now had _____ whereas before they did not.

2. The kit you are using, or your instructor, will identify which of the fossils are invertebrate animals. List the names of these fossils in Table 12.2, and give a description of the fossil. Use the Geologic Timescale (Table 12.1) to tell the era and period they were most abundant; list the fossils from the latest (*top*) to earliest (*bottom*).

### Table 12.2 Invertebrate Fossils

| Type of Fossil | Era, Period | Description |
|---|---|---|
|  |  |  |
|  |  |  |
|  |  |  |
|  |  |  |
|  |  |  |
|  |  |  |

## Observation: Vertebrate Fossils

The various groups of vertebrates are shown in Figure 12.3. Today it is generally agreed that birds are reptiles rather than being a separate group. That means that the major groups of vertebrates are (1) various types of fishes (jawless, jawed, cartilaginous, and bony fishes); (2) amphibians such as frogs and salamanders; (3) reptiles such as lizards, crocodiles, and birds; and (4) mammals. There are many types of mammals, from whales to mice to humans. The fossil record can rely on skeletal differences, such as limb structure, to tell if an animal is a mammal.

In order to familiarize yourself with possible vertebrate fossils, a few of the most common fossilized vertebrate groups are depicted in Figure 12.3. Which of the fossils available to you are vertebrates? _____

_____

Use Table 12.1 to associate each fossil with the particular era and period when this type of animal was most abundant. Fill in Table 12.3 according to sequence of the time frames from the latest (*top*) to earliest (*bottom*).

Fossilized bird, a reptile

Fossilized bony fish

Fossilized deerlike mammal

Fossilized frog, an amphibian

Fossilized turtle, a reptile

Fossilized duckbill dinosaur, a reptile

**Figure 12.3 Vertebrate fossils.**

(bird) WaterFrame/Alamy Stock Photo; (deerlike mammal): Gary Ombler/Getty Images; (duckbill dinosaur) Kevin Schafer/The Image Bank/Getty Images; (fish) Jason Edwards/National Geographic/Getty Images; (frog) G. CIGOLINI/DEA/De Agostini/Getty Images; (turtle) Kevin Schafer/The Image Bank/Getty Images

### Table 12.3 Vertebrate Fossils

| Type of Fossil | Era, Period | Description |
|---|---|---|
| | | |
| | | |
| | | |
| | | |
| | | |
| | | |
| | | |

## Observation: Plant Fossils

See Figure 12.4, which shows the evolution of plants, including the bryophytes, ferns and their allies, gymnosperms, and angiosperms. The fossil record for plants is not as good as that for invertebrates and vertebrates because plants have no hard parts that are easily fossilized. In order to familiarize you with possible plant fossils, a few of the most common fossilized plant groups are depicted in Figure 12.4.

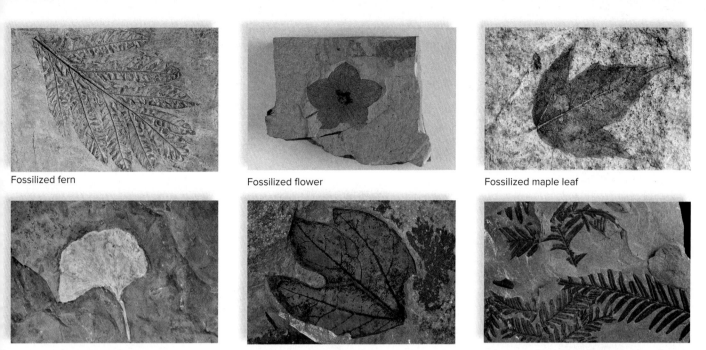

Fossilized fern

Fossilized flower

Fossilized maple leaf

Fossilized gingko leaf

Fossilized sassafras leaf

Fossilized early seed plant leaves

**Figure 12.4   Plant fossils.**

(fern) Andreeva Ekaterina Evgenevna/123RF; (flower) Barbara Strnadova/Science Source; (maple leaf) Biophoto Associates/Science Source; (gingko leaf) Corbin17/ Alamy Stock Photo; (sassafras leaf) Jonathan Blair/Getty Images; (seed plant leaves) Sinclair Stammers/Science Source

Plants that have no hard parts become fossils when their impressions are filled in by minerals. Use Table 12.1 to associate each fossilized plant in your kit with a particular era and period. Assume trees are flowering plants and associate them with the era and period when flowering plants were most abundant. Fill in Table 12.4 with a description of each fossil in sequence from the latest (*top*) to earliest (*bottom*).

| Table 12.4   Plant Fossils | | |
|---|---|---|
| Type of Fossil | Era, Period | Description |
| | | |
| | | |
| | | |
| | | |
| | | |
| | | |

## 12.2   Evidence from Comparative Anatomy

**Pre-Lab**

4. Which four structures will be examined throughout this section using comparative anatomy? _____

_____

Parts of organisms are said to be **homologous** if they exhibit similar basic structures and embryonic origins. Homologous structures indicate an evolutionary relationship and are used to classify organisms.

## Comparison of Adult Vertebrate Forelimbs

The limbs of vertebrates are homologous structures because of common descent (Fig. 12.5).

**Figure 12.5   Vertebrate forelimbs.**
Because all vertebrates evolved from a common ancestor, their forelimbs share homologous structures.

(bird) Holly Hildreth/McGraw Hill; (human) J.W. Ramsey/McGraw Hill; (lizard) Mauricio Handler/Photodisc/Getty Images; (bat) Jack Michanowski/age fotostock/Getty Images; (cat) Marc Henrie/Dorling Kindersley/Getty Images

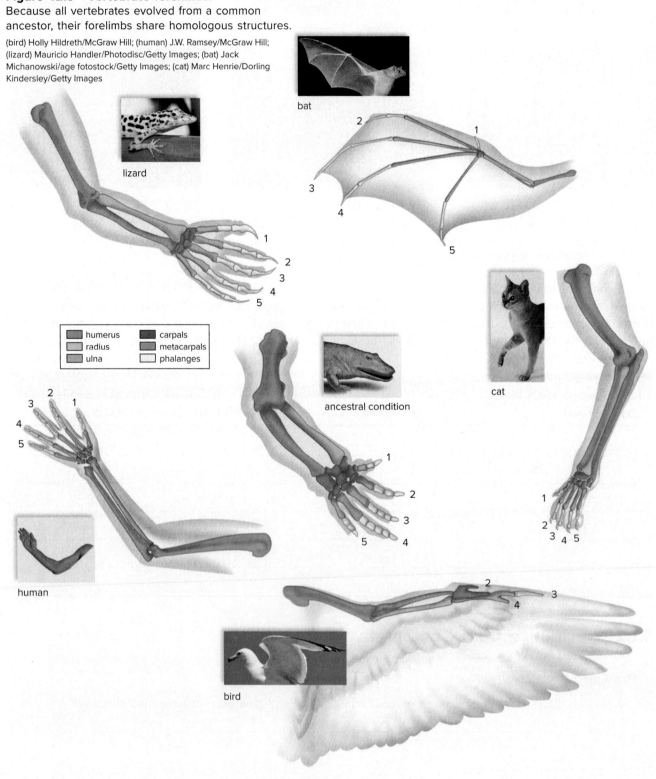

bat

lizard

humerus    carpals
radius     metacarpals
ulna       phalanges

ancestral condition

cat

human

bird

1. Find the forelimb bones of the ancestral vertebrate in Figure 12.5. The basic components are the humerus (h), ulna (u), radius (r), carpals (c), metacarpals (m), and phalanges (p) in the five digits.
2. *Label the corresponding forelimb bones of the lizard, the bird, the bat, the cat, and the human.*
3. Fill in Table 12.5 to indicate which bones in each specimen appear to most resemble the ancestral condition and which most differ from the ancestral condition.
4. Adaptation to a way of life can explain the modifications that have occurred. Relate the change in bone structure to mode of locomotion in two examples.

Example 1: _____

Example 2: _____

| Table 12.5 Comparison of Vertebrate Forelimbs | | |
|---|---|---|
| **Animal** | **Bones That Resemble Common Ancestor** | **Bones That Differ from Common Ancestor** |
| Lizard | | |
| Bird | | |
| Bat | | |
| Cat | | |
| Human | | |

## Comparison of Vertebrate Embryos

The anatomy shared by vertebrates extends to their embryological development. During early developmental stages, all animal embryos resemble each other closely, but as development proceeds the different types of vertebrates take on their own shape and form.

In this observation, you will see that as embryos all vertebrates have a postanal tail, a dorsal spinal cord, pharyngeal pouches, and various organs. In aquatic animals, pharyngeal pouches become functional gills (Fig. 12.6). In humans, the first pair of pouches becomes the cavity of the middle ear and auditory tube, the second pair becomes the tonsils, and the third and fourth pairs become the thymus and parathyroid glands.

**Figure 12.6 Vertebrate embryos.**

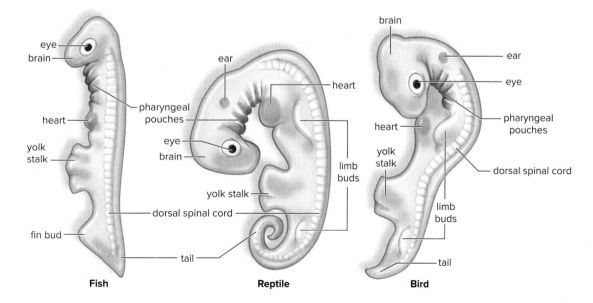

Fish          Reptile          Bird

## Observation: Vertebrate Embryos

1. Obtain prepared slides of vertebrate embryos at comparable stages of development. Observe each of the embryos using a stereomicroscope.
2. List five similarities of the embryos:

    a. _____

    b. _____

    c. _____

    d. _____

    e. _____

## Comparison of Chimpanzee and Human Skeletons

Chimpanzees and humans are closely related, as is apparent from an examination of their skeletons. However, they are adapted to different ways of life. Chimpanzees are adapted to living in trees and are herbivores—they eat mainly plants. Humans are adapted to walking on the ground and are omnivores—they eat both plants and meat.

### Observation: Chimpanzee and Human Skeletons

#### Posture

Chimpanzees are arboreal and climb in trees. While on the ground, they tend to knuckle-walk, with their hands bent. Humans are terrestrial and walk erect. In Table 12.6, compare:

1. **Head and torso:** Where are the head and trunk with relation to the hips and legs—thrust forward over

    the hips and legs or balanced over the hips and legs (Fig. 12.7)? _____

2. **Spine:** Which animal has a long and curved lumbar region, and which has a short and stiff lumbar region?

    How does this contribute to an erect posture in humans? _____

    _____

3. **Pelvis:** Chimpanzees sway when they walk because lifting one leg throws them off balance. Which animal has a narrow and long pelvis, and which has a broad and short pelvis? Record your observations in Table 12.6.

| Table 12.6 | Comparison of Chimpanzee and Human Postures | |
|---|---|---|
| **Skeletal Part** | **Chimpanzee** | **Human** |
| 1. Head and torso | | |
| 2. Spine | | |
| 3. Pelvis | | |
| 4. Femur | | |
| 5. Knee joint | | |
| 6. Foot:<br>    Opposable toe | | |
|     Arch | | |

4. **Femur:** In humans, the femur better supports the trunk. In which animal is the femur angled between articulations with the pelvic girdle and the knee? In which animal is the femur straight with no angle? Record your observations in Table 12.6.

5. **Knee joint:** In humans, the knee joint is modified to support the body's weight. In which animal is the femur larger at the bottom and the tibia larger at the top? Record your observations in Table 12.6.

6. **Foot:** In humans, the foot is adapted for walking long distances and running with less chance of injury.

   In which animal is the big toe opposable? _____ How does an opposable toe assist chimpanzees?

   _____

   Which foot has an arch? _____ How does an arch assist humans? _____
   Record your observations in Table 12.6.

7. How does the difference in the position of the foramen magnum, a large opening in the base of the skull for the spinal cord, correlate with the posture and stance of the two organisms?

   _____

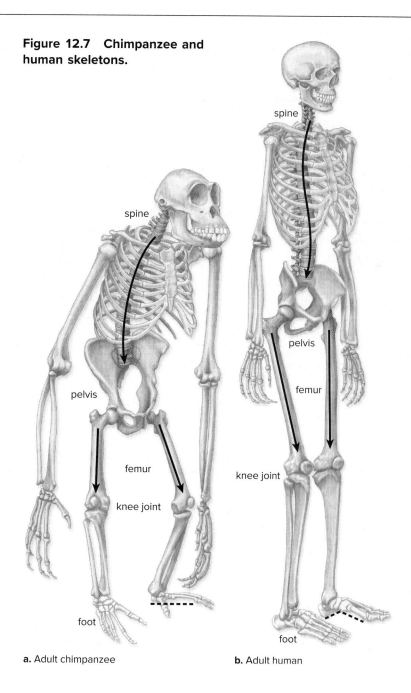

**Figure 12.7 Chimpanzee and human skeletons.**

**a.** Adult chimpanzee      **b.** Adult human

## Skull Features

Humans are omnivorous. A diet rich in meat does not require strong grinding teeth or well-developed facial muscles. Chimpanzees are herbivores, and a vegetarian diet requires strong teeth and strong facial muscles that attach to bony projections. Compare the skulls of the chimpanzee and the human in Figure 12.8 and answer the following questions:

1. **Supraorbital ridge:** For which skull is the supraorbital ridge (the region of frontal bone just above the eye socket) thicker? Record your observations in Table 12.7.
2. **Sagittal crest:** Which skull has a sagittal crest, a projection for muscle attachments that runs along the top of the skull? Record your observation in Table 12.7.
3. **Frontal bone:** Compare the slope of the frontal bones of the chimpanzee and human skulls. How are they different? Record your observations in Table 12.7.
4. **Teeth:** Examine the teeth of the adult chimpanzee and adult human skulls. Are the incisors (two front teeth) vertical or angled? Do the canines overlap the other teeth? Are the molars larger or moderate in size? Record your observations in Table 12.7.
5. **Chin:** What is the position of the mouth and chin in relation to the profile for each skull? Record your observations in Table 12.7.

**Figure 12.8   Chimpanzee and human skulls.**

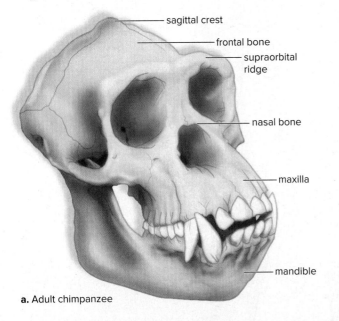

**a.** Adult chimpanzee

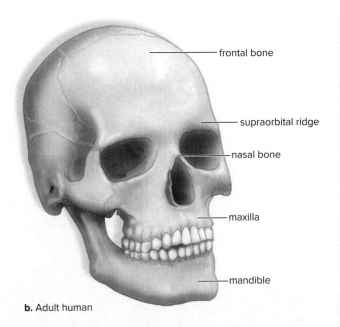

**b.** Adult human

### Table 12.7   Skull Features of Chimpanzees and Humans

| Feature | Chimpanzee | Human |
|---|---|---|
| 1.  Supraorbital ridge | | |
| 2.  Sagittal crest | | |
| 3.  Slope of frontal bone | | |
| 4.  Teeth | | |
| 5.  Chin | | |

## Conclusions: Chimpanzee and Human Skeletons

- Do your observations show that the skeletal differences between chimpanzees and humans can be related to posture? _____ Explain your answer. _____

- Do your observations show that diet can be related to the skull features of chimpanzees and humans? __ Explain your answer. _____

## Comparison of Hominid Skulls

The designation *hominid* includes humans and primates that are humanlike. Paleontologists have uncovered several fossils dated from 7.5 MYA to 30,000 years BP (before present), when humans called Cro-Magnons arose that are virtually identical to modern humans (*Homo sapiens*). (Fig. 12.9).

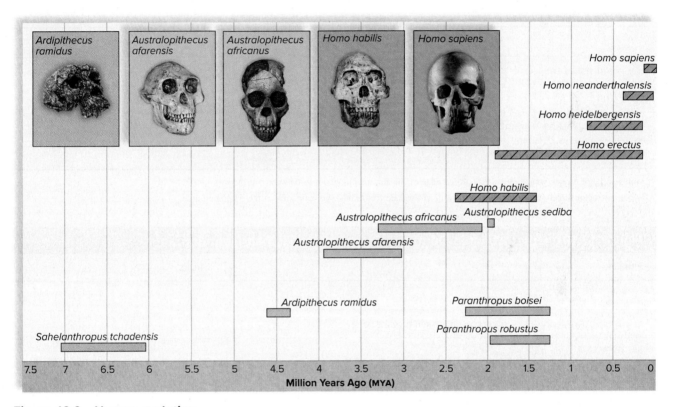

**Figure 12.9  Human evolution.**

*(A. ramidus)* Puwadol Jaturawutthichai/123RF; *(A. afarensis)* Scott Camazine/Alamy Stock Photo; *(A. africanus)* Philippe Plailly/Science Source; *(H. habilis)* Kike Calvo/Visual & Written/SuperStock; *(H. sapiens)* Kenneth Garrett/Getty Images

## Observation of Hominid Skulls

Several of the skulls noted in Figure 12.9 may be on display. Use Tables 12.8, 12.9, and 12.10 to record data pertaining to the cranium (or braincase), the face, and the teeth. Compare the early skulls on display with a modern human skull. For example, is the forehead like or more flat than the human skull? List at least three skulls you examined for each feature.

## Table 12.8  Other Hominid Craniums Compared to Human Cranium

| Feature | Skulls | | |
|---|---|---|---|
| | 1. | 2. | 3. |
| a. Frontal bone (like or more flat?) | | | |
| b. Supraorbital ridge (divided or continuous?) | | | |
| c. Sagittal crest (present?) | | | |
| d. Mastoid process (flat or projecting?) | | | |

## Table 12.9  Other Hominid Faces Compared to Human Face

| Feature | Skulls | | |
|---|---|---|---|
| | 1. | 2. | 3. |
| a. Nasal bones (raised or flat?) | | | |
| b. Nasal opening (larger?) | | | |
| c. Chin (projecting forward?)* | | | |
| d. Width of face (wider?)** | | | |

*If your instructor directs you to measure from the edge of the foramen magnum to between the incisors.
**If your instructor directs you to measure the width of the face from mid-zygomatic arch to the other mid-arch.

## Figure 12.10  Other Hominid Dentition Compared to Human Dentition

| Feature | Skulls | | |
|---|---|---|---|
| | 1. | 2. | 3. |
| a. Teeth rows (parallel or diverging from each other?) | | | |
| b. Incisors (vertical or angled?) | | | |
| c. Canine teeth (overlapping other teeth?) | | | |
| d. Molars (more massive?) | | | |

### Conclusions

- Do your data appear to be consistent with the evolutionary sequence of the hominids in Figure 12.9? Explain your answer. _____
- Report here any data you collected that would indicate a particular hominid was closer in time to humans than indicated in Figure 12.9. _____

  _____

- Report here any data you collected that would indicate a particular hominid was more distant in time from humans than indicated in Figure 12.9. _____

  _____

## 12.3 Molecular Evidence

**Pre-Lab**

5. What is cytochrome *c*? _____

_____

6. Compared to humans, how different is cytochrome *c* in monkeys and fish? _____

Molecular data substantiate the comparative and developmental data that biologists have accumulated over the years. *Hox* genes are developmental genes. Their activities vary among different groups of animals. In vertebrates these genes account, for why they have a dorsally placed nerve cord while it is ventrally placed in invertebrates. Sequencing DNA data show which organisms are closely related. Molecular data among primates are of extreme interest because they can help determine which of the primates humans are most closely related to. Chromosomal and genetic data allow us to conclude that we are more closely related to chimpanzees than to other types of apes.

In this section, we note that scientists can compare the amino acid sequence in proteins to determine the degree to which any two groups of organisms are related. The sequence of amino acids in cytochrome *c*, a carrier of electrons in the electron transport chain found in mitochondria, has been determined per a variety of organisms. On the basis of the number of amino acid *differences* reported in Figure 12.10, it is concluded that the evolutionary relationship between humans and these organisms decreases in the order stated: monkeys, pigs, ducks, turtles, fishes, moths, and yeast. This conclusion agrees with the sequence of dates these organisms are found in the fossil record.

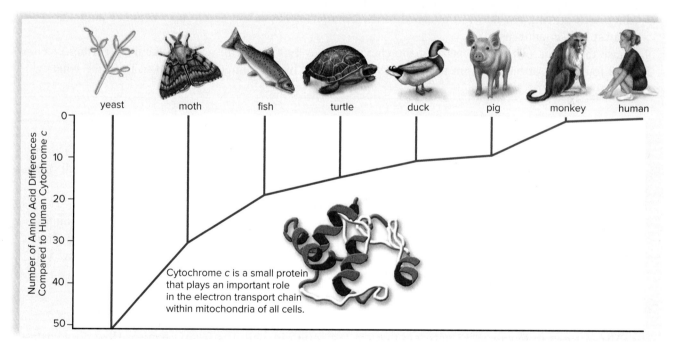

**Figure 12.10  Amino acid differences in cytochrome *c*.**
The few differences found in cytochrome *c* between monkeys and humans show that of these organisms, humans are most closely related to monkeys.

## Protein Similarities

The immune system makes **antibodies** (proteins) that react with foreign proteins, called **antigens.** Antigen-antibody reactions are specific. An antibody will react only with a particular antigen. In today's procedure, it is assumed rabbit antibodies to human antigens are in rabbit serum (Fig. 12.11). When these antibodies are allowed to react against the antigens of other animals, the stronger the antibody-antigen reaction (determined by the amount of precipitate), the more closely related the animal is to humans.

**Figure 12.11  Antigen-antibody reaction.**
When antibodies react to antigens, a precipitate appears.

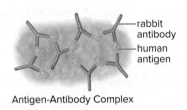

Antigen-Antibody Complex

### Experimental Procedure: Protein Similarities

1. Obtain a chemplate (a clear glass tray with wells), one bottle of synthetic human blood serum, one bottle of synthetic rabbit blood serum, and five bottles (I–V) of blood serum test solution.
2. Put two drops of synthetic rabbit blood serum in each of the six wells in the chemplate. *Label the wells 1–6.* See yellow circles in Figure 12.12.
3. Add two drops of synthetic human blood serum to each well. See red circles in Figure 12.12. Stir with the plastic stirring rod that was attached to the chemplate. The rabbit serum now contains antibodies against human antigens.
4. Rinse the stirrer. (The large cavity of the chemplate may be filled with water to facilitate rinsing.)
5. Add four drops of blood serum test solution III (contains human antigens) to well 6. Describe what you see. _____

   _____

   This well will serve as the basis by which to compare all the other samples of test blood serum.
6. Now add four drops of blood serum test solution I to well 1. Stir and observe. Rinse the stirrer. Do the same for each of the remaining blood serum test solutions (II–V)—adding II to well 2, III to well 3, and so on. Be sure to rinse the stirrer after each use.
7. At the end of 10 and 20 minutes, record the amount of precipitate in each of the six wells in Figure 12.12. Well 6 is recorded as having ++++ amount of precipitate after both 10 and 20 minutes. Compare the other wells with this well (+ = trace amount; 0 = none). Holding the plate slightly above your head at arm's length and looking at the underside toward an overhead light source will allow you to more clearly determine the amount of precipitate.

### Conclusions: Protein Similarities

- The last row in Figure 12.12 tells you that the test serum in well 3 is from a human. How do your test results confirm this? _____

- Judging by the amount of precipitate, complete the last row in Figure 12.12 by indicating which serum you believe came from which animal. _____

   _____

## Figure 12.12 Protein similarity.

The greater the amount of precipitate, the more closely related an animal is to humans.

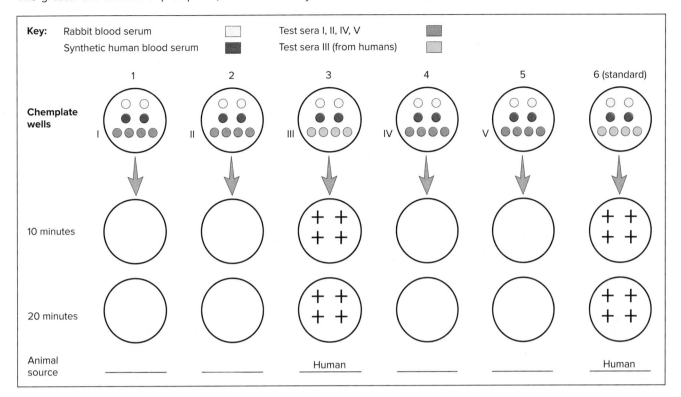

## Laboratory Review 12

_____ 1. What process has taken place when a group of bacteria becomes resistant to a specific antibiotic?

_____ 2. In addition to comparative anatomy and molecular evidence, what shows that all organisms are related to one another?

_____ 3. What era of geologic time are we in currently?

_____ 4. During what period of geologic time did the coal-forming forests evolve and flourish?

_____ 5. During what era of geologic time did the first cells evolve?

_____ 6. How many significant mass extinctions have occurred over time?

_____ 7. What body parts of an organism are most likely to become fossilized?

_____ 8. For which group of organisms—invertebrates, vertebrates, or plants—is the fossil record the least substantial?

_____ 9. What term describes parts of organisms that exhibit similar basic structures and embryonic origins?

_____ 10. In addition to a postanal tail and a dorsal nerve cord, what structures are observed in vertebrate embryos?

_____ 11. Do the skulls of chimpanzees or humans have a sagittal crest?

_____ 12. What hominid genus flourished 2.8 MYA?

**Laboratory 12** Evidences of Evolution

_____ **13.** The amino acid sequence of what electron carrier protein has been determined for a variety of organisms?

_____ **14.** Are two species closely or distantly related if the antibodies to the antigens of one species form very little precipitate when mixed with the antigens of the other species?

## Thought Questions

**15.** Analogous structures perform similar functions but do not arise from common ancestry. What explains the existence of analogous structures?

**16.** Why are there differences in the axial skeletons and skulls of chimpanzees and humans?

**17.** People with sickle-cell trait (heterozygotes $Hb^A Hb^S$) are immune to malaria, but people who are $Hb^S Hb^S$ will have sickle-cell disease.

**a.** If two people with sickle-cell trait have offspring, what are the chances their offspring will have sickle-cell disease?

**b.** Explain where, in the world, the $Hb^S$ allele is mostly likely to be prevalent and why the allele persists in spite of the danger associated with having two $Hb^S$ alleles.

# 13

# Microbiology

## Introduction

### Pre-Lab

**1.** What are the differences between prokaryotes and eukaryotes? _____

_____

The history of life began with the origin and evolution of the prokaryotic cell. Although the prokaryotic cell contains genetic material, it is not located in a nucleus, and the cell also lacks any other type of membranous organelle. At one time prokaryotes were believed to be a unified group, but based on molecular data, they are now divided into two major groups—**domain Bacteria** and **domain Archaea.** Eukaryotes in **domain Eukarya** have a membrane-bound nucleus and membranous organelles. Eukaryotes are more closely related to archaea than bacteria. Prokaryotes resemble each other structurally but are metabolically diverse. Eukaryotes, on the other hand, are structurally diverse and exist as protists, fungi, plants, and animals (Fig. 13.1).

**Figure 13.1 The world of organisms.**

Prokaryotes are represented in this illustration by the bacteria. The protists, fungi, plants, and animals are all eukaryotes.

(bacteria) Source: Janice Haney Carr/CDC; (*Paramecium*) Michael Abbey/Science Source; (morel) Robert Marien/Corbis/Getty Images; (sunflower) Apollon/Iconotec/ Glow Images; (woodpecker) MDParr/Flickr Open/Moment Open/Getty Images

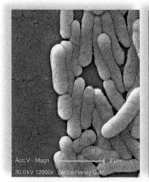    

*Salmonella,* a bacterium  *Paramecium,* a protist  Morel, a fungus  Sunflower, a plant  Great spotted woodpecker, an animal

# 13.1 Bacteria

### Pre-Lab

2. *Label Figure 13.2 with the help of your textbook.*

3. List three common shapes of bacterial cells. _____

In this laboratory, you will first relate the general structure of a bacterium to its ability to cause disease. The specific shape, growth habit, and staining characteristics of bacteria are often used to identify them. Therefore, you will observe a variety of bacteria using the microscope. Aside from their medical importance, bacteria are essential in ecosystems because, along with fungi, they are decomposers that break down dead organic remains and thereby return inorganic nutrients to plants.

## Pathogenic Bacteria

Pathogenic bacteria are infectious agents that cause disease. Infectious bacteria are able to invade and multiply within a host. Some also produce a toxin. Antibiotic therapy is often an effective treatment against a bacterial infection.

We will explore how it is possible to relate the structure of a bacterium to its ability to avoid detection and destruction by the host's immune system. We will also consider what characteristics allow bacteria to be resistant to antibiotics and to pass on the necessary genes to other bacteria.

### Observation: Structure of a Bacterium

1. Study the generalized structure of a bacterium in Figure 13.2 and, if available, examine a model of a bacterium.
2. Identify the following:

   **capsule:** a gel-like coating outside the cell wall. Capsules often allow bacteria to stick to surfaces such as teeth. They also prevent phagocytic white blood cells from engulfing and destroying them.

   **fimbriae:** hairlike bristles that allow adhesion to surfaces. This can be how a bacterium clings to and gains access to the body prior to an infection.

**conjugation pilus:** an elongated, hollow appendage used to transfer DNA to other cells. Genes that allow bacteria to be resistant to antibiotics can be passed in this manner.

**flagellum:** a rotating filament that propels the cell.

**cell wall:** a structure that provides support and shapes the cell. Antibiotics that prevent the formation of a cell wall are most effective against Gram-positive bacteria because they have more peptidoglycan in the cell wall than Gram-negative bacteria.

**plasma membrane:** a sheet that surrounds the cytoplasm and regulates the entrance and exit of molecules. Resistance to antibiotics can be due to plasma membrane alterations that prevent the drug from binding to or crossing the membrane as well as alterations that enable the plasma membrane to eliminate the drug from the bacteria.

**ribosomes:** site of protein synthesis. Some bacteria possess antibiotic-inactivating enzymes that make them resistant to antibiotics.

**nucleoid:** the location of the bacterial chromosome.

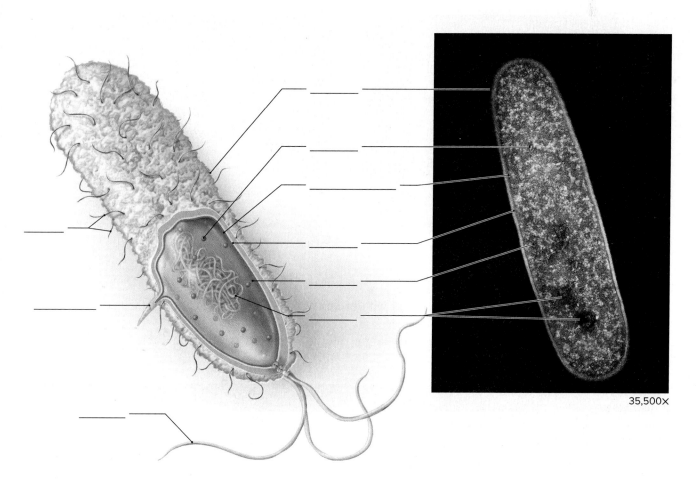

35,500×

**Figure 13.2 Generalized structure of a bacterium.**
*Complete the labels with the help of your textbook.*

Dr. Kari Lounatmaa/Science Source

Also, some bacteria contain plasmids, small rings of DNA that replicate independently of the chromosomes and can be passed to other bacteria. Genes that allow bacteria to be resistant to antibiotics are often located in a plasmid.

## Conclusions: Structure of a Bacterium

- Which portions of a bacterial cell aid the ability of a bacterium to cause infections? _____

  _____

- Which portions of a bacterial cell aid the ability of a bacterium to be resistant to antibiotics? _____

  _____

## Colony Morphology

On a nutrient material called agar, bacteria grow as colonies. A **colony** contains cells descended from one original cell. Sometimes, it is possible to identify the type of bacterium by the appearance of the colony (Fig. 13.3).

**Figure 13.3  Colony morphology.**

Colonies of bacteria on agar plates. Note the variation in features like size and color that help researchers identify the type of bacterium.

(a–b) Kathy Park Talaro

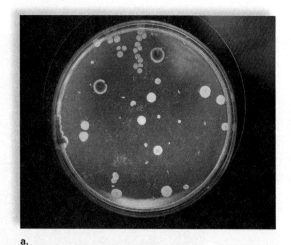
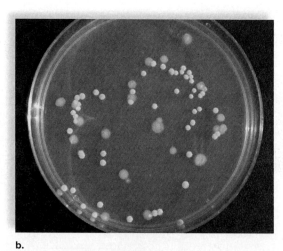

a.                                b.

### *Observation: Colony Morphology*

1. View agar plates that have been inoculated with bacteria and then incubated. Notice the "colonies" of bacteria growing on the plates.
2. Compare the colonies' color, surface, and margin, and note your observations in Table 13.1. It is not necessary to identify the type of bacteria.

| Table 13.1  Agar Plates | |
|---|---|
| **Plate Number** | **Description of Colonies** |
| | |
| | |
| | |

3. If available, obtain a sterile agar plate, and inoculate the plate with your thumbprint, or use a sterile swab and inoculate the plate with material from around your teeth or inside your nose. Put your name on the plate, and place it where directed by your instructor. Remember to view the plate next laboratory period. Describe your plate:

_____

_____

4. If available, obtain a sterile agar plate, and expose it briefly (at most for 10 minutes) anywhere you choose, such as in the library, your room, or your car. No matter where the plate is exposed, it subsequently will show bacterial colonies. Describe your plate:

_____

_____

## Shape of Bacterial Cell

Most bacteria are found in three basic shapes: **spirillum** (spiral or helical), **bacillus** (rod), and **coccus** (round or spherical) (Fig. 13.4). Bacilli may form long filaments, and cocci may form clusters or chains. Some bacteria form endospores. An **endospore** contains a copy of the genetic material encased by heavy protective spore coats. Spores survive unfavorable conditions and germinate to form vegetative cells when conditions improve.

### Observation: Shape of Bacterial Cell

1. View the microscope slides of bacteria on display. What magnification is required to view bacteria?

_____

2. Using Figure 13.4 as a guide, identify the three different shapes of bacteria.

3. Do any of the slides on display show bacterial cells with endospores? _____

**Figure 13.4  Diversity of bacteria.**
**a.** Spirillum, a spiral-shaped bacterium. **b.** Bacilli, rod-shaped bacteria. **c.** Cocci, round bacteria.

(a) Ed Reschke/Stone/Getty Images; (b) Eye of Science/Science Source; (c) Scimat/Science Source

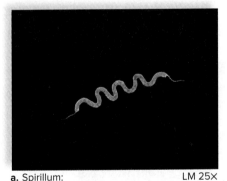
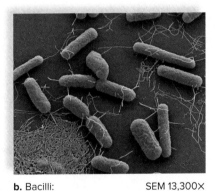
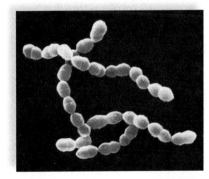

**a.** Spirillum:            LM 25×
*Spirillum volutans*

**b.** Bacilli:            SEM 13,300×
*Escherichia coli*

**c.** Cocci:            SEM 6,250×
*Streptococcus thermophilus*

## Cyanobacteria

**Cyanobacteria** were formerly called blue-green algae because their general growth habit and appearance through a compound light microscope are similar to green algae. Electron microscopic study of cyanobacteria, however, revealed that they are structurally similar to other bacteria, particularly other photosynthetic bacteria. Although cyanobacteria do not have chloroplasts, they do have thylakoid membranes, where photosynthesis occurs.

**Gloeocapsa**

1. Prepare a wet mount of a *Gloeocapsa* culture, if available, or examine a prepared slide, using high power (45×) or oil immersion (if available). The single cells adhere together because each is surrounded by a sticky, gelatinous sheath (Fig. 13.5).

2. What is the estimated size of a single cell? _____

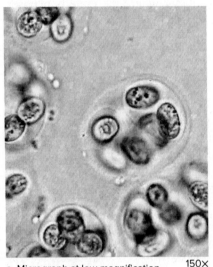

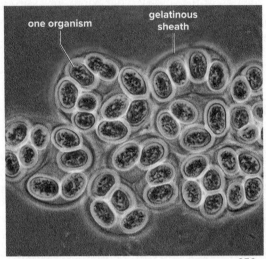

one organism

gelatinous sheath

**a.** Micrograph at low magnification    150×        **b.** Micrograph at high magnification    250×

**Figure 13.5    *Gloeocapsa.***

(a) Steven R. Spilatro; (b) Michael Abbey/Science Source

**Oscillatoria**

1. Prepare a wet mount of an *Oscillatoria* culture, if available, or examine a prepared slide, using high power (45×) or oil immersion (if available). This is a filamentous cyanobacterium with individual cells that resemble a stack of pennies (Fig. 13.6).

2. *Oscillatoria* takes its name from the characteristic oscillations that you may be able to see if your sample is alive. If you have a living culture, are oscillations visible? _____

**Figure 13.6    *Oscillatoria.***

M.I. Walker/Science Source

250×

### Anabaena

1. Prepare a wet mount of an *Anabaena* culture, if available, or examine a prepared slide, using high power (45×) or oil immersion (if available). This is also a filamentous cyanobacterium, although its individual cells are barrel-shaped (Fig. 13.7).
2. Note the thin nature of this strand. If you have a living culture, what is its color? _____

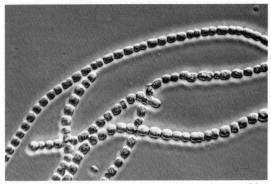

160×

**Figure 13.7  *Anabaena*.**
Robert Knauft/Biology Pics/Science Source

## 13.2  Protists

### Pre-Lab

4. What are the two ways groups of protists acquire food? _____

5. Why are algae most like plants? _____

6. List the three ways protozoans can move. _____

Protists were the first eukaryotes to evolve. Their diversity and complexity make it difficult to categorize them, but on the basis of molecular data, they are now placed in supergroups. Animals, fungi, and land plants are also included in supergroups (Fig. 13.8). However, the diagram in Figure 13.8 may be helpful.

Traditionally, protists were organized according to their mode of nutrition. This section will continue to do so while also noting the supergroup to which each type of protist belongs. The algae will run photosynthesis, similar to what we see in the land plants. Some green algae share a common ancestor with land plants; both groups possess chlorophylls *a* and *b,* store reserve food as starch, and have cell walls containing cellulose. Heterotrophic protists, commonly called protozoans, must acquire food from an external source by ingesting it as animals do, or by absorbing it as fungi do. Some of the protozoans may be ancestral to animals; the relationship of any of the protists to fungi is currently unclear.

### Photosynthetic Protists

*Algae* is a term that has been used for aquatic organisms that photosynthesize in the same manner as land plants. All photosynthetic protists contain green chlorophyll, but they also may contain other pigments that mask the chlorophyll color, and this accounts for their common names—the green algae, red algae, brown algae, and golden-brown algae.

#### *Observation: Green Algae*

If available, view a video showing the many forms of green algae. Notice that green algae can be single cells, filaments, colonies, or multicellular sheets. You will examine a filamentous form (*Spirogyra*) and a colonial form (*Volvox*). A **colony** is a loose association of cells.

1. Make a wet mount of live *Spirogyra* or observe a prepared slide. *Spirogyra* is a filamentous alga, lives in fresh water, and often is seen as a green scum on the surface of ponds and lakes. The most prominent feature of the cells is the spiral, ribbonlike chloroplast (Fig. 13.9). How do you think *Spirogyra* got its

   name? _____

   _____

**Figure 13.8   Protist diversity.**

Supergroups into which protists are classified along with animals, fungi, and land plants.

(*Volvox*): Stephen Durr; (*diatoms*): Lester V. Bergman/Corbis Documentary/Getty Images; (*radiolarians*): Eye of Science/Science Source; (*Giardia*): Dr. Stan Erlandsen and Dr. Dennis Feely/CDC; (*amoeba*): Micro photo/iStockphoto/Getty Images; (*choanoflagellates*): David J. Patterson

| Supergroup | | Members | Distinguishing Features |
|---|---|---|---|
| Archaeplastida | 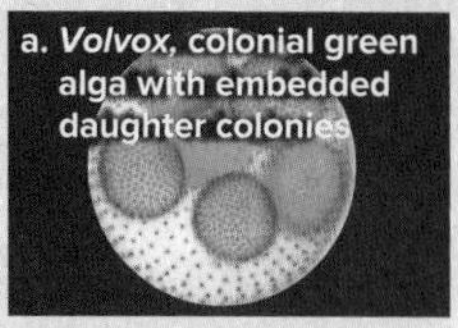 a. *Volvox*, colonial green alga with embedded daughter colonies | Green algae, red algae, charophytes<br><br>**Other Members**<br><br>Plants | Plastids; unicellular, colonial, and multicellular |
| Chromalveolata |  b. Assorted fossilized diatoms | Stramenopiles: brown algae, diatoms, golden algae, water molds<br><br>Alveolates: ciliates, apicomplexans, dinoflagellates | Most with plastids; unicellular and multicellular<br><br>Alveoli support plasma membrane; unicellular |
| Rhizaria | 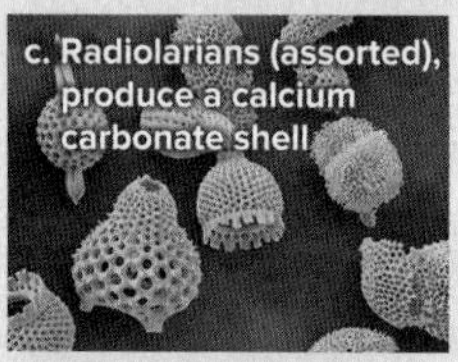 c. Radiolarians (assorted), produce a calcium carbonate shell | Cercozoans, foraminiferans, radiolarians | Thin pseudopods; some with tests; unicellular |
| Excavata | 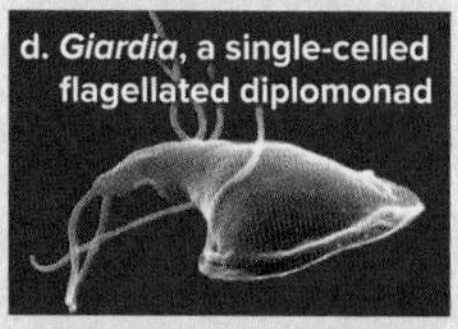 d. *Giardia*, a single-celled flagellated diplomonad | Euglenids, kinetoplastids, parabasalids, diplomonads | Feeding groove; unique flagella; unicellular |
| Amoebozoa | 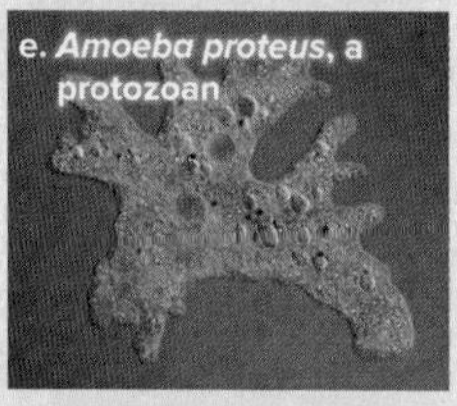 e. *Amoeba proteus*, a protozoan | Amoeboids, plasmodial and cellular slime molds | Pseudopods; unicellular |
| Opisthokonta | 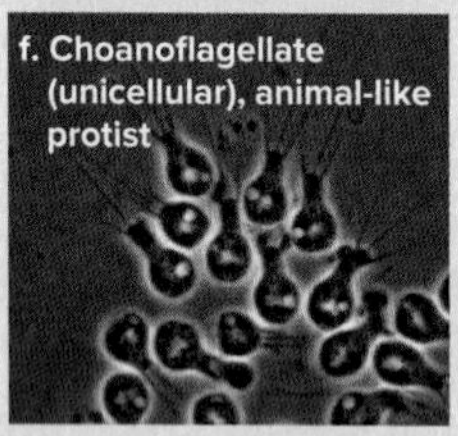 f. Choanoflagellate (unicellular), animal-like protist | Choanoflagellates, *nucleariids*<br><br>**Other Members**<br><br>Animals, fungi | Some with flagella; unicellular and colonial |

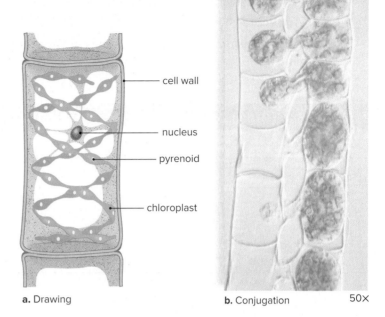

**Figure 13.9  *Spirogyra.***
**a.** *Spirogyra* is a filamentous green alga, in which each cell has a ribbonlike chloroplast. **b.** During conjugation, the cell contents of one filament enter the cells of another filament. Zygote formation follows.

(a) M.I. Walker/Science Source

cell wall

nucleus

pyrenoid

chloroplast

**a.** Drawing

**b.** Conjugation                    50×

*Spirogyra*'s chloroplast contains a number of circular bodies, the **pyrenoids,** centers of starch polymerization. The nucleus is in the center of the cell, anchored by cytoplasmic strands.

2.  Your slide may show **conjugation,** a sexual means of reproduction illustrated in Figure 13.9*b*. If it does not, obtain a slide that does show this process. Conjugation tubes form between two adjacent filaments, and the contents of one set of cells enter the other set. As the nuclei fuse, a zygote is formed. The zygote overwinters, and in the spring, meiosis and, subsequently, germination occur. The resulting adult protist is therefore haploid.

### Volvox

1.  Using a depression slide, prepare a wet mount of live *Volvox,* or study a prepared slide. *Volvox* is a green algal colony. It is motile (capable of locomotion) because the thousands of cells that make up the colony have flagella. These cells are connected by delicate cytoplasmic extensions (Fig. 13.10).

    *Volvox* is capable of both asexual and sexual reproduction. Certain cells of the adult colony can divide to produce **daughter colonies** (Fig. 13.10) that reside for a time within the parental colony. A daughter colony escapes the parental colony by releasing an enzyme that dissolves away a portion of the matrix of the parental colony. During sexual reproduction, some colonies of *Volvox* have cells that

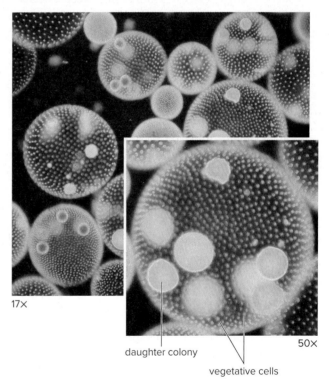

17×

daughter colony

50×

vegetative cells

**Figure 13.10  *Volvox.***
*Volvox* is a colonial green alga. The adult *Volvox* colony often contains daughter colonies, asexually produced by special cells.

(both) Manfred Kage/Science Source

produce sperm, and others have cells that produce eggs. The resulting zygote undergoes meiosis, and the adult *Volvox* is haploid.

2. In Table 13.2, list the genus names of each of the green algae specimens available, and give a brief description.

| Table 13.2 Green Algae Diversity | |
| --- | --- |
| Specimen | Description |
| | |
| | |
| | |
| | |

## Observation: Brown Algae and Red Algae

Brown algae and red algae are commonly called *seaweed,* along with the multicellular green algae.

1. If available, study preserved specimens of brown algae (Fig. 13.11), which have brown pigments that mask chlorophyll's green color. These algae are large and have specialized parts. *Laminaria* algae are called kelps. *Fucus* is called rockweed because it is seen attached to rocks at the seashore when the tide is out (Fig. 13.11). If available, view a preserved specimen. Note the dichotomously branched body plan, so called because the **stipe** repeatedly divides into two branches (Fig. 13.11). Note also the **holdfast** by which the alga anchors itself to the rock; the **air vesicles,** or bladders, that help hold the thallus erect in the water; and the **receptacles,** or swollen tips. The receptacles are covered by small raised areas, each with a hole in the center. These areas are cavities in which the sex organs are located, with the gametes escaping to the outside through the holes. *Fucus* is unique among algae in that as an adult it is diploid (2n) and always reproduces sexually.

2. If available, study preserved specimens of red algae. Like most brown algae, the red algae (Fig. 13.12) are multicellular, but they occur chiefly in warmer seawater, growing both in shallow waters and as deep as light penetrates. Some forms of red algae are filamentous, but more often they are complexly branched with a feathery, flat, and expanded or ribbonlike appearance. Coralline algae are red algae that have cell walls impregnated with calcium carbonate ($CaCO_3$). Coralline algae act like a glue, holding pieces of the coral reef together.

3. In Table 13.3, list the genus names of each of the brown and red algae specimens available, and give a brief description.

## Figure 13.11 Brown algae.

*Laminaria* and *Fucus* are seaweeds known as kelps. They live along rocky coasts of the north temperate zone. The other brown algae featured, *Macrocystis* and *Nereocystis,* form spectacular underwater "forests" at sea.

(rockweed) Dr. D. P. Wilson/Science Source

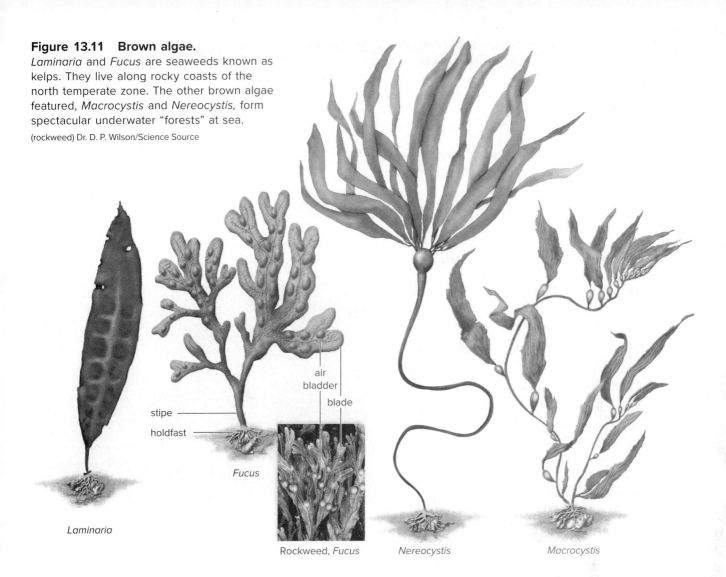

stipe

holdfast

air bladder

blade

*Fucus*

*Laminaria*

Rockweed, *Fucus*

*Nereocystis*

*Macrocystis*

## Figure 13.12 Red algae.

Generally, red algae are smaller and more delicate than brown algae. **a.** *Sebdenia* has a pronounced filamentous structure. **b.** Calcium carbonate is deposited in the walls of the red alga *Corallina*.

(a) Steven P. Lynch; (b) Jeff Rotman Photography/Corbis/Getty Images

a.

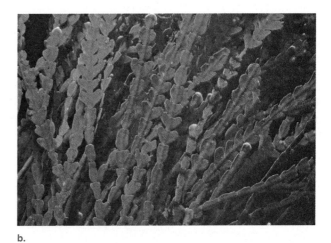

b.

## Table 13.3  Brown and Red Algae

| Specimen | Genus | Description |
|----------|-------|-------------|
| 1 | | |
| 2 | | |
| 3 | | |
| 4 | | |

### Observation: Diatoms and Dinoflagellates

**Diatoms** (golden-brown algae) have a yellow-brown pigment that, in addition to chlorophyll, gives them their color.

1. Make a wet mount of live diatoms, or view a prepared slide (Fig. 13.13). Describe what you see: _____

_____

_____

_____

**Figure 13.13  Diatoms.**
Diatoms, photosynthetic protists.

Jan Hinsch/Science Source

130×

The cell wall of diatoms is in two sections, with the larger one fitting over the smaller as a lid fits over a box. Since the cell wall is impregnated with silica, diatoms are said to "live in glass houses." The glass cell walls of diatoms do not decompose, so they accumulate in thick layers subsequently mined as diatomaceous earth and used in filters and as a natural insecticide. Diatoms, being photosynthetic and extremely abundant, are important food sources for the small heterotrophs in both marine and freshwater environments.

2. Make a wet mount of live dinoflagellates or view a prepared slide (Fig. 13.14). Describe what you see:

_____

_____

_____

_____

**Dinoflagellates** are photosynthetic, but they have two flagella; one is free, but the other is located in a transverse groove that encircles the organism. The beating of these flagella causes the organism to spin like a top. The cell wall, when present, is frequently divided into closely joined polygonal plates of cellulose. At times there are so many of these organisms in the ocean that they cause a condition called "red tide." The toxins given off in these red tides cause widespread fish kills and can cause paralysis in humans who eat shellfishes that have fed on the dinoflagellates.

**Figure 13.14   Dinoflagellates.**
Dinoflagellates such as *Gonyaulax* have cellulose plates.

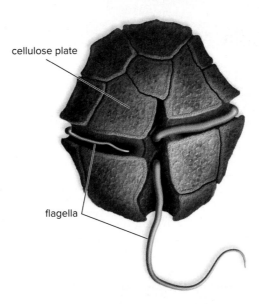

cellulose plate

flagella

## Heterotrophic Protists

The term *protozoan* refers to single-celled eukaryotes and is often restricted to heterotrophic organisms that ingest food by forming **food vacuoles.** Other vacuoles, such as **contractile vacuoles** that rid the cell of excess water, can be found among many of the protozoans. Usually protozoans have some form of locomotion, and as shown in Figure 13.15, some use **pseudopodia,** some move by **cilia,** and some use flagella.

*Plasmodium vivax,* which causes malaria, is an apicomplexan. Apicomplexans have a special organelle called an apicoplast that assists in penetrating host cells. *Plasmodium vivax* spends a portion of its life cycle in mosquitoes (sexual phase) and the other part in human hosts (asexual phase). During the asexual phase of their life cycle, all apicomplexans exist as particulate spores, which is why they are also called sporozoans. Malaria is caused when the *Plasmodium* spores multiply in and rupture red blood cells. In general, how do

sporozoans differ from the other protozoans shown in Figure 13.15? _____

_____

_____

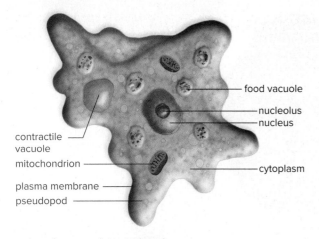

food vacuole
nucleolus
nucleus
contractile vacuole
mitochondrion
plasma membrane
pseudopod
cytoplasm

a. *Amoeba* moves by pseudopods.

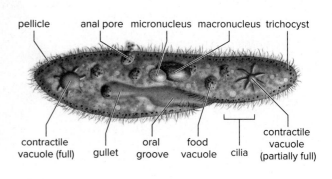

pellicle   anal pore   micronucleus   macronucleus   trichocyst

contractile vacuole (full)   gullet   oral groove   food vacuole   cilia   contractile vacuole (partially full)

b. *Paramecium* moves by cilia.

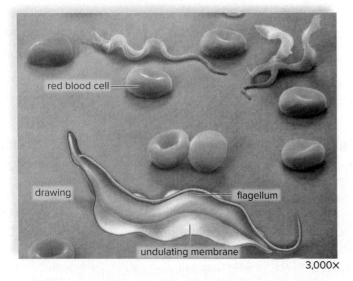

red blood cell

drawing

flagellum

undulating membrane

3,000×

c. *Trypanosoma*, which lives in bloodstream of host, moves by flagella.

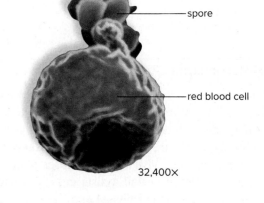

spore

red blood cell

32,400×

d. *Plasmodium* bursts forth as spores from infected red blood cells.

**Figure 13.15   Protozoan diversity.**
Many protozoans are motile. **a.** *Amoeba* moves by pseudopodia. **b.** *Paramecium* moves by cilia. **c.** *Trypanosoma* move by flagella. A blood infection is the cause of African sleeping sickness. **d.** *Plasmodium*, a cause of malarias, exists as non-motile spores inside the red blood cells of humans.

(c) Eye of Science/Science Source; (d) Dr. Gopal Murti/Science Source

### *Observation: Heterotrophic Protists*

#### *Individual Protozoans*

1. Watch a video if available, and note the various forms of protozoans.
2. Prepare wet mounts or examine prepared slides of protozoans as directed by your instructor.
3. Complete Table 13.4, listing the structures for locomotion in the types of protozoans you have observed.

## Table 13.4 Heterotrophic Protists

| Name | Structures for Locomotion | Observations |
|------|---------------------------|--------------|
|      |                           |              |
|      |                           |              |
|      |                           |              |
|      |                           |              |
|      |                           |              |

### *Observation:* Euglena

*Euglena* (Fig. 13.16) typifies the problem of classifying protists. One-third of all *Euglena* genera have chloroplasts; the rest do not. Scientists hypothesize that the chloroplasts were green algae taken up by phagocytosis (engulfing them) and instead of digesting them, they were retained to provide energy for the cell through photosynthesis. A pyrenoid is a region of the chloroplast where a special type of carbohydrate is formed.

    *Euglena* have a long flagellum that projects out of a vaselike indentation and a much shorter one that does not project out. It moves very quickly, and you will be advised to add Protoslo to your wet mount to slow it down. Like some of the protozoans discussed previously, *Euglena* is bounded by a flexible pellicle made of protein. This means it can also assume all sorts of shapes. *Euglena* lives in fresh water and contains a contractile vacuole that collects water and then contracts, ridding the body of excess water.

    Make a wet mount of *Euglena* by using a drop of a *Euglena* culture and adding a drop of Protoslo (methyl cellulose solution) onto the slide to slow it down. Describe what you see: _____

_____

_____

**Figure 13.16** *Euglena.*
*Euglena* is a unicellular, flagellated protist.
(b) M I (Spike) Walker/Alamy Stock Photo

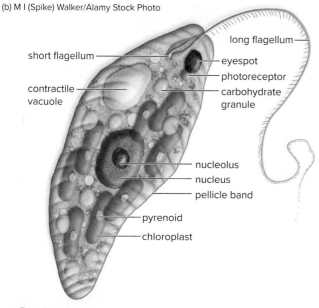

short flagellum — 
contractile vacuole — 
long flagellum — 
eyespot — 
photoreceptor — 
carbohydrate granule — 
nucleolus — 
nucleus — 
pellicle band — 
pyrenoid — 
chloroplast — 

**a.** Drawing

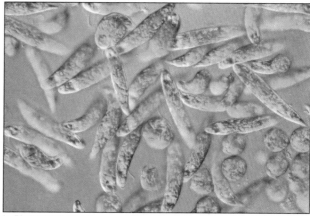

**b.** Photomicrograph 150×

Pond water typically contains various examples of protists studied in this laboratory. Your instructor may have an illustrated manual that will help you identify the ones you are not familiar with.

1. Make a wet mount of pond water by taking a drop from the bottom of a container of pond water. Be sure to select some of the sediment or strands of filamentous algae if present.
2. Scan the slide for organisms: Start at the upper left-hand corner, and move the slide forward and back as you work across the slide from left to right.
3. Experiment by using all available objective lenses, by focusing up and down with the fine-adjustment knob, and by adjusting the light so that it is not too bright.
4. Identify the organisms you see by consulting Figure 3.9 on page 27, and use any pictorial guides provided by your instructor.

## 13.3 Fungi

**Pre-Lab**

7. Why are fungi beneficial? _____

_____

8. Which group of fungi includes mushrooms? _____

**Fungi** (kingdom Fungi) (Fig. 13.17) are multicellular heterotrophs that release digestive juices into the environment and then absorb the resulting nutrients. Many fungi and heterotrophic bacteria are decomposers that break down waste and dead organisms, making inorganic nutrients available to plants. A fungal body, called a **mycelium,** is composed of many strands, called **hyphae** (Fig. 13.18). Sometimes, the nuclei within a hypha are separated by septae.

Fungi produce windblown **spores** (small, haploid bodies with a protective covering) when they reproduce sexually or asexually. The types of sexual spores produced by fungi enable biologists to further classify the various fungi. In this part of the lab exercise, you'll examine some fungi most familiar to students.

**Figure 13.17 Diversity of fungi.**
**a.** Scarlet hood, an inedible mushroom. **b.** Spores exploding from a puffball. **c.** Common bread mold.
**d.** Morel, an edible fungus.

(a) Biophoto Associates/Science Source; (b) RF Company/Alamy Stock Photo; (c) imagebroker/Alamy Stock Photo; (d) Carol Wolfe

## Figure 13.18  Body of a fungus.

**a.** The body of a fungus is called a mycelium.
**b.** A mycelium contains many individual chains of cells, and each chain is called a hypha.

(a) Chanus/Shutterstock

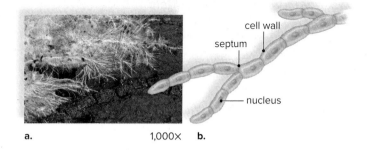

cell wall
septum
nucleus

a. 1,000× b.

## Black Bread Mold

In keeping with its name, black bread mold grows on bread and any other type of baked goods. Notice in Figure 13.19 the sporangia at the tips of aerial hyphae that produce spores in the asexual portion of the life cycle. A **zygospore** is diploid (2n). The other structures associated with the asexual and sexual portions of bread mold's life cycle are haploid (n).

### Figure 13.19  Black bread mold.

The mycelium of this mold (1) uses sporangia to produce windblown spores. (2) During sexual reproduction, the ends of plus and minus hyphae fuse as (3) fertilization occurs. (4) The resulting zygospore undergoes meiosis to produce spores that germinate on the bread to produce a new mycelium (5).

Garry DeLong/Oxford Scientific/Getty Images

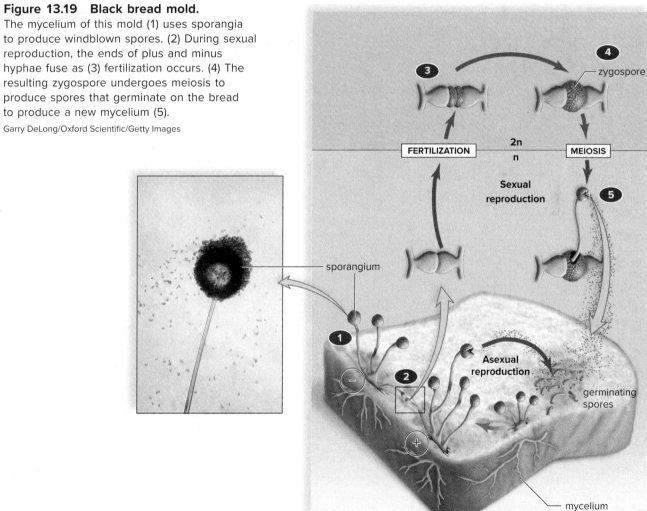

zygospore

FERTILIZATION   2n
                 n

MEIOSIS

Sexual
reproduction

sporangium

Asexual
reproduction

germinating
spores

mycelium

1. If available, examine bread that has become moldy. Do you recognize black bread mold on the bread?
2. Obtain a petri dish that contains living black bread mold. Observe with a stereomicroscope. *Label the mycelium and a sporangium in Figure 13.20a.*
3. View a prepared slide of *Rhizopus,* using both a stereomicroscope and the low-power setting of a light microscope. The absence of cross walls in the hyphae is an identifying feature of zygospore fungi. *Label the mycelium and zygospore in Figure 13.20b.*

**Figure 13.20   Microscope slides of black bread mold.**
**a.** Asexual life cycle structures. **b.** Sexual life cycle structures.

(a) Aldona Griskeviciene/Alamy Stock Photo; (b) Biophoto Associates/Science Source

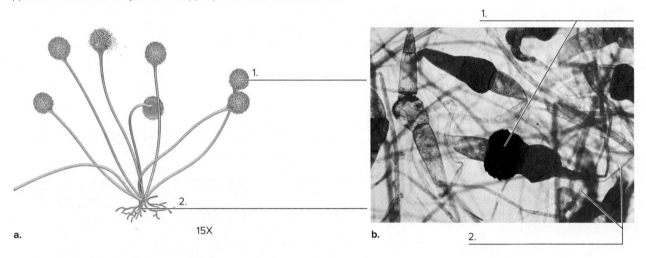

a.          15X          b.

## Club Fungi

**Club fungi** are just as familiar as black bread mold to most laypeople because they include the mushrooms. A gill mushroom consists of a stalk and a terminal cap with gills on the underside (Fig. 13.21). The cap, called a **basidiocarp,** is a fruiting body that arises following the union of + and – hyphae. The gills bear basidia, club-shaped structures where nuclei fuse, and meiosis occurs during spore production. The spores are called **basidiospores.**

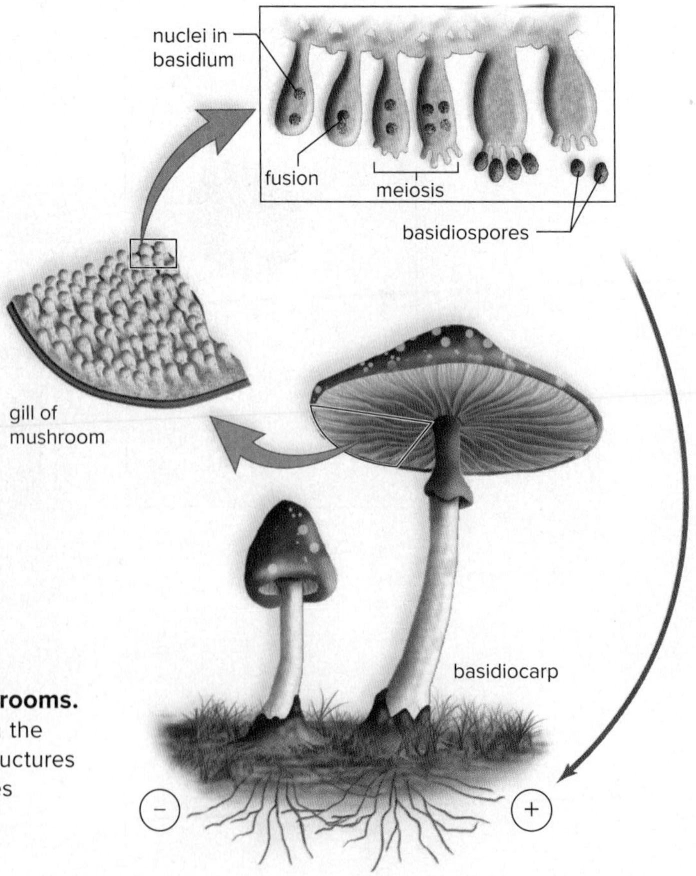

nuclei in basidium

fusion    meiosis

basidiospores

gill of mushroom

basidiocarp

**Figure 13.21   Sexual reproduction produces mushrooms.**
Fusion of + and – hyphae tips results in hyphae that form the mushroom (a fruiting body). The nuclei fuse in clublike structures attached to the gills of a mushroom, and meiosis produces basidiospores.

–          +

1. Obtain an edible mushroom—for example, *Agaricus*—and identify as many of the following structures as possible:
   a. **Stalk:** the upright portion that supports the cap.
   b. **Annulus:** a membrane surrounding the stalk where the immature (button-shaped) mushroom was attached.
   c. **Cap:** the umbrella-shaped basidiocarp of the mushroom.
   d. **Gills:** on the underside of the cap, radiating lamellae on which the basidia are located.
   e. **Basidia:** on the gills, club-shaped structures where basidiospores are produced.
   f. **Basidiospores:** spores produced by basidia.

2. View a prepared slide of a cross section of *Coprinus*. Using all three microscope objectives, look for the gills, basidia, and basidiospores.

3. Can you see individual hyphae in the gills? _____

4. Are the basidiospores inside or outside the basidia? _____

5. What type of nuclear division does the zygote undergo to produce the basidiospores? _____

   _____

6. Can you suggest a reason for some of the basidia having fewer than four basidiospores? _____

   _____

7. What happens to the basidiospores after they are released? _____

   _____

## Fungi and Human Diseases

Fungi cause a number of human diseases. Oral thrush is a yeast infection of the mouth common in newborns and AIDS patients (Fig. 13.22*a*). Ringworm is a group of related diseases caused by the fungus *Tinea*. The fungal colony grows outward, forming a ring of inflammation (Fig. 13.22*b*). Athlete's foot is a form of *Tinea* that affects the foot, mainly causing itching and peeling of the skin between the toes (Fig. 13.22*c*).

**Figure 13.22 Human fungal diseases.**
**a.** Thrush, or oral candidiasis, is characterized by the formation of white patches on the tongue.
**b.** Ringworm and (**c**) athlete's foot are caused by *Tinea* spp.

(a) Dr M.A. Ansary/Science Source; (b) John Hadfield/Science Source; (c) Dr P. Marazzi/Science Source

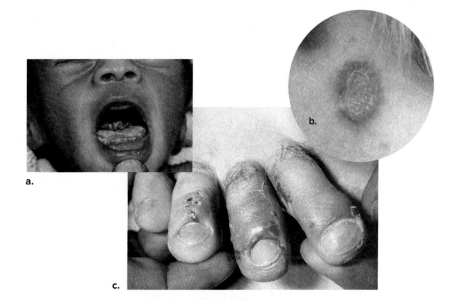

_____ 1. Into which domain are protists, fungi, plants, and animals classified?

_____ 2. What kind of cell lacks a nucleus and other membrane-bound organelles?

_____ 3. What structure synthesizes proteins in bacteria?

_____ 4. What shape are the bacteria classified as *Bacillus subtilis*?

_____ 5. Where does photosynthesis occur in cyanobacteria?

_____ 6. Where does photosynthesis occur in green algae?

_____ 7. What kind of algae has silica in its cell walls?

_____ 8. What type of protist spins like a top when its flagella beat?

_____ 9. What word refers to single-celled eukaryotes that are typically heterotrophic consumers?

_____ 10. What kind of protozoan moves using pseudopods?

_____ 11. Which organisms studied in this lab exercise are eukaryotic, multicellular, and heterotrophic?

_____ 12. What is an important ecological "job" done by many fungi and bacteria that ensures plants have a source of inorganic nutrients?

_____ 13. When black bread mold sexually reproduces, what kind of spore does it make?

_____ 14. Where are the spore-producing structures located on a mushroom?

## Thought Questions

15. Compare and contrast cyanobacteria and algae with regard to their importance to other organisms and their cellular characteristics.

16. Many people have the misconception that all bacteria are "germs." Briefly make a case that many bacteria do good work; be sure to mention the group of organisms that rely on this good work.

17. When fungi reproduce (asexually or sexually) they produce many, many spores. Speculate about why it is beneficial/essential for them to produce so many spores and how a large number of spores can be detrimental to many people.

# 14

# Nonvascular Plants and Seedless Vascular Plants

## Learning Outcomes

**14.1 The Evolution and Diversity of Land Plants**
- Describe the main events in the evolution of plants as they became adapted to life on land.
- Describe the plant life cycle and the concept of the dominant generation.

**14.2 Nonvascular Plants**
- Describe, in general, the nonvascular plants and how they are adapted to reproducing on land.
- Describe the life cycle of a moss, including the appearance of its two generations.

**14.3 Seedless Vascular Plants**
- Describe, in general, the seedless vascular plants and how they have adapted to reproducing on land.
- Describe the lycophytes, whisk ferns, and horsetails.
- Describe the life cycle of a fern and the appearance of its two generations.

## Introduction

### Pre-Lab

1. What reproductive and structural adaptations allowed plants to flourish on land? _____

_____

Plants are multicellar photosynthetic eukaryotes; their evolution is marked by adaptations to a land existence. Among aquatic green algae, the **charophytes** are most closely related to the plants that now live on land. The charophytes have several features that would have promoted the evolution of land plants, including retention of and care of the zygote. The most successful land plants are those that protect all phases of reproduction (sperm, egg, zygote, and embryo) from drying out and that have an efficient means of dispersing offspring on land.

In this laboratory, you will have the opportunity to examine the various adaptations of plants to living on land. Although this lab will concentrate on land plant reproduction, you will also see that much of a land plant's body is covered by a **waxy cuticle** that prevents water loss, and that land plants require structural support to oppose the force of gravity and to lift their leaves up toward the sun. **Vascular tissue** offers this support and transports water to, and nutrients from, the leaves.

Also, you will examine the adaptations of nonvascular plants and the seedless vascular plants to a land-based existence, and you will see that reproduction in these plants requires an outside source of water.

# 14.1   The Evolution and Diversity of Land Plants

**Pre-Lab**

   **2.** What is the closest algal relative to land plants? _____

   **3.** Meiosis produces _____ in plants and _____ in animals (Fig. 14.3).

Figure 14.1 shows the evolution of land plants.

**Figure 14.1   Evolutionary relationships among the plants.**

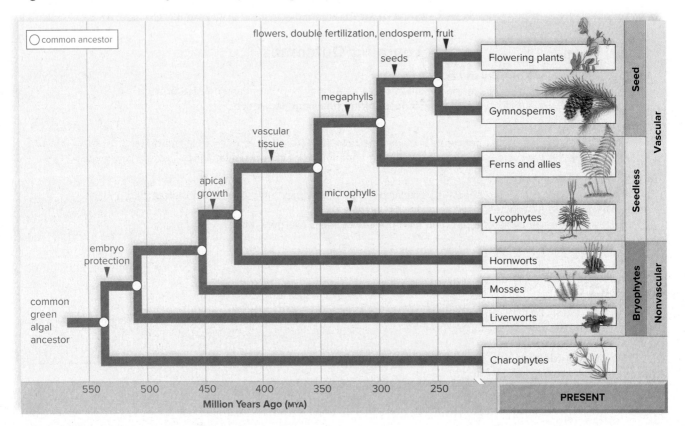

## Algal Ancestor of Land Plants

In the evolutionary tree, the charophytes are green algae that share a common ancestor with land plants. This common ancestor may have resembled a Charales, such as *Chara,* which can live in warm, shallow ponds that occasionally dry up. Adaptation to such periodic desiccation may have facilitated the ability of certain members of the common ancestor population to invade land, and with time, become the first land plants.

### *Observation:* Chara

Examine a living *Chara* (Fig. 14.2). How does it superficially resemble a land plant? _____

_____

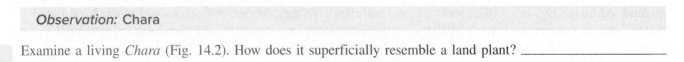

   *Chara* is a filamentous green alga that consists of a primary branch and many side branches (Fig. 14.2). Each branch has a series of very long cells. Note where one cell ends and the other begins. Chara is also called stonewort because its cell walls are covered with calcium carbonate deposits.

**Figure 14.2 _Chara_.**

_Chara_ is an example of a stonewort, the type of green alga believed to be most closely related to the land plants.

(left) Bob Gibbons/Alamy Stock Photo; (right) Kingsley Stern

_Chara_, several individuals     One individual

— branch

— main axis

— node

## Alternation of Generations

Land plants have a two-generation life cycle called **alternation of generations.**

1. The **sporophyte** (diploid) **generation** produces haploid spores by meiosis. Spores develop into a haploid generation by mitosis. _Label the appropriate arrow "mitosis" in the first part of Figure 14.3._
2. The **gametophyte** (haploid) **generation** produces **gametes** (eggs and sperm) by mitosis. The gametes then unite to form a diploid zygote. The zygote becomes the sporophyte by mitosis. _Label the appropriate arrow "mitosis" in Figure 14.3._

In this life cycle, the two generations are dissimilar, and one dominates the other. The dominant generation is larger and exists for a longer period. Figure 14.3 contrasts the plant life cycle (alternation of generations) with the animal life cycle (**diploid life cycle**).

1. In the plant life cycle, meiosis occurs during the production of _____.
2. In the human life cycle, meiosis occurs during the production of _____.
3. In the plant life cycle, the generation that produces gametes is (n or 2n) _____.
4. In the human life cycle, the individual that produces gametes is (n or 2n) _____.

**Figure 14.3 Plant and animal life cycles.**

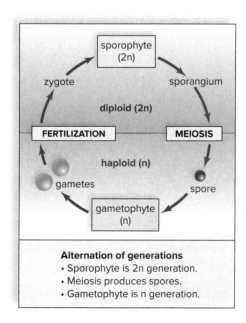

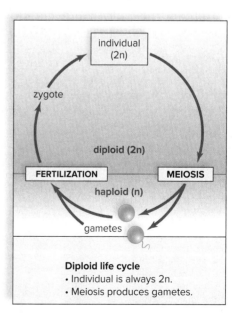

## 14.2 Nonvascular Plants

**Pre-Lab**

  **4.** What are some examples of nonvascular plants? _____

The nonvascular plants include the **mosses** (phylum Bryophyta) and **liverworts** (phylum Hepaticophyta). The gametophyte is dominant in all nonvascular plants. The gametophyte produces eggs within the archegonia and flagellated sperm in the **antheridia.** The sperm swim to the egg, and the embryo develops within the **archegonia.** The nonvascular sporophyte grows out of the archegonium. The sporophyte is dependent on the female gametophyte (Fig. 14.4) for nourishment.

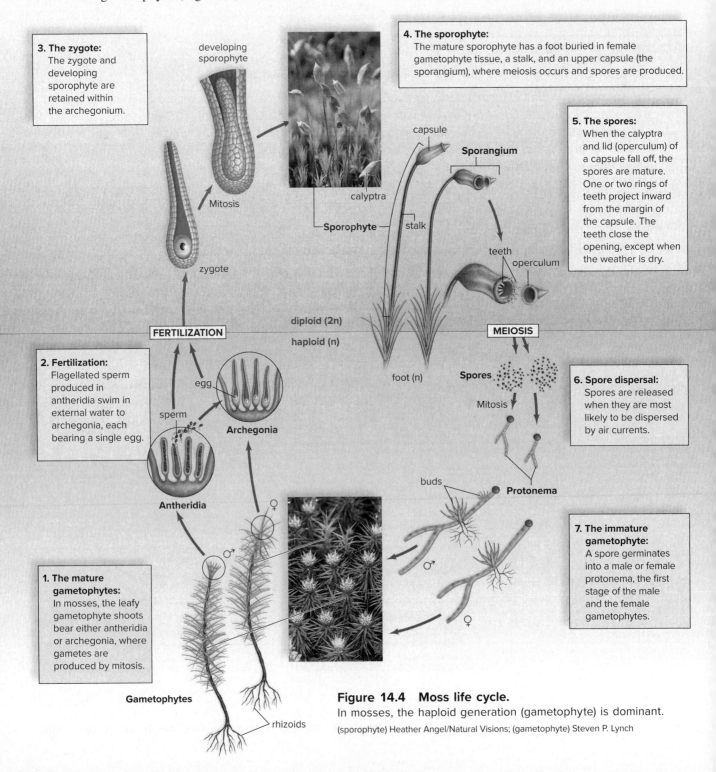

**3. The zygote:**
The zygote and developing sporophyte are retained within the archegonium.

developing sporophyte

Mitosis

zygote

**4. The sporophyte:**
The mature sporophyte has a foot buried in female gametophyte tissue, a stalk, and an upper capsule (the sporangium), where meiosis occurs and spores are produced.

capsule

calyptra

**Sporangium**

**Sporophyte**

stalk

teeth

operculum

**5. The spores:**
When the calyptra and lid (operculum) of a capsule fall off, the spores are mature. One or two rings of teeth project inward from the margin of the capsule. The teeth close the opening, except when the weather is dry.

diploid (2n)

**FERTILIZATION**

haploid (n)

**MEIOSIS**

**2. Fertilization:**
Flagellated sperm produced in antheridia swim in external water to archegonia, each bearing a single egg.

sperm

egg

**Archegonia**

**Antheridia**

foot (n)

**Spores**

Mitosis

**6. Spore dispersal:**
Spores are released when they are most likely to be dispersed by air currents.

buds

**Protonema**

**1. The mature gametophytes:**
In mosses, the leafy gametophyte shoots bear either antheridia or archegonia, where gametes are produced by mitosis.

**7. The immature gametophyte:**
A spore germinates into a male or female protonema, the first stage of the male and the female gametophytes.

**Gametophytes**

rhizoids

**Figure 14.4  Moss life cycle.**
In mosses, the haploid generation (gametophyte) is dominant.
(sporophyte) Heather Angel/Natural Visions; (gametophyte) Steven P. Lynch

1. Put a check mark beside the phrases that describe nonvascular plants:

I

_____ No vascular tissue to transport water

_____ Flagellated sperm that swim to egg

_____ Dominant gametophyte

II

_____ Vascular tissue to transport water

_____ Sperm protected from drying out

_____ Dominant sporophyte

2. Which listing of features (**I** or **II**) would you expect to find in a plant fully adapted to a land environment? _____ Explain. _____

3. In nonvascular plants, windblown spores are dispersal agents, and some species forcefully expel their spores. How are windblown spores an adaptation to reproduction on land? _____
_____

## Observation: Moss Gametophyte

### Living or Plastomount

Obtain a living moss gametophyte or a plastomount of this generation. Describe its appearance. _____
_____

The leafy green shoots of a moss are said to lack true roots, stems, and leaves because, by definition, roots, stems, and leaves are structures that contain vascular tissue.

### Microscope Slide

1. Study a slide of the top of a male moss shoot that contains antheridia, the reproductive structures where sperm are produced (Fig. 14.5). What is the chromosome number (choose 2n or n) of the sperm (see Fig. 14.4)? _____ Are the surrounding cells haploid or diploid? _____

2. Study a slide of the top of a female moss shoot that contains archegonia, the reproductive structures where eggs are produced (Fig. 14.6). What is the chromosome number of the egg? _____
Are the surrounding cells haploid or diploid? _____ When sperm swim from the antheridia to the archegonia, a zygote results. The zygote develops into the sporophyte. Is the sporophyte haploid or diploid? _____

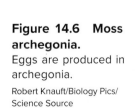

sperm

**Figure 14.5 Moss antheridia.**
Flagellated sperm are produced in antheridia.

Ed Reschke/Stone/Getty Images

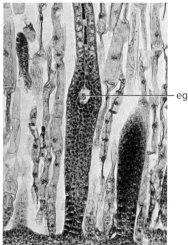

egg

**Figure 14.6 Moss archegonia.**
Eggs are produced in archegonia.

Robert Knauft/Biology Pics/ Science Source

### Living Sporophyte

1. Examine the living sporophyte of a moss in a minimarsh, or obtain a plastomount of a female shoot with the sporophyte attached. Identify the **sporangium,** a capsule where spores are produced and released.

2. *Bracket and label the gametophyte and sporophyte in Figure 14.7a. Place an* n *beside the gametophyte and a* 2n *beside the sporophyte.*

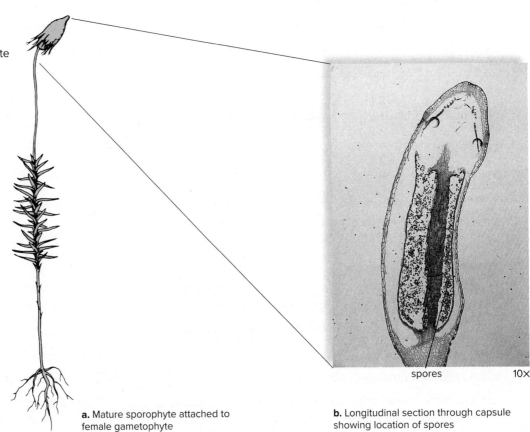

**Figure 14.7 Moss sporophyte.**
**a.** The moss sporophyte is dependent on the female gametophyte.
**b.** The sporophyte produces spores by meiosis.

(b) Ed Reschke

spores                     10✕

**a.** Mature sporophyte attached to female gametophyte

**b.** Longitudinal section through capsule showing location of spores

### Microscope Slide

Examine a slide of a longitudinal section through a moss sporophyte (Fig. 14.7*b*). Identify the stalk and the sporangium, where spores are being produced. By what process are the spores being produced? _____

When spores germinate, what generation begins to develop? _____

Why is it proper to say that spores are dispersal agents? _____

_____

### Observation: Liverworts

Obtain a living sample of the liverwort *Marchantia* (Fig. 14.8) thallus, body of the plant. Examine it under the stereomicroscope. What generation is this sample? _____ Identify the following:

1. The gametophyte consisting of lobes. Each lobe is about a centimeter or so in length; the upper surface is smooth, and the lower surface bears numerous **rhizoids** (rootlike hairs).

2. **Gemma cups** on the upper surface of the thallus. These contain groups of cells called **gemmae** that can asexually start new plants.
3. Disk-headed stalks bear antheridia, where flagellated sperm are produced.
4. Umbrella-headed stalks bear archegonia, where eggs are produced. Following fertilization, tiny sporophytes arise from the archegonia.

**Figure 14.8 Liverwort, *Marchantia*.**
**a.** Gemmae can detach and start new plants. **b.** Antheridia are present in disk-shaped structures. **c.** Archegonia are present in umbrella-shaped structures.

(a) Ed Reschke; (b) J. M. Conrader/Science Source; (c) Ed Reschke/Stone/Getty Images

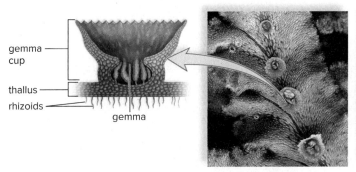

**a.** Thallus with gemma cups    **b.** Male gametophytes bear antheridia.    **c.** Female gametophytes bear archegonia.

# 14.3 Seedless Vascular Plants

**Pre-Lab**

5. What is the function of vascular tissue in plants? _____

_____

6. List some examples of seedless vascular plants. _____

Seedless vascular plants include the **lycophytes** (club mosses) and **pteridophytes**—ferns and their allies, the whisk ferns and horsetails. These plants were prevalent and quite large, forming the forests of the Carboniferous period. It was during this time period that plants began being formed into the coal deposits we use today.

The sporophyte is dominant in the seedless vascular plants. The dominant sporophyte has adaptations for living on land; it has vascular tissue and produces windblown spores. The spores develop into a separate gametophyte generation that is very small (less than 1 cm). The gametophyte lacks vascular tissue and produces flagellated sperm.

1. Place a check mark beside the phrases that describe seedless vascular plants:

_____ Independent gametophyte    _____ Gametophyte dependent on sporophyte, which has vascular tissue

_____ Flagellated sperm    _____ Sperm protected from drying out

2. Are seedless vascular plants fully adapted to living on land? Explain your answer. _____

_____

**Lycophytes** (phylum Lycophyta) are commonly called **club mosses.** Lycophytes are representative of the first vascular plants. They have an aerial stem and a horizontal root (rhizome with attached rhizoids) both of which have vascular tissue. The leaves are called **microphylls** because they have only one strand of vascular tissue.

### Ground Pines

1. Examine a living or preserved specimen of *Lycopodium* (Fig. 14.9).
2. Note the shape and the size of the microphylls and the branches of the stems.
3. Note the terminal clusters of leaves, called **strobili,** that are club-shaped and bear sporangia.
4. *Label strobili, leaves, stem, and rhizoids in Figure 14.9.*
5. Examine a prepared slide of a *Lycopodium* that shows the sporangia with spores inside. The spore develops into a tiny microscopic gametophyte that remains in the soil.

**Figure 14.9** *Lycopodium.*

In the club moss *Lycopodium,* green photosynthetic stems are covered by scalelike leaves, and spore-bearing leaves are clustered in strobili.

Steve Solum/Avalon

Molecular (DNA) studies tell us that the whisk ferns, horsetails, and ferns are closely related.

### Whisk Ferns

*Psilotum* is representative of **whisk ferns,** named for their resemblance to whisk brooms.

1. Examine a preserved specimen of *Psilotum,* and note that it has no leaves. The underground stem, called a **rhizome,** gives off upright, aerial stems with a dichotomous branching pattern, where bulbous sporangia are located (Fig. 14.10).

   What generation are you examining? _____

2. *Label a sporangium, the stem, and the rhizome in Figure 14.10b.*

## Figure 14.10  Whisk fern, *Psilotum.*

This whisk fern has no roots or leaves—the branches carry on photosynthesis. The sporangia are yellow.

(a) Rattiya Thongdumhyu/Shutterstock

a.

b.

### Horsetails

In **horsetails,** a rhizome produces aerial stems that stand about 1.3 meters tall.

1. Examine *Equisetum,* a horsetail, and note the minute, scalelike leaves (Fig. 14.11).
2. Feel the stem. *Equisetum* contains a large amount of silica in its stem. For this reason, these plants are sometimes called scouring rushes and may be used by campers for scouring pots.
3. Strobili appear at the tips of the stems, or else special buff-colored stems bear the strobili. Sporangia are in the strobili.
4. *Label the strobilus, branches, leaves, and rhizome of the horsetail in Figure 14.11.*

## Figure 14.11  Horsetail.

In the horsetail *Equisetum,* whorls of branches or tiny leaves appear in the joints of the stem. The sporangia are borne in strobili.

Robert P. Carr/Avalon

# Ferns

**Ferns** are quite diverse and range in size from those that are low growing and resemble mosses to those that are as tall as trees. The rhizome grows horizontally underground, which allows ferns to spread without sexual reproduction (Fig. 14.12). The gametophyte (called a **prothallus**) is small (about 0.5 cm) and usually heart-shaped. The prothallus contains both archegonia and antheridia. Ferns are largely restricted to moist, shady habitats because sexual reproduction requires adequate moisture.

### Figure 14.12  Fern life cycle.
The sporophyte is the frond, and the gametophyte is the heart-shaped prothallus.

(*Dryopterus*) Matt Meadows/Photolibrary/Getty Images

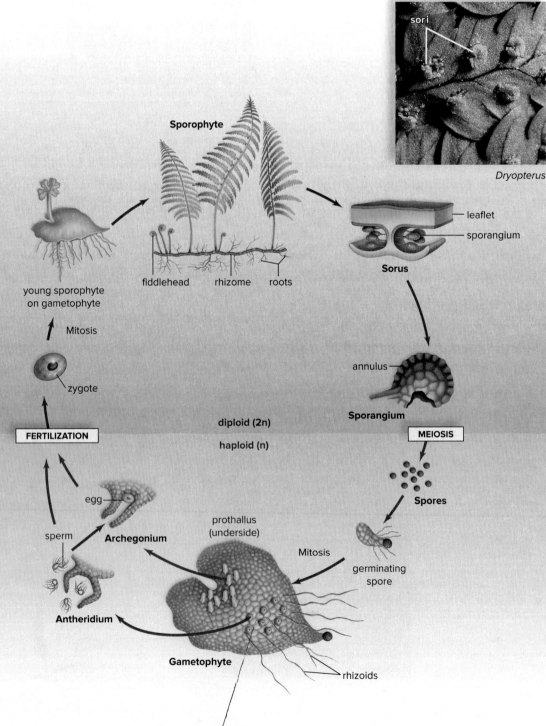

Fern spores are produced by meiosis in sporangia, which in many species occur on the underside of large leaves called **fronds.** These leaves are called **megaphylls** because they are broad leaves with several strands of vascular tissue. The spores are released when special cells of the **annulus** (a line or ring of thickened cells on the outside of the sporangium) dry out, and the sporangium opens. How do ferns disperse to new locations? _____

Observe the ferns on display, and then complete Table 14.1.

| Table 14.1 Fern Diversity | |
| --- | --- |
| **Type of Fern** | **Description of Frond** |
| | |
| | |
| | |

## Observation: Fern Sporophyte

Study the life cycle of the fern (Fig. 14.12), and find the sporophyte generation. This large, complexly divided leaf is known as a frond. Fronds arise from an underground stem called a rhizome.

### Living or Preserved Frond

Examine a living or preserved specimen of a frond, and on the underside, notice a brownish clump called a **sorus** (pl., sori), each a cluster of many sporangia (Fig. 14.13).

What is being produced in the sporangia? _____

Given that this is the generation we call the fern, what generation is dominant in ferns? _____

**Figure 14.13 Underside of frond leaflets.**
Sori occur on the underside of frond leaflets.

Carlyn Iverson/McGraw Hill

sorus

## Microscope Slide of Sorus

1. Examine a prepared slide of a cross section of a frond leaflet. Using Figure 14.14 as a guide, locate the fern leaf above and the sorus below.

**Figure 14.14   Micrograph of cross section of a frond leaflet.**
Micrograph of the internal anatomy of a sorus depicts many sporangia, where spores are produced.
Ed Reschke/Stone/Getty Images

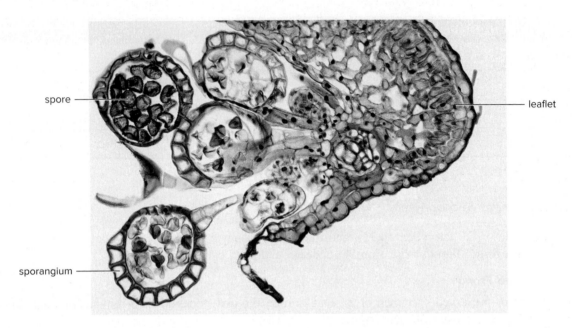

2. Within the sorus, find the sporangia and spores. Look for an **indusium** (not present in all species), a shelflike structure that protects the sporangia until they are mature.

### Observation: Fern Gametophyte

**Plastomount**

1. Examine a plastomount showing the fern life cycle.

2. Notice the prothallus, a small, heart-shaped structure. The prothallus is the gametophyte generation of the fern. What is the function of this structure? _____

_____

## Microscope Slide

1. Examine whole-mount slides of fern prothallium-archegonia and fern prothallium-antheridia (Fig. 14.15).

2. If you focus up and down very carefully on an archegonium, you may be able to see an egg inside. What is being produced inside an antheridium? _____ When sperm produced in an antheridium swim to the archegonium in a film of water, what results? _____ This structure develops into what generation? _____

## Conclusions: Ferns

- How are ferns dispersed from one area to another? _____
- Is either generation in the fern dependent for any length of time on the other generation? _____ Explain. _____

### Figure 14.15 Fern prothallus.
The underside of the heart-shaped fern prothallus contains archegonia, where eggs are produced, and antheridia, where flagellated sperm are produced.

(right side, top) Robert Knauft/Biology Pics/Science Source; (right side, bottom) Rattiya Thongdumhyu/Shutterstock

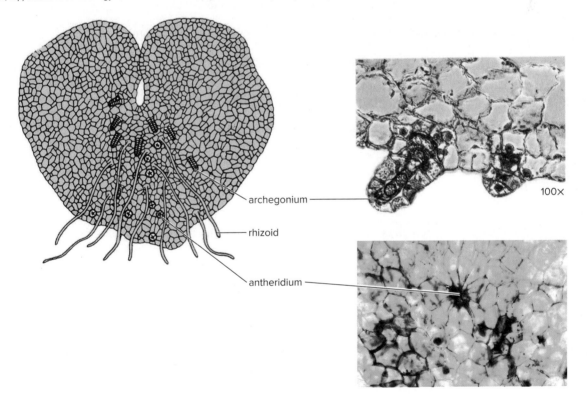

archegonium

rhizoid

antheridium

100×

_____  **1.** What is the closest living relative to the land plants?

_____  **2.** What generation of the plant life cycle undergoes meiosis, sporophyte or gametophyte?

_____  **3.** What process causes the spores to develop into the gametophytes?

_____  **4.** When two gametes fuse, what plant generation is formed?

_____  **5.** When a plant identified as moss is observed, what generation of the moss life cycle is it?

_____  **6.** What kind of gamete is produced by moss antheridia?

_____  **7.** What key features enabled the earliest plants to adapt to life on land?

_____  **8.** How are mosses dispersed to new locations?

_____  **9.** Do seedless vascular plants have dominant sporophytes or gametophytes?

_____  **10.** Which generation in the life cycle of ferns lacks vascular tissue?

_____  **11.** Are fern zygotes dispersed to new locations?

_____  **12.** Do fern roots and rhizomes contain vascular tissue?

_____  **13.** What are the brownish clumps on the underside of a fern frond called?

_____  **14.** What structure produces eggs in the life cycle of ferns?

## Thought Questions

**15.** Is there anything comparable to a spore in the life cycle of animals? Justify your reply.

**16.** Compare and contrast the sporophyte generation of mosses and ferns. Your reply could take into consideration vascular tissue, dominant form, diploid or haploid, and what happens to it as the life cycle advances.

**17.** During the life cycle of animals, gametes are produced when meiosis occurs. Why must gametes be produced by mitosis during the plant life cycle?

# 15
# Seed Plants

## Learning Outcomes

**15.1 Life Cycle of Seed Plants**
- Compare the life cycles of seedless and seed plants.
- List significant innovations of seed plants.

**15.2 Gymnosperms**
- Name three groups of gymnosperms, and give the distinguishing features of each group.
- Distinguish between pollination and fertilization in the life cycle of a pine tree.

**15.3 Angiosperms**
- Identify the parts of a flower and the two major groups of flowering plants.
- Identify the male and female gametophyte and their roles in the life cycle of a flowering plant.

**15.4 Comparison of Gymnosperms and Angiosperms**
- Compare the adaptations of flowering plants to those of conifers.

## Introduction

### Pre-Lab

1. What are the unique adaptations of seed plants compared to seedless plants? _____
_____

2. How are angiosperms unique compared to gymnosperms? _____
_____

Among plants, **gymnosperms** and **angiosperms** are seed plants. Review the plant evolutionary tree (see Fig. 13.1) and note that the gymnosperms and angiosperms share a common ancestor, which produced **seeds,** structures that contain the next sporophyte generation. We shall see that the use of seeds to disperse the next generation requires a major overhaul of the alternation-of-generations life cycle. Seedless plants disperse the gametophyte, the n generation, when spores are carried by the wind to a new location. Seed plants disperse the sporophyte, the 2n generation.

Seedless plants utilize an archegonium to protect the egg. After fertilization, the sporophyte develops within and remains attached to the archegonium. Seed plants protect the entire female gametophyte within an **ovule.** The ovule develops into a sporophyte-containing seed following fertilization. You will see that the formation of the ovule and also the **pollen grains** (contain male gametophytes) are radical innovations in the life cycle of seed plants. In gymnosperms, pollen grains are windblown, but many flowering plants have a mutualistic relationship with animals, particularly flying insects, which disperse pollen within the species. Animals also disperse the seeds of many flowering plants. Relationships with animals help explain why flowering plants are so diverse and widespread today.

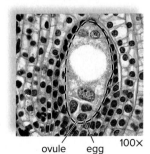

ovule    egg    100×

©Ed Reschke

# 15.1  Life Cycle of Seed Plants

## Pre-Lab

3. How are pollen grains in seed plants advantageous over flagellated sperm in seedless plants? _____
_____

Figure 15.1 shows alternation of generations as it occurs in seedless plants, and Figure 15.2 shows alternation of generations as it occurs in seed plants. Note which structures are haploid and which are diploid in these diagrams.

1. In which life cycle, seedless or seed, do you note pollen sacs (microsporangia) and ovules

   (megasporangia)? _____ In which life cycle, seedless or seed, do you note two types of spores,

   microspores and megaspores? _____ The formation of **heterospores** (unlike spores) is an innovation

   that leads to the production of pollen grains and the formation of ovules in seed plants. *Label*

   *heterospores where appropriate in Figure 15.2.* In which life cycle do you note male gametophyte

   (pollen grain) and female gametophyte (embryo sac in ovule)? _____

2. In seedless plants, flagellated sperm must swim in external water to the egg. In seed plants, **pollination** (the transport of male gametophytes by wind or pollen carrier to the vicinity of female gametophytes) does not require external water. The production of pollen grains is an innovation in seed plants.

3. Formation of a seed from an embryo sac in the ovule is an innovation in seeds. What generation is

   present in a seed? _____

**Figure 15.2  Alternation of generations in flowering plants that produce seeds.**

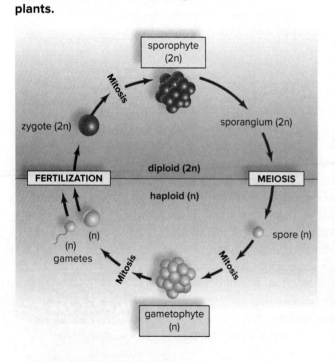

**Figure 15.1  Alternation of generations in seedless plants.**

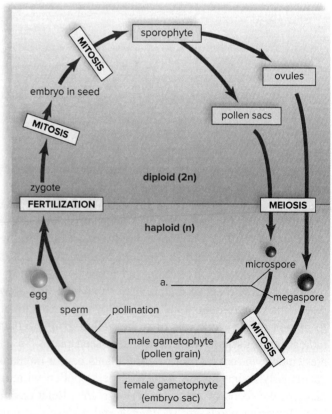

# 15.2 Gymnosperms

## Pre-Lab

4. Which group of gymnosperms is most common? _____

5. Where are seeds held on gymnosperms? _____

6. How does pollination occur in gymnosperms? _____

The term *gymnosperm* means "naked seed." The ovules and seeds of gymnosperms are exposed on a cone scale or some other comparable structure. Most gymnosperms produce cones. Three groups of gymnosperms are especially familiar: cycads, ginkgoes, and conifers (Fig. 15.3).

**Figure 15.3 Gymnosperms.**
Three divisions of gymnosperms are well known: **(a)** cycads, **(b)** ginkgoes, and **(c)** conifers.

(a) Hoberman Collection/Alamy Stock Photo; (a, inset) PlazaCameraman/iStock/Getty Images; (b) ©Kingsley Stern; (c) Michael Giannechini/Science Source

**a.** An African cycad

**b.** Ginkgo, *Ginkgo biloba.* Female maidenhair tree with seeds.

**c.** Conifer, *Picea.* Spruce with pollen cones and seed cones.

# Conifers

The **conifers** (phylum Pinophyta) are by far the most abundant group of gymnosperms. Pines, hemlocks, and spruces are evergreen conifers because their leaves remain on the tree through all seasons. The cypress tree and larch (tamarack) are examples of conifers that are not evergreen.

# Pine Trees

The pine tree is the dominant sporophyte generation (Fig. 15.4) in the life cycle of a pine. Vascular tissue extends from the roots, through the stem, to the leaves. Pine tree leaves are needlelike, leathery, and covered

## Figure 15.4   Pine life cycle.

The sporophyte is the tree. The male gametophytes are windblown pollen grains shed by pollen cones. The female gametophytes are retained within ovules on seed cones. The ovules develop into windblown seeds.

J M Barres/agefotostock/Alamy Stock Photo

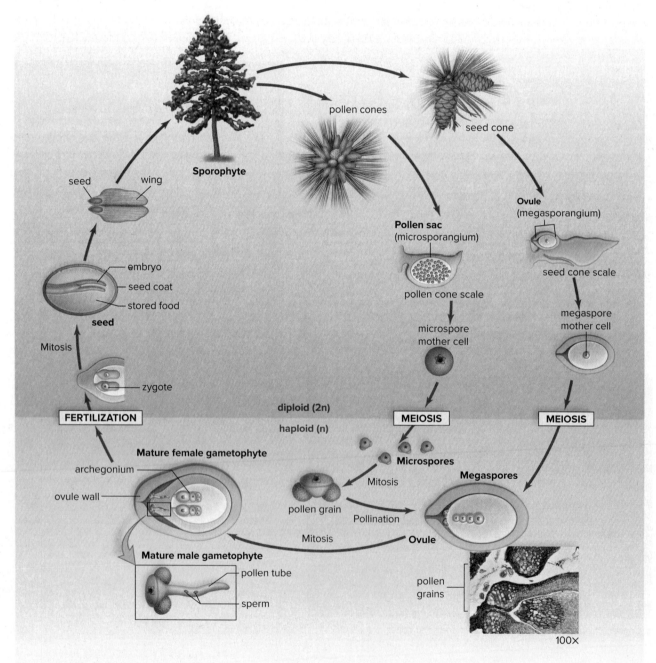

with a waxy, resinous cuticle. The **stomata,** openings in the leaves for gas exchange, are sunken. The structure of the leaf and the leaf's internal anatomy are adaptive to a drier climate.

In pine trees, the **pollen sacs** and ovules are located in cones. Pollen sacs on pollen cone scales contain cells that undergo meiosis to produce **microspores.** Microspores (n) become sperm-bearing **male gameto-phytes** (pollen grains). Seed cone scales bear ovules where cells undergo meiosis to produce **megaspores.** A megaspore (n) becomes an egg-bearing **female gametophyte.**

**Pollination** occurs when pollen grains are windblown to the seed cone on, ideally, a different plant. After **fertilization,** the egg becomes a sporophyte (2n) embryo enclosed within the ovule, which develops a seed coat. The seeds are winged and are dispersed by the wind.

1. Which part of the pine life cycle is the sporophyte? _____

2. Which part of the pine life cycle is the male gametophyte? The female gametophyte? _____

_____

_____

3. Where does fertilization occur and what structure becomes a seed? _____

_____

### *Observation: Pine Leaf*

Obtain a cluster of pine leaves (needles). A very short woody stem is at its base. Each type of pine has a typical number of leaves in a cluster (Fig. 15.5). How many leaves are in the cluster you are examining? _____

What is the common name of your specimen? _____

b. Pitch pine

a. White pine

c. Red pine

**Figure 15.5   Pine leaves (needles).**
**a.** The needles of white pines are in clusters of five. **b.** The needles of pitch pines are in clusters of three. **c.** The needles of red pines are in clusters of two.

*Preserved Cones*

1. Compare a pine pollen cone to a pine seed
   cone. _____
   _____
   _____

   Note the size and texture of the pollen cones
   relative to the larger, woody, seed cones.

2. Pollen cones. Remove a single scale (sporophyll)
   from the pollen cone and examine with a
   stereomicroscope. Note the two pollen sacs on the
   lower surface of each scale (Fig. 15.6a). What do
   the pollen sacs produce? _____
   _____

3. Seed cones.
   a. Seed cones of three distinct ages are present.
      First-year cones are about 1 cm wide. Second-
      year cones are about 10 cm long. They are
      green and the scales are tightly closed together.
      In third-year cones the scales have opened,
      revealing the mature seeds at the base of the
      cone scale.
   b. If available, examine a first-year cone cut
      lengthwise. The ovules are visible as small,
      milky-white domes at the base of the cone
      scale. Ovules will hold the female gametophyte
      generation, and then they become seeds.

   c. Examine mature seed cones. See if any seeds are present (Fig. 15.6b). Where are they located? _____
   _____ What is the function of the seed coat?
   _____

   If instructed to do so, use tweezers to pull out a seed and note the wing. What is the wing for? _____
   _____

4. Pine seeds. If available, examine a pine seed (called a pine nut) that has the seed coat removed. These
   seeds can be used for cooking foods such as pesto. Carefully cut the seed lengthwise and examine. Can
   you find an embryo inside? _____

**Figure 15.6  Pine cones.**
**a.** The scales of pollen cones bear pollen sacs where
microspores become pollen grains. **b.** The scales of seed
cones bear ovules that develop into winged seeds.

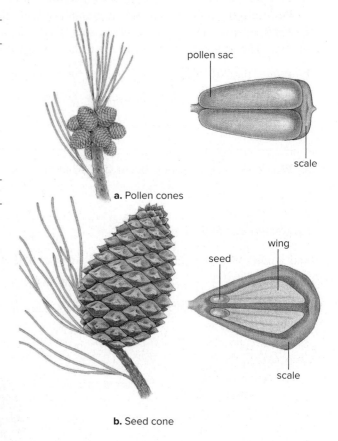

pollen sac

scale

**a.** Pollen cones

wing

seed

scale

**b.** Seed cone

1. Examine a prepared slide of a longitudinal section through a mature pine pollen cone. *Label a pollen sac in Figure 15.7a and a pollen grain in Figure 15.7b.* A pollen grain has a central body and two attached hollow bladders. How do these help in the dispersal of pine pollen? _____

_____

One pollen grain cell will divide to become two nonflagellated sperm, one of which fertilizes the egg after pollination. The other cell forms the **pollen tube** through which a sperm travels to the egg.

**Figure 15.7  Pine pollen cone.**
**a.** Pollen cones bear pollen sacs (microsporangia) in which microspores develop into pollen grains. **b.** Enlargement of pollen grains.

(a) Biophoto Associates/Science Source; (b) ©Ed Reschke

1. _____

2. _____

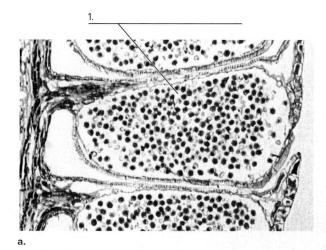

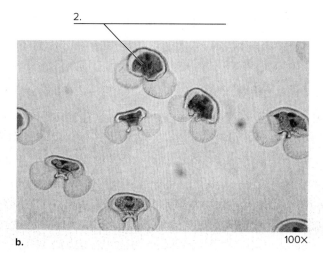

a.

b.                                              100×

2. Examine a prepared slide of a longitudinal section through an immature pine seed cone. Seed cone scales bear ovules. The ovule contains a megaspore mother cell, which undergoes meiosis to produce four megaspores, three of which disintegrate. This megaspore (n) becomes an egg-bearing female gametophyte. *Label the ovule and the megaspore mother cell in Figure 15.8. Also, label the pollen grains that you can see just outside the ovule.*

a. _____          c. _____

**Figure 15.8  Seed cone.**
Seed cones bear ovules, each of which will contain a female gametophyte. Note pollen grains near the entrance.

J M Barres/agefotostock/Alamy Stock Photo

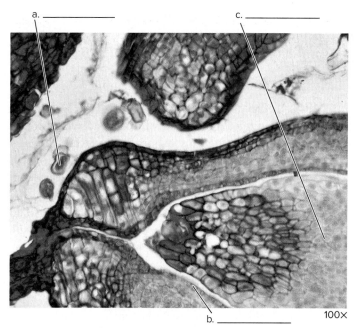

b. _____          100×

## 15.3 Angiosperms

### Pre-Lab

7. What is the unique feature of angiosperms? _____

8. What does the term *angiosperm* mean? _____

9. How does pollination occur in angiosperms? _____

Flowering plants (phylum Magnoliophyta) are the dominant plants today. Some varieties occur as trees, shrubs, vines, and garden plants. At some point in their life cycle, all angiosperms bear flowers.

**Figure 15.9   Generalized flower.**
A flower has four main kinds of parts: sepals, petals, stamens, and a carpel. A stamen has an anther and a filament. A carpel has a stigma, a style, and an ovary. An ovary contains ovules.

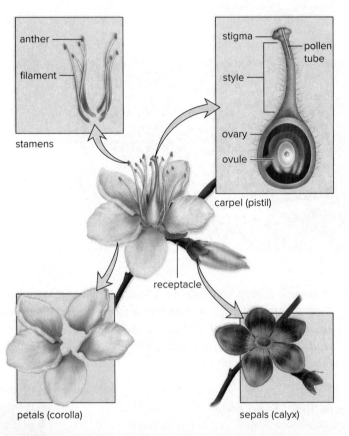

### Observation: A Flower

1. With the help of Figure 15.9, identify the following structures on a model of a flower:
   a. **Receptacle:** the portion of a stalk to which the flower parts are attached.
   b. **Sepals:** an outermost whorl of modified leaves, collectively called the **calyx.** Sepals are green in most flowers. They protect a bud before it opens.
   c. **Petals:** usually colored leaves that collectively constitute the **corolla.** These are generally used to attract a pollinator.
   d. **Stamen:** a swollen terminal **anther** and the slender supporting **filament.** The anther contains two pollen sacs, where microspores develop into male gametophytes (pollen grains).
   e. **Carpel:** a modified sporophyll consisting of a swollen basal ovary; a long, slender **style** (stalk); and a terminal **stigma** (sticky knob).
   f. **Ovary:** the enlarged part of the carpel that develops into a fruit.
   g. **Ovule:** the structure within the ovary where a megaspore develops into a female gametophyte (embryo sac). The ovule becomes a seed.

## Table 15.1 Monocots and Eudicots

Flower parts in threes and multiples of three

| Monocots | Eudicots |
|---|---|
| One cotyledon | Two cotyledons |
| Flower parts in threes or multiples of three | Flower parts in fours or fives or multiples of four or five |
| Usually herbaceous | Woody or herbaceous |
| Usually parallel venation | Usually net venation |
| Scattered bundles in stem | Vascular bundles in a ring |
| Fibrous root system | Taproot system |
| Pollen grain with one pore | Pollen grain with three pores |

Flower parts in fours or fives and their multiples

2. Inspect a fresh flower. Remove the sepals and petals by breaking them off at the base. How many sepals and petals are there? _____

3. Remove a stamen, and touch the anther to a drop of water on a slide. If nothing comes off in the water, crush the anther a little to release some of its contents. Place a coverslip on the drop, and observe with low- and high-power magnification. What are you observing? _____

4. Remove the carpel by cutting it free just below the base. Make a series of thin cross sections through the ovary. The ovary is hollow, and you can see nearly spherical bodies inside. What are these bodies? _____

_____

_____

5. Flowering plants are divided into two classes, called **monocots** and **eudicots.** Table 15.1 lists significant differences between the two classes of plants. Is your flower a monocot or eudicot? _____

## Life Cycle of Flowering Plants

The life cycle of a flowering plant is like that of the pine tree except for these innovations:

- In angiosperms, the often brightly colored flower contains the pollen sacs and ovules. Locate these structures in Figure 15.10.
- Pollination in flowering plants—when pollen is delivered to the stigma of the carpel—is sometimes accomplished by wind but more often by the assistance of an animal pollinator. The pollinator acquires nutrients (e.g., nectar) from the flower and inadvertently collects pollen, which it takes to the next flower.
- Notice that flowering plants practice **double fertilization.** A mature pollen grain contains two sperm; one fertilizes the egg, and the other joins with the two polar nuclei to form **endosperm** (3n), which serves as food for the developing embryo.
- Also, flowering plants have seeds enclosed in fruits. Fruits protect the seeds and aid in seed dispersal. Sometimes, animals eat the fruits, and after the digestion process, the seeds are deposited far away from the parent plant. The term *angiosperm* means "covered seeds." The seeds of angiosperms are found in fruits, which develop from parts of the flower, particularly the ovary.

# Figure 15.10 Flowering plant life cycle.

The parts of the flower involved in reproduction are the anthers of stamens and the ovules in the ovary of a carpel.

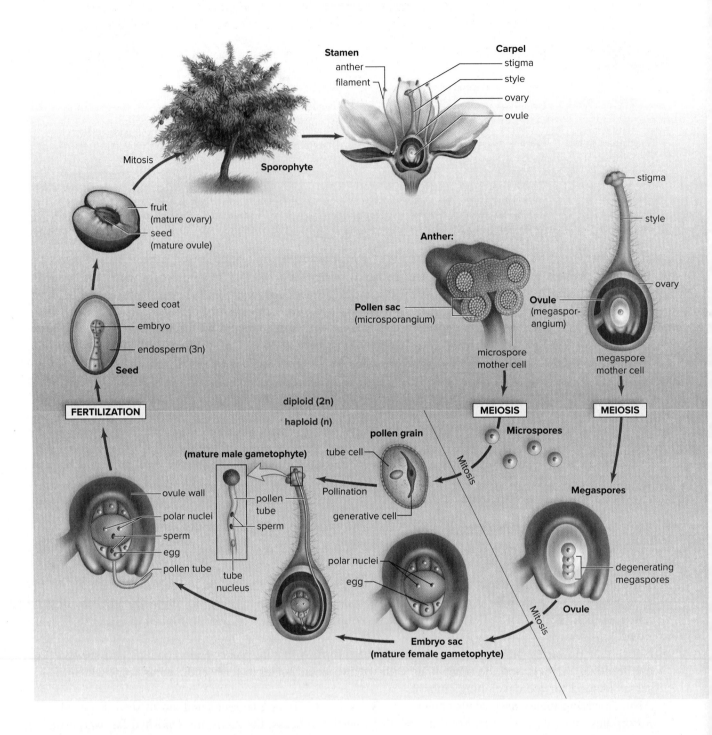

## The Male Gametophyte

Notice that in the life cycle of flowering plants (Fig. 15.10), meiosis in pollen sacs produces four microspores, each of which will become a two-celled pollen grain. In flowering plants, **pollination** is the transfer of the pollen grain from the anther to the stigma, where the pollen grain germinates and becomes the mature male gametophyte. How is pollination accomplished in the flowering plant life cycle? _____

_____

### Observation: Pollen Grain Slide

1. A nongerminated pollen grain newly released from the anther has two cells. The larger of the two cells is the **tube cell,** and the smaller is the **generative cell.** Examine a prepared slide of pollen grains. Identify the tube cell and the generative cell. Sketch your observation here.

2. During germination, the pollen grain's tube cell gives rise to the pollen tube. As it grows, the pollen tube passes through the stigma and grows through the style and into the ovary. Two sperm cells produced by division of the pollen grain's generative cell migrate through the pollen tube into the embryo sac. Examine a prepared slide of germinated pollen grains with pollen tubes. You should be able to see the tube cell nucleus and two sperm cells. What portion of a germinated pollen grain tells you it is the mature male gametophyte?

_____

### Experimental Procedure: Pollen Grains

Inoculate an agar-coated microscope slide with pollen grains. Invert the slide, and place it onto a pair of wooden supports in a covered petri dish. Leave the inverted slide in the covered petri dish for 1 hour. Then remove the slide, and examine it with the compound microscope. Have any of the pollen grains germinated?

Describe what you see. _____

_____

_____

## The Female Gametophyte

In the ovule, a megaspore undergoes three mitotic divisions to produce a seven-celled (eight-nuclei) female gametophyte called an **embryo sac** (Fig. 15.11). One of these cells is an egg cell. The largest cell contains two polar nuclei.

**Figure 15.11   Embryo sac in a lily ovule.**

An embryo sac is the female gametophyte of flowering plants. It contains seven cells, one of which is the egg. The fate of each cell is noted.

W. P. Armstrong 2004

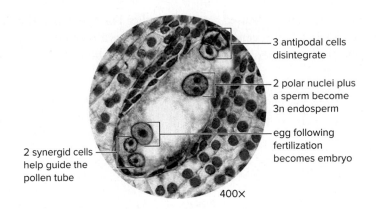

3 antipodal cells disintegrate

2 polar nuclei plus a sperm become 3n endosperm

egg following fertilization becomes embryo

2 synergid cells help guide the pollen tube

400×

### Observation: Embryo Sac Slide

Examine the demonstration slide of the mature embryo sac of *Lilium*. Identify the egg labeled in Figure 15.11.

Answer these questions.

1. What portion of the embryo sac tells you that it is the female gametophyte? _____

_____

2. When the pollen tube delivers sperm to the embryo sac, double fertilization occurs. Because of double fertilization, what happens to the egg? _____

_____

What happens to the polar nuclei? _____

_____

3. Following fertilization, the ovules develop into seeds, and the ovary develops into the fruit.

What are the three parts of a seed (as shown in Figure 15.10)? _____

*Label the flower remnants, the fruit, and the seeds in the drawings of a pea pod and apple provided.* The flesh of an apple comes from the enlarged receptacle that grows up and around the ovary, while the ovary largely consists of the core.

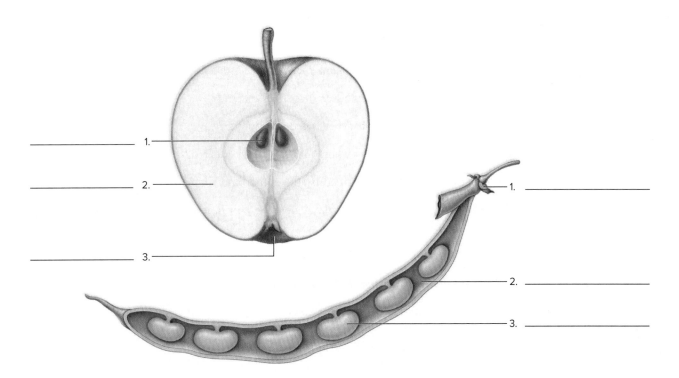

## 15.4   Comparison of Gymnosperms and Angiosperms

**1.** Complete Table 15.2 using "yes" or "no" to compare adaptations of gymnosperms and angiosperms.

| Table 15.2   Comparison of Gymnosperms and Angiosperms | | | | | |
|---|---|---|---|---|---|
| | Heterospores | Pollen grains/ Ovule | Cones | Flower | Fruit |
| Gymnosperms | | | | | |
| Angiosperms | | | | | |

**2.** What structure in gymnosperms and angiosperms delivers sperm to the vicinity of the egg? _____

**3.** What structure in gymnosperms and angiosperms becomes a seed? _____

**4.** The embryo of what generation, sporophyte or gametophyte, is in a seed? _____

**5.** What innovation in angiosperms led to the seeds being covered by fruit? _____

_____ 1. Which generation, gametophyte or sporophyte, is dispersed by seed plants?

_____ 2. What structure protects the female gametophyte in seed plants?

_____ 3. When meiosis occurs in a seed plant, what results?

_____ 4. What is the transfer of the seed plants' male gametophyte by wind or a carrier called?

_____ 5. What is the name given to plants that produce "naked seeds"?

_____ 6. How are the seeds of a pine tree dispersed?

_____ 7. In the life cycle of a pine, what is the sporophyte?

_____ 8. In the life cycle of a pine, where are the ovules located?

_____ 9. What flower structure is formed by the anther and filament?

_____ 10. What does mitotic division of the generative cell in a pollen grain produce?

_____ 11. What structure do the sperm move through to reach an angiosperm's female gametophyte?

_____ 12. What results from double fertilization in angiosperms?

_____ 13. What is the name of the female gametophyte of flowering plants?

_____ 14. What flower structure forms the covering around the seed of an angiosperm?

## Thought Questions

15. **a.** How are the gametophytes of mosses and flowering plants different from one another?

**b.** How are the gametophytes of mosses and flowering plants similar to one another?

16. Why do pines produce clusters of pollen cones, and why are the pollen cones located at the tips of branches instead of closer to the trunk of the tree?

17. *Angiosperm* means "covered seed." What covers the seeds of angiosperms, and what role does this covering have in ensuring the success of the embryos in the seeds?

# 16

# Organization of Flowering Plants

## Learning Outcomes

**16.1 External Anatomy of a Flowering Plant**
- Distinguish between the root system and the shoot system of a plant.
- Identify the external anatomical features of a flowering plant.

**16.2 Major Tissues of Roots, Stems, and Leaves**
- List and give functions for the major tissues in the organs of a plant.

**16.3 Root System**
- Identify a cross section and the specific tissues of a eudicot root and a monocot root.
- Identify various types of root diversity.

**16.4 Stems**
- Identify a cross section and the specific tissues of eudicot herbaceous and monocot herbaceous stems.
- Identify various types of stem diversity.
- Explain the occurrence of annual rings, and determine the age of a tree from a trunk cross section.

**16.5 Leaves**
- Identify a cross section and the specific tissues of a leaf.
- Distinguish between various arrangements of leaves on a stem.

**16.6 Xylem Transport**
- Explain the continuous water column in xylem.

## Introduction

### Pre-Lab

1. What are the three vegetative organs of a plant? _____

Despite their great diversity in size and shape, all flowering plants have three vegetative organs that have nothing to do with reproduction: the root, the stem, and the leaf (Fig. 16.1). Roots anchor a plant and absorb water and minerals from the soil. A stem usually supports the leaves so that they are exposed to sunlight. Leaves carry on photosynthesis and thereby produce the nutrients that sustain a plant and allow it to grow.

**Figure 16.1  Organization of plant body.**
Roots, stems, and leaves—the vegetative organs of a plant—are shown in this photo of an onion plant.
Dwight Kuhn

# 16.1   External Anatomy of a Flowering Plant

Figure 16.2 shows that a plant has a root system and a shoot system. The **root system** consists of the roots. The **shoot system** consists of the stem and the leaves.

*With the help of your textbook, label the following structures in Figure 16.2: axillary bud, blade, branch root, internode, leaf, node (2×), petiole, primary root, root hairs, root tip, stem, terminal bud, vascular tissues, vein.*

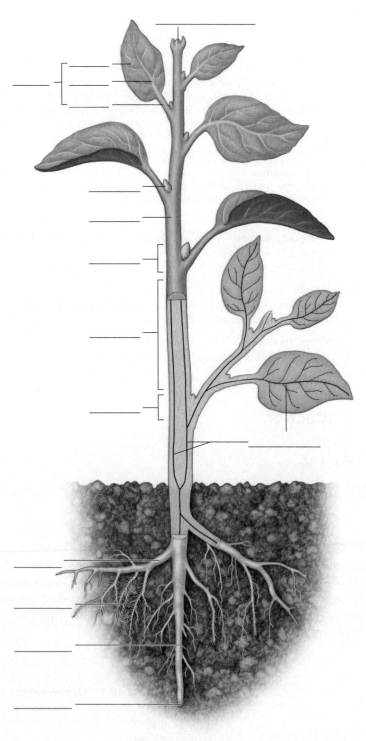

**Figure 16.2   Organization of a plant.**
Roots, stems, and leaves are vegetative organs.

# 16.2 Major Tissues of Roots, Stems, and Leaves

Unlike humans, flowering plants continue to grow in size their entire life due to **meristematic tissue** which is composed of cells that continuously divide. **Apical meristem** is located at the terminal end of the stem, the branches, and at the root tip and the root branches. When apical meristem cells divide, some of the cells differentiate into the mature tissues of a plant:

   **Dermal tissue:** forms the outer protective covering of a plant organ

   **Ground tissue:** fills the interior of a plant organ; photosynthesizes and stores the products of photosynthesis

   **Vascular tissue:** transports water and sugar, the product of photosynthesis, in a plant and provides support

Note in Table 16.1 that roots, stems, and leaves have all three tissues but they are given different specific names.

### Table 16.1   Mature Tissues of Vegetative Organs

| Tissue Type | Roots | Stems | Leaves |
|---|---|---|---|
| 1. Dermal tissue (epidermis) | Protects inner tissues Root hairs absorb water and minerals. | Protects inner tissues | Protects inner tissues Cuticle prevents $H_2O$ loss. Stomata carry on gas exchange. |
| 2. Ground tissue | Cortex: Stores products of photosynthesis<br><br>Pith: Stores products of photosynthesis | Cortex: Carries on photosynthesis, if green<br><br>Pith: Stores products of photosynthesis | Mesophyll: Photosynthesis |
| 3. Vascular tissue (xylem and phloem) | Vascular cylinder: Transports water and nutrients | Vascular bundle: Transports water and nutrients | Leaf vein: Transports water and nutrients |

### Observation: Tissues of Roots, Stems, and Leaves

Meristematic tissue. A slide showing the apical meristem in a shoot tip and another showing the apical meristem in a root tip are on demonstration. Meristematic cells are spherical and stain well when they are dense and have thin cell walls (Fig. 16.3).

**Figure 16.3**
**Apical meristem.**
A shoot tip and a root tip contain meristem tissue, which allows them to grow longer the entire life of a plant.

(left) Steven P. Lynch; (right) ©Ray F. Evert, University of Wisconsin

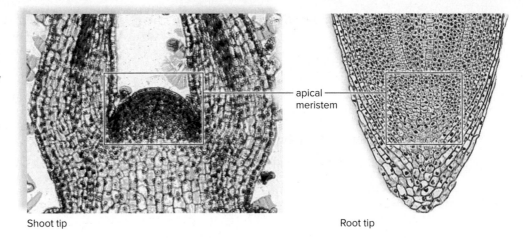

apical meristem

Shoot tip

Root tip

## Mature Tissues

1. Dermal tissue. In a cross section of a leaf (Fig. 16.4), for example, focus only on the upper or lower epidermis. Epidermal cells tend to be square or rectangular in shape. In a leaf, the epidermis is interrupted by openings called **stomata** (sing., stoma). Later you will have an opportunity to see the epidermis in roots, stems, and leaves. What is a function of epidermis in all

   three organs (see Table 16.1)? _____

   _____

   _____

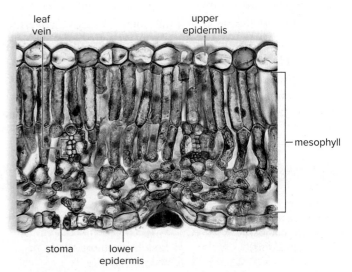

leaf vein

upper epidermis

mesophyll

stoma

lower epidermis

**Figure 16.4   Microscopic leaf structure.**
Like the stem and root, a leaf contains epidermal tissue, vascular tissue (leaf vein), and ground tissue (mesophyll).

©Ray F. Evert, University of Wisconsin

2. Ground tissue. The ground tissue fills the space between epidermis in roots, stems, and leaves. In leaves, for example, ground tissue is called mesophyll (Fig. 16.4). Ground tissue largely contains parenchyma cells and sclerenchyma cells. **Parenchyma cells** can be of different sizes and vary from fairly circular to oval. Those that contain chloroplasts carry on photosynthesis. Those that contain leucoplasts store starches and oils. **Sclerenchyma cells** are usually elongate and have thick walls impregnated with lignin. These dead cells appear hollow, and the presence of lignin means that they stain a red color. Sclerenchyma cells are strong and provide support. Which type of cell (parenchyma or sclerenchyma) carries on photosynthesis or stores the products of photosynthesis

   in a leaf? _____ Which one lends strength to ground tissue in roots and

   stems? _____

**3.** Vascular tissue. In a leaf, strands of vascular tissue are called leaf veins (Fig. 16.4). There are two types of vascular tissue, called xylem and phloem. **Xylem** contains hollow dead cells that transport water. The presence of lignin makes the cell walls strong, stains red, and makes xylem easy to spot. **Phloem** contains thin-walled, smaller living cells that transport sugars in a plant. Phloem is harder to locate than xylem, but it is always found in association with xylem. Which type of tissue (xylem or phloem) transports sugars in a plant? _____ Which type of tissue transports water? _____

## Monocots Versus Eudicots

Flowering plants are classified into two major groups: **monocots** and **eudicots.** In this laboratory, you will be studying the differences between monocots and eudicots as noted in Figure 16.5. The arrangement of tissues is distinctive enough that you should be able to identify the plant as a monocot or eudicot when examining a slide of a root, stem, or leaf.

### Experimental Procedure: Monocot Versus Eudicot

**1.** Examine the live plant again (see Fig. 16.2). The leaf vein pattern—that is, whether the veins run parallel to one another or whether the veins spread out from a central location (called the net pattern)—indicates that a plant is either a monocot or a eudicot. Is the plant in Figure 16.2 a monocot or eudicot? _____ The leaf pattern in Figure 16.4 appears to be parallel. Is this the leaf of a monocot or eudicot? _____

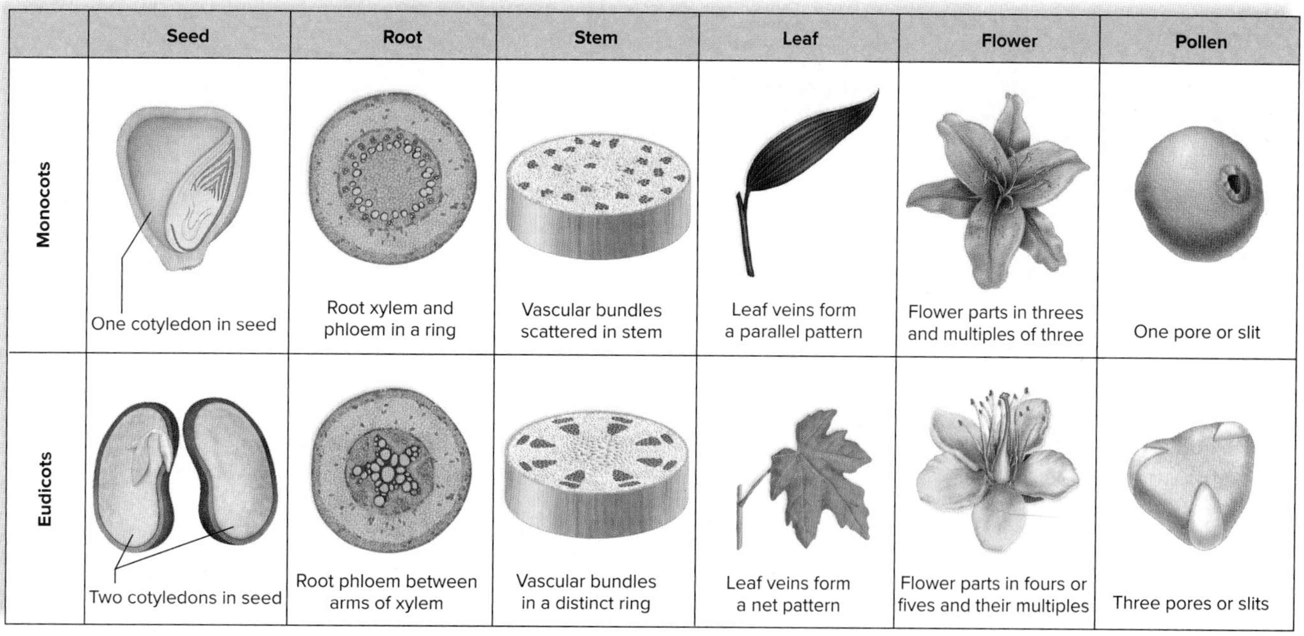

| | Seed | Root | Stem | Leaf | Flower | Pollen |
|---|---|---|---|---|---|---|
| **Monocots** | One cotyledon in seed | Root xylem and phloem in a ring | Vascular bundles scattered in stem | Leaf veins form a parallel pattern | Flower parts in threes and multiples of three | One pore or slit |
| **Eudicots** | Two cotyledons in seed | Root phloem between arms of xylem | Vascular bundles in a distinct ring | Leaf veins form a net pattern | Flower parts in fours or fives and their multiples | Three pores or slits |

**Figure 16.5  Monocots versus eudicots.**
The six features illustrated here are used to distinguish monocots from eudicots.

2. Observe any other available plants, and record in Table 16.2 the name of the plant and note the leaf vein pattern. On the basis of the leaf vein pattern, decide if this plant is a monocot or a eudicot.

3. Aside from leaf vein pattern, other external features indicate whether a plant is a monocot or a eudicot. For example, open a peanut if available. The two halves you see are cotyledons. Is the plant that produced the peanut a monocot or eudicot? _____
If available, examine a flower. If a flower has three petals or six petals, or any multiple of three, is the plant that produced this flower a monocot or a eudicot? _____

In this laboratory, you will have the opportunity to examine the cross sections of roots and stems microscopically; the arrangement of vascular tissue roots and stems also indicates whether a plant is a monocot or a eudicot.

| Table 16.2 Monocots Versus Eudicots | | |
|---|---|---|
| **Name of Plant** | **Organization of Leaf Veins** | **Monocot or Eudicot?** |
| | | |
| | | |
| | | |

## 16.3 Root System

### Pre-Lab

5. What are the functions of xylem and phloem? _____
_____

The **root system** anchors the plant in the soil, absorbs water and minerals from the soil, and stores the products of photosynthesis received from the leaves.

### Anatomy of a Root Tip

**Primary growth** of a plant increases its length. Note the location of the root apical meristem in Figure 16.6. As primary growth occurs, root cells enter zones that correspond to various stages of differentiation and specialization.

# Figure 16.6 Eudicot root tip.

In longitudinal section, the root cap is followed by the zone of cell division, zone of elongation, and zone of maturation.

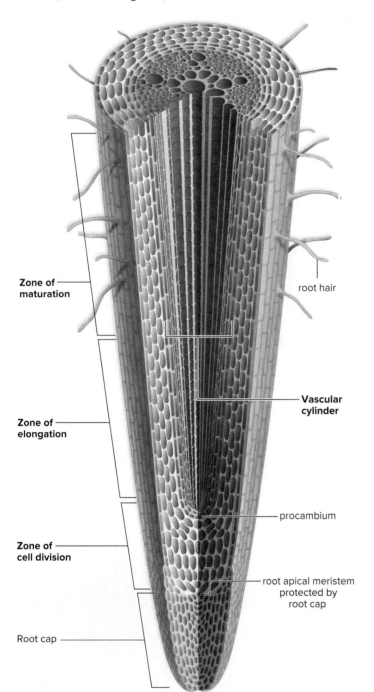

Zone of maturation

root hair

Vascular cylinder

Zone of elongation

procambium

Zone of cell division

root apical meristem protected by root cap

Root cap

1. Examine a model and/or a slide of a root tip (Fig. 16.6).
2. Identify the **root cap** (dead cells at the tip of a plant that provide protection as the root grows).
3. Locate the **zone of cell division.** Apical meristem is found in this zone. As mentioned previously, meristematic tissue is composed of embryonic cells that continually divide, providing new cells for root growth.
4. Find the **zone of elongation.** In this zone, the newly produced cells elongate as they begin to grow larger.
5. Identify the **zone of maturation.** In this zone, the cells become differentiated into particular cell types. When epidermal cells differentiate, they produce **root hairs,** small extensions that absorb water and minerals. Also noticeable are the cells that make up the xylem and phloem of vascular tissue.

## Anatomy of Eudicot and Monocot Roots

Eudicot and monocot roots differ in the arrangement of their vascular tissue.

### Observation: Cross-Section Anatomy of Eudicot and Monocot Roots

#### Eudicot Root

1. Obtain a prepared cross-section slide of a buttercup (*Ranunculus*) root. Use both low power and high power to identify the **epidermis** (the outermost layer of small cells that gives rise to root hairs). The epidermis protects inner tissues and absorbs water and minerals.
2. Locate the **cortex,** which consists of several layers of thin-walled cells (Fig. 16.7a, b). In Figure 16.7b, note the many stained starch grains in the cortex cells. The cortex is ground tissue that functions in food storage.
3. Find the **endodermis,** a single layer of cells whose walls are thickened by a layer of waxy material known as the **Casparian strip.** (It is as though these cells are glued together with a waxy glue.) Because of the Casparian strip, the only access to the xylem is through the living endodermal cells. What materials that enter the plant through it's root, will the endodermis regulate? _____

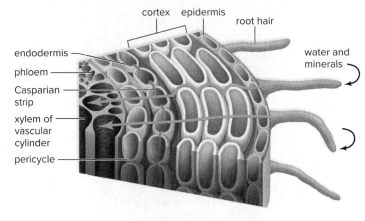

Use this illustration to trace the path of water and minerals from the root hairs to xylem. _____

4. Identify the **pericycle,** a layer one or two cells thick just inside the endodermis. Branch roots originate from this tissue.
5. Locate the **xylem** in the vascular cylinder of the root. Xylem has several "arms" that extend like the spokes of a wheel. This tissue conducts water and minerals from the roots to the stem.
6. Find the **phloem,** located between the arms of the xylem. Phloem conducts organic nutrients from the leaves to the roots and other parts of the plant.

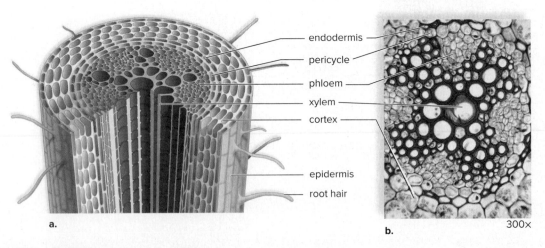

**Figure 16.7  Eudicot root cross section.**
The vascular cylinder of a dicot root contains the vascular tissue. Xylem is typically star-shaped, and phloem lies between the points of the star. **a.** Drawing. **b.** Micrograph.

(b) Richard Gross/McGraw Hill

### Monocot Root

1. Obtain a prepared cross-section slide of a corn (*Zea mays*) root (Fig. 16.8*a*, *b*). Use both low power and high power to identify the six tissues mentioned for the eudicot root.

2. In addition, identify the **pith,** a centrally located ground tissue that functions in food storage.

**Figure 16.8  Monocot root cross section.**
**a.** Micrograph of a monocot root cross section. **b.** An enlarged portion.

(a) Dr. Keith Wheeler/Science Source; (b) George Ellmore, Tufts University

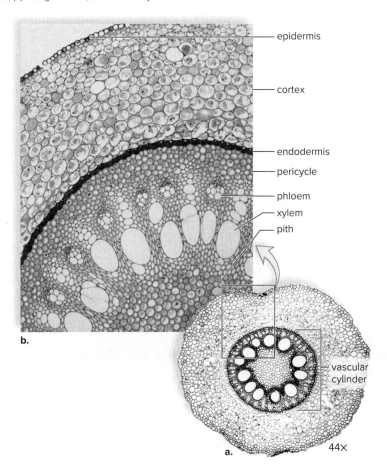

### Comparison

Contrast the arrangement of vascular tissue (xylem and phloem) in the vascular cylinder of monocot roots and eudicot roots by *writing "monocot" or "eudicot" on the appropriate line.*

_____ Xylem has the appearance of a wheel. Phloem is between the spokes of the wheel.

_____ Ring of xylem (inside) and phloem (outside) surrounds pith.

## Root Diversity

Roots are quite diverse, and we will take this opportunity to become acquainted with only a few select types.

Taproots and fibrous roots. Most plants have either a taproot or a fibrous root. Note in Figure 16.9a that carrots have a **taproot.** The main root is many times larger than the branch roots. Grasses such as the type shown in Figure 16.9b have a **fibrous root:** All the roots are approximately the same size.

Examine the taproots on display, and name one or two in which the taproot is enlarged for storage. _____

_____

The dandelion on display has a taproot. Describe. _____

_____

Taproots often function in food storage.

Adventitious roots. Some plants have **adventitious** roots. Roots that develop from nonroot tissues, such as nodes of stems, are called adventitious roots. Examples include the prop roots of corn (Fig. 16.9c) and the aerial roots of ivy that attach this plant to structures such as stone walls.

Which plants on display have adventitious roots? _____

Other types of roots. Mangroves and other swamp-dwelling trees have roots called pneumatophores that rise above the waterline. Pneumatophores have numerous lenticels, which are openings that allow gas exchange to occur.

What root modifications not noted here are on display in your laboratory? _____

_____

**Figure 16.9  Root specializations.**
**a.** Potato plants have food storage roots. **b.** Prop roots are specialized for support. **c.** The pneumatophores of this tree allow it to acquire oxygen even though it lives in water. **d.** The aerial roots of orchids offer physical support, water, and nutrient uptake, and in some cases even photosynthesis.
(a) Madlen/Shutterstock; (b) NokHoOkNoi/iStock/Getty Images; (c) Pat Canova/Photographer's Choice/Getty Images; (d) STUDIO75/Alamy Stock Photo

a.        b.        c.        d.

# 16.4  Stems

**Pre-Lab**

6. Are the majority of trees monocots or eudicots? _____

7. How is growth in herbaceous stems different from woody stems? _____

_____

Stems are usually found aboveground where they provide support for leaves and flowers. Vascular tissue extends from the roots through the stem and its branches to the leaves.

Stems that do not contain wood are called **herbaceous,** or nonwoody, stems. Usually, monocots remain herbaceous throughout their lives. Some eudicots, such as those that live a single season, are also herbaceous. Other eudicots, namely trees, become woody as they mature.

## Anatomy of Herbaceous Stems

Herbaceous stems undergo primary growth. **Primary growth** results in an increase in length due to the activity of the apical meristem located in the terminal bud (see Fig. 16.3) of the shoot system.

### Observation: Anatomy of Eudicot and Monocot Herbaceous Stems

#### Eudicot Herbaceous Stem

1. Examine a prepared slide of a eudicot herbaceous stem (Fig. 16.10), and identify the **epidermis** (the outer protective layer). *Label the epidermis in Figure 16.10a.*
2. Locate the **cortex,** which may photosynthesize or store nutrients.
3. Find a **vascular bundle,** which transports water and organic nutrients. The vascular bundles in a eudicot herbaceous stem occur in a ring pattern. *Label the vascular bundle in Figure 16.10a.* Which vascular tissue (xylem or phloem) is closer to the surface?_____
4. *Label the central **pith,** which stores organic nutrients.* Both cortex and pith are composed of which tissue type listed in Table 16.1? _____

**Figure 16.10   Eudicot herbaceous stem.**
The vascular bundles are in a definite ring in this photomicrograph of a eudicot herbaceous stem. *Complete the labeling as directed by the Observation.*

(a) Ed Reschke; (b) Ray F. Evert, University of Wisconsin

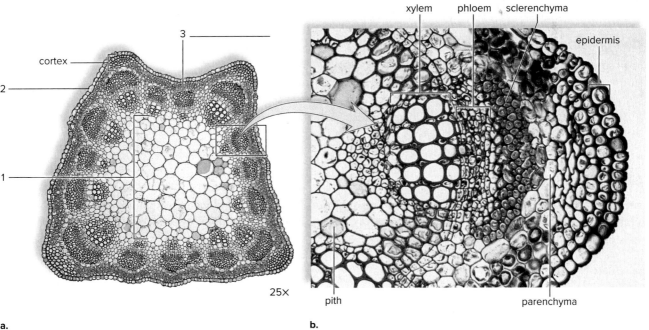

a.                              b.

## Monocot Herbaceous Stem

1. Examine a prepared slide of a monocot herbaceous stem (Fig. 16.11). Locate the epidermal, ground, and vascular tissues in the stem.

**Figure 16.11   Monocot stem.**
The vascular bundles, one of which is enlarged, are scattered in this photomicrograph of a monocot herbaceous stem.

(left) Artem Povarov/123RF; (right) Kingsley Stern

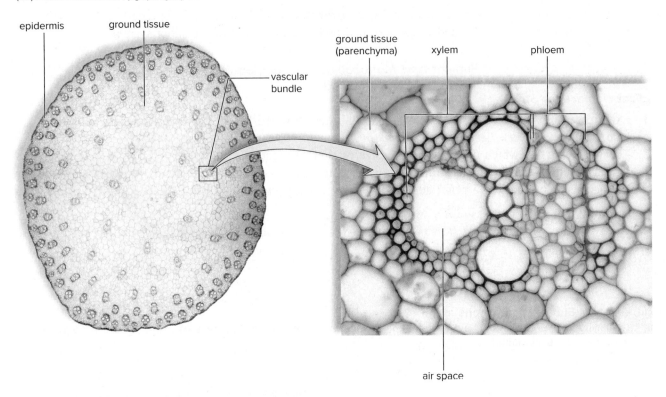

### Comparison

1. Compare the arrangement of ground tissue in eudicot and monocot stems. _____

_____

2. Compare the arrangement of vascular bundles in the stems of eudicots and monocots. _____

_____

### Stem Diversity

Stems can be quite diverse as well. We take this opportunity to become acquainted with those that allow a plant to accomplish vegetative reproduction and/or function in food storage. Several of these and other types of stems may be on display in the laboratory.

Stolons. The strawberry plant in Figure 16.12a has a horizontal aboveground stem called a runner or **stolon.** The stolon produces adventitious roots and new shoots at nodes. *Label an adventitious root and a new shoot in Figure 16.12a.*

List any other plants on display that spread and produce new shoots by sending out stolons. _____
_____

Rhizomes. An iris has a belowground horizontal stem called a **rhizome,** which functions as a fleshy food storage organ (Fig. 16.12b). New plants can grow from a single piece of the rhizome.

List any other plants on display that have the same belowground horizontal stems (rhizomes) as the iris.
_____

Tubers. A white potato has a belowground rhizome that gives off food storage **tubers** (Fig. 16.12c). Each eye is a node that can produce new plants.

List any other types of plants on display whose belowground stems have tubers. _____
_____

Corms. A gladiolus has a belowground vertical stem called a **corm,** which functions in food storage and has thin, papery leaves (Fig. 16.12d).

List any other types of plants on display that have a vertical stem called a corm. _____
_____

**Figure 16.12   Stem diversity.**
**a.** Stolon of a strawberry plant. **b.** Rhizome of an iris. **c.** Tuber of a potato. **d.** Corm of a gladiolus.

(a) Evelyn Jo Johnson/McGraw Hill; (b) jelloyd/Shutterstock; (c) Carlyn Iverson/McGraw Hill; (d) LanaSweet/Shutterstock

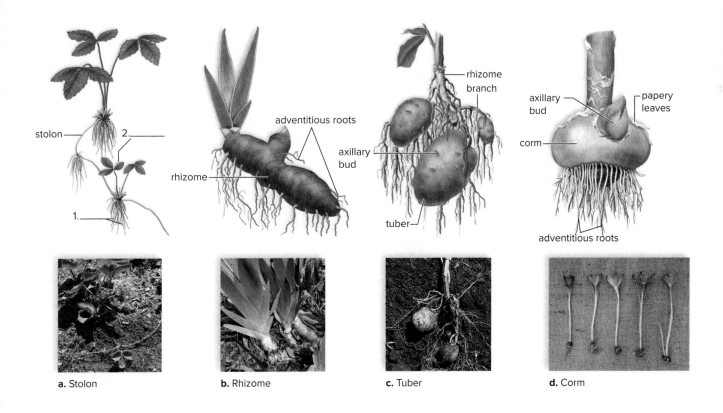

a. Stolon          b. Rhizome          c. Tuber          d. Corm

## Anatomy of Woody Stems

Woody stems undergo both primary growth (increase in length) and secondary growth (increase in girth). When *primary growth* occurs, the apical meristem within a terminal bud is active. When *secondary growth* occurs, the vascular cambium is active. **Vascular cambium** is meristem tissue, which produces new xylem and phloem called **secondary xylem** and **phloem** each year. The buildup of secondary xylem year after year is called **wood.** Complete Table 16.3 to distinguish between primary growth and secondary growth of a stem.

| Table 16.3  Primary Growth Versus Secondary Growth | | |
|---|---|---|
| | **Primary Growth** | **Secondary Growth** |
| Active meristem | | |
| Result | | |

### Observation: Anatomy of a Winter Twig

1. A winter twig typically shows several past years' primary growth. Examine several examples of winter twigs (Fig. 16.13), and identify the **terminal bud** located at the tip of the twig. This is where new primary growth will originate. During the next growing season, the terminal bud produces new tissues including vascular bundles and leaves.
2. Locate a **terminal bud scar.** These are marks left on a stem from terminal bud scales (modified leaves protecting the bud). The distance between two adjacent terminal bud scars equals one year's primary growth.
3. Find a **leaf scar.** Mark where a leaf was attached to the stem.
4. Note the **bundle scars.** Complete this sentence: Marks left in the leaf scar where the vascular tissue

   _____.
5. Identify a **node.** This is the region where you find leaf scars and bundle scars. The region between nodes

   is called an _____.
6. Locate an **axillary bud.** This is where new branch growth can occur.
7. Note the numerous lenticels, breaks in the outer surface where gas exchange can occur.

**Figure 16.13  External structure of a winter twig.**
Counting the terminal bud scars tells the age of a particular branch.

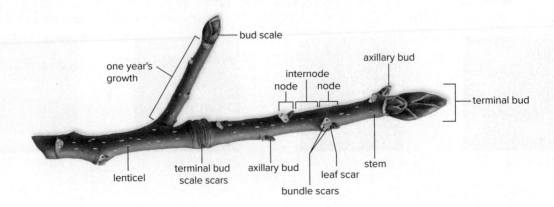

1. Examine a prepared slide of a cross section of a woody stem (Fig. 16.14), and identify the **bark** (the dark outer area), which contains **cork,** a protective outer layer; **cortex,** which stores nutrients; and **phloem,** which transports organic nutrients.

**Figure 16.14   Woody eudicot stem cross section.**
Because xylem builds up year after year, it is possible to count the annual rings to determine the age of a tree. This tree is three years old.

Ed Reschke/Stone/Getty Images

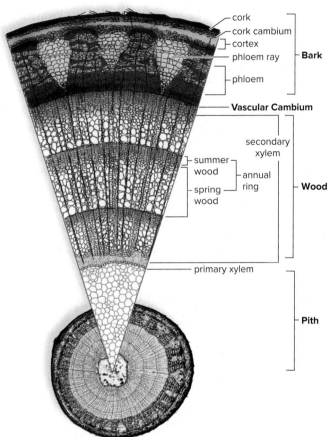

2. Locate the **vascular cambium** at the inner edge of the bark, between the bark and the wood. Vascular cambium is meristem tissue whose activity accounts for secondary growth, which causes increased girth of a tree. Secondary phloem (which disappears) and secondary xylem (which builds up) are produced by vascular cambium each growing season.
3. Find the **wood,** which contains annual rings. An **annual ring** is the amount of xylem added to the plant during one growing season. Rings appear to be present because spring wood has large xylem vessels and looks light in color, while summer wood has much smaller vessels and appears much darker. How old is the stem you are observing? _____ Are all the rings the same width? _____
4. Identify the **pith,** a ground tissue at the center of a woody stem that stores organic nutrients and may disappear.
5. Locate **rays,** groups of small, almost cuboid cells that extend out from the pith laterally.

## 16.5   Leaves

**Pre-Lab**

8. Why are leaves typically the topmost and outermost structures on a plant? _____
_____

A **leaf** is the organ that produces food for the plant by carrying on photosynthesis. Leaves are generally broad and quite thin. An expansive surface facilitates the capture of solar energy and gas exchange. Water and nutrients are transported to the cells of a leaf by leaf veins, extensions of the vascular bundles from the stem.

## Anatomy of Leaves

### Observation: Anatomy of Leaves

1. Examine a model of a leaf. With the help of Figure 16.15, identify the waxy **cuticle,** the outermost layer that protects the leaf and prevents water loss.

2. Locate the **upper epidermis** and **lower epidermis,** single layers of cells at the upper and lower surfaces. Trichomes are hairs that grow from the upper epidermis and help protect the leaf from insects and water loss.

3. Find the leaf veins in your model. The bundle sheath is the outer boundary of a vein; its cells surround and protect the vascular tissue. If this is a model of a monocot, all the leaf veins will be _____ _____. If this is a model of a eudicot, some leaf veins will be circular and some will be oval. Why? _____

4. Identify the **palisade mesophyll,** located near the upper epidermis. These cells contain chloroplasts and carry on most of the plant's photosynthesis. Locate the **spongy mesophyll,** located near the lower epidermis. These cells have air spaces that facilitate the exchange of gases across the plasma membrane. *Label the layers of mesophyll in Figure 16.15.* Collectively, the mesophyll represents which of the three types of tissue found in all parts of a plant (see Table 16.1)? _____

5. *Label the two layers of epidermis in Figure 16.15.* Find a **stoma** (pl., stomata), an opening through which gas exchange occurs and water escapes. Stomata are more numerous in the lower epidermis. A stoma has two guard cells that regulate its opening and closing.

**Figure 16.15   Leaf anatomy.**

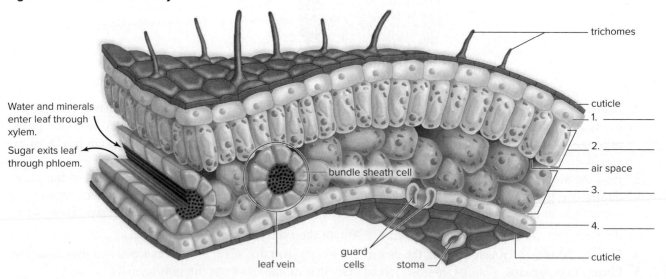

Water and minerals enter leaf through xylem.

Sugar exits leaf through phloem.

bundle sheath cell

leaf vein

guard cells

stoma

trichomes

cuticle
1. _____

2. _____

air space

3. _____

4. _____

cuticle

## Leaf Diversity

A eudicot leaf consists of a flat blade and a stalk, called the petiole. An axillary bud appears at the point where the petiole attaches a leaf to the stem. In other words, an axillary bud is a tip-off that you are looking at a single leaf.

## Observation: Leaf Diversity

Several types of eudicot leaves will be on display in the laboratory. Examine them using these directions.

- A leaf may be **simple,** in which case it consists of a single blade; or a leaf may be **compound,** meaning its single blade is divided into leaflets. *In Figure 16.16a write* simple *or* compound *before the word* leaf *in 1–3.* (Among the leaves on display, find one that is simple and one that is compound.)
- A compound leaf can be **palmately** compound, meaning the leaflets are spread out from one point. Which leaf in Figure 16.16a is palmately compound? *Add* palmately *in front of* compound *where appropriate.* (See if you can find a palmately compound leaf among those on display.)
- A compound leaf can be **pinnately** compound, meaning the leaflets are attached at intervals along the petiole. Which leaf in Figure 16.16a is pinnately compound? *Add* pinnately *in front of* compound *where appropriate.* (See if you can find a pinnately compound leaf among those on display.)
- As shown in Figure 16.16b, leaves can be in various positions on a stem. On which stem in Figure 16.16b, 4–6, are the leaves **opposite** one another? *Write the word* opposite *where appropriate.* On which stem do the leaves **alternate** along the stem? *Write the word* alternate *where appropriate.* On which stem do the leaves **whorl** about a node? *Add the word* whorl *where appropriate in Figure 16.16b.* (See if you can find different arrangements of leaves among the stems on display.)

**Figure 16.16 Classification of leaves.**

1. _____ leaf, magnolia

**a.** Simple versus compound leaves

axillary bud

2. _____

leaf, buckeye

3. _____

leaf, black walnut

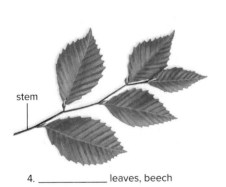

stem

4. _____ leaves, beech

stem

5. _____ leaves, bedstraw

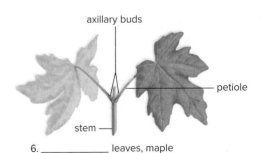

axillary buds

petiole

stem

6. _____ leaves, maple

**b.** Arrangement of leaves on stem

## 16.6  Xylem Transport

### Pre-Lab

9. Is water pushed or pulled up a plant from roots to leaves? _____

Xylem (Fig. 16.17), transports water from the roots to the leaves and contains two types of conducting cells: tracheids and vessel elements. Both types of conducting cells are hollow and nonliving; the vessel elements are larger, lack transverse end walls, and are arranged to form a continuous pipeline for water and mineral transport.

## Water Column

The water column in xylem is continuous because water molecules are cohesive (they cling together) and because water molecules adhere to the sides of xylem cells. Evaporation of water from the leaves occurs when the stomata are open and this process is called transpiration. This water loss creates a force that pulls the water column upward from the roots to the leaves.

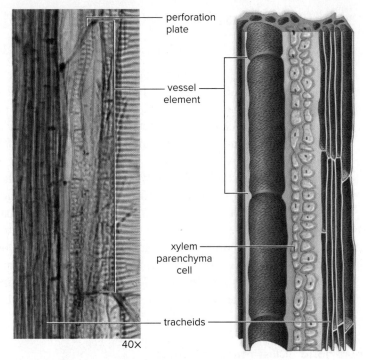

perforation plate

vessel element

xylem parenchyma cell

tracheids

40×

**Figure 16.17 Xylem structure.**
Xylem contains two types of conducting cells: tracheids and vessel elements. Tracheids have pitted walls, but vessel elements are larger and form a continuous pipeline from the roots to the leaves.

(left) Garry Delong/Science Source

### *Experimental Procedure: The Water Column*

1. Place a small amount of red-colored water in two beakers. *Label one beaker "wet" and the other beaker "dry."*
2. Transfer a stalk of celery (which was cut and then immediately placed in a container of water) into the "wet" beaker so that the large end is in the colored water.
3. Transfer a stalk of celery of approximately the same length and width (that was kept in the air after being cut) into the "dry" beaker so that the large end is in the colored water.
4. With scissors, cut off the top end of each stalk, leaving about 10 cm total length.
5. Time how long it takes for the red-colored water to reach the top of each stalk, and record these data in Table 16.4.
6. In which celery stalk was the water column broken? _____
   Use this information to write a conclusion in Table 16.4.
7. Make a cross-sectional wet mount of the stalk in the "wet" beaker. Observe this slide under the microscope. What type of tissue has been stained by the dye?

| Table 16.4 Celery Stalk Experiment | | |
|---|---|---|
| **Stalk** | **Speed of Dye (Minutes)** | **Conclusion** |
| Cut end placed in water prior to experiment | | |
| Cut end kept in air prior to experiment | | |

_____ **1.** What kind of plant tissue is composed of cells that divide?

_____ **2.** What term is used to identify ground tissue in a leaf?

_____ **3.** What kind of plant, monocot or eudicot, has the xylem and phloem in its leaves arranged in a parallel pattern?

_____ **4.** In what zone of a eudicot root will tracheids and vessel elements be found?

_____ **5.** What type of tissue gives rise to root hairs?

_____ **6.** How is xylem typically arranged in a eudicot root cross section?

_____ **7.** What term refers to stems that are nonwoody?

_____ **8.** What kind of plant, monocot or eudicot, has stem vascular bundles that occur in a ring?

_____ **9.** What kind of growth produces an increase in the diameter of a woody stem?

_____ **10.** Where does new primary growth occur on a winter twig?

_____ **11.** What type of vascular tissue forms the annual rings in a woody stem?

_____ **12.** What is the outermost layer of a leaf that protects and prevents water loss?

_____ **13.** What cells are on either side of a stoma and regulate whether it is open or closed?

_____ **14.** If its leaf veins form a net pattern and there is one blade attached to a petiole, is the leaf simple or compound, and is it from a monocot or a eudicot?

## Thought Questions

**15.** Where is glucose produced in a leaf? What type of tissue carries glucose away from a leaf, and where is the glucose stored in a plant like a beet?

**16.** Stomata are more numerous in the lower epidermis of leaves. Explain why it would be detrimental to the plant to have more stomata in the upper epidermis of leaves.

**17.** Compare and contrast monocots and eudicots by identifying two similarities and two differences between them.

# 17
# Invertebrates

## Pre-Lab

**1.** List the organisms that will be studied during this lab, and describe their main similarities. _____

_____

_____

_____

## Introduction

This laboratory focuses on invertebrates in the animal kingdom. Animals are all multicellular, heterotrophic and have varying degrees of motility at some point of their lifecycle. **Invertebrates** were the first animals to evolve, and by contrast with vertebrates, they lack an internal skeleton of bone or cartilage.

One of the themes of today's laboratory will be the similarities and differences between the invertebrate animals studied. All of the animals in today's laboratory have true tissues, bilateral symmetry, a complete digestive tract, and a **coelom** or body cavity. A significant difference between them is that molluscs are nonsegmented, while the annelids and arthropods are segmented animals. Arthropods also have jointed appendages and are particularly adapted for locomotion on land.

Each animal in this lab is a **coelomate,** meaning they have a true internal body cavity. However, molluscs, annelids, and arthropods are protostomes, or coelomate animals in which the first (*proto*) embryonic opening becomes the mouth (*stoma*). Echinoderms and vertebrates are examples of animals that are deuterostomes. In coelomates that are deuterostomes, the first embryonic opening becomes the anus, and the second (*deutero*) opening becomes the mouth.

**Pre-Lab**

2. Why do all animals share similar characteristics? _____

_____

_____

_____

## 17.1   Evolution of Animals

Today, molecular data are used in addition to comparative anatomy to trace the evolutionary history of animals. All animals share a common ancestor (Fig. 17.1). This common ancestor was most likely a choanoflagellate. All but one of the phyla depicted in the tree consist of only invertebrates—the chordates contain a few invertebrates as well as the vertebrates.

Certain anatomical features of animals are used in the tree. The first feature of interest is formation of tissue layers. Sponges have no true tissue layers. The other phyla have two or three tissue layers.

Another feature of interest is **symmetry. Asymmetry** means the animal has no particular symmetry, as in sponges. Some phyla have **radial symmetry** and most have **bilateral symmetry.**

Finally, complex animals are either **protostomes** (first opening during development is the mouth) or **deuterostomes** (first opening during development is the anus, the second opening forms the mouth).

## Figure 17.1 Evolution of animals.

All animal phyla living today are most likely descended from a colonial flagellated protist living about 600 MYA. This phylogenetic (evolutionary) tree is a hypothesis and uses morphological and molecular data (rRNA sequencing) to determine which phyla are most closely related to one another. While many relationships are still not clear, placement of Ctenophora and relationships among Spiralia are most uncertain (marked with red dots).

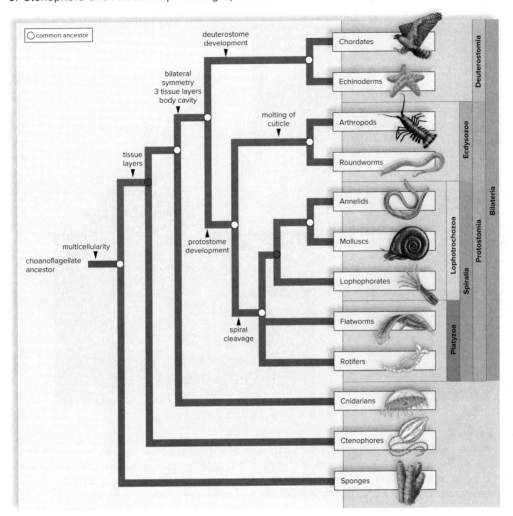

## 17.2 Molluscs

### Pre-Lab

3. *With the help of your textbook, complete the labeling of Figures 17.4 and 17.5.*

Most **molluscs** (phylum Mollusca) are marine, but there are also some freshwater and terrestrial molluscs (Fig. 17.2). Among molluscs, the grazing marine herbivores, known as **chitons,** have a body flattened dorsoventrally covered by a shell consisting of eight plates (Fig. 17.2a). The **bivalves** contain marine and freshwater sessile filter feeders, such as clams and scallops, with a body enclosed by a shell consisting of two valves (Fig. 17.2b). The **gastropods** contain marine, freshwater, and terrestrial species. In snails, the shell, if present, is coiled (Fig. 17.2c). The **cephalopods** contain active marine predators, such as squids and nautiluses. Tentacles are used to capture their prey (Fig. 17.1d).

All molluscs have a three-part body consisting of (1) a muscular **foot** specialized for various means of locomotion; (2) **visceral mass** that includes the internal organs; and (3) a **mantle,** a thin tissue that encloses the visceral mass and may secrete a shell. **Cephalization** is the development of a head region.

**a.** Chiton, *Tonicella*

**c.** Snail, *Helix* is a gastropod.

**b.** Scallop, *Aequipecten* is a bivalve.

**d.** Nautilus, *Nautilus,* is a cephalopod.

**Figure 17.2   Molluscan diversity.**

**a.** You can see the exoskeleton of this chiton but not its dorsally flattened foot. **b.** A scallop doesn't have a foot but it does have strong adductor muscles to close the shell. In this specimen, the edge of the mantle bears tentacles and many blue eyes. **c.** A gastropod, such as a snail, is named for the location of its large foot beneath the visceral mass. **d.** In a cephalopod, such as this nautilus, a funnel (its foot) opens in the area of the tentacles and allows it to move by jet propulsion.

(a) Randimal/Shutterstock; (b) NHPA/Photoshot; (c) Rosemary Calvert/Getty Images; (d) Douglas Faulkner/Science Source

## Anatomy of a Clam

Clams are bivalved because they have right and left shells secreted by the mantle. Clams lack a distinct head, and can burrow into sand by extending a **muscular foot** between the valves. Clams are **filter feeders** and feed on debris that enters the mantle cavity. In the visceral mass, the blood leaves the heart and enters sinuses (cavities) by way of anterior and posterior aortas. There are many different types of clams. The one examined here is the freshwater clam *Venus*.

### External Anatomy

1. Examine the external shell (Fig. 17.3) of a preserved clam (*Venus*). The shell is an **exoskeleton.**
2. Find the posterior and anterior ends. The more pointed end of the **valves** (the halves of the shell) is the posterior end.
3. Determine the clam's dorsal and ventral regions. The valves are hinged together dorsally.
4. What is the function of a heavy shell? _____

**Figure 17.3   External view of the clam shell.**

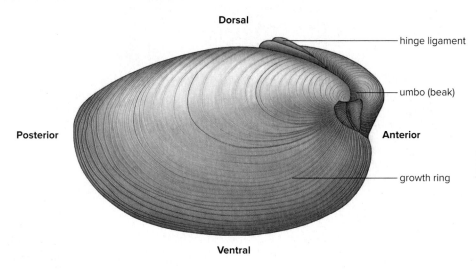

### Internal Anatomy

1. Place the clam in the dissecting pan, with the **hinge ligament** and **umbo** (blunt dorsal protrusion) down. Carefully separate the **mantle** from the right valve by inserting a scalpel into the slight opening of the valves. What is a mantle? _____

   _____

2. Insert the scalpel between the mantle and the valve you just loosened.
3. The **adductor muscles** hold the valves together. Cut the adductor muscles at the anterior and posterior ends by pressing the scalpel toward the dissecting pan. After these muscles are cut, the valve can be carefully lifted away. What is the advantage of powerful adductor muscles? _____

   _____

4. Examine the inside of the valve you removed. Note the concentric lines of growth on the outside, the hinge teeth that interlock with the other valve, the adductor muscle scars, and the mantle line. The inner layer of the shell is mother-of-pearl.
5. Examine the rest of the clam (Fig. 17.4) attached to the other valve. Notice the adductor muscles and the mantle, which lies over the visceral mass and foot.
6. Bring the two halves of the mantle together. Explain the term *mantle cavity.* _____

   _____

7. Identify the **incurrent** (more ventral) and **excurrent siphons** at the posterior end (Fig. 17.4).

   Explain how water enters and exits the mantle cavity. _____

   _____

# Figure 17.4  Anatomy of a bivalve.

The mantle has been removed to reveal the internal organs. **a.** Drawing. *Fill in the missing labels with the help of your textbook.* **b.** Dissected specimen.

(b) Ken Taylor/Wildlife Images

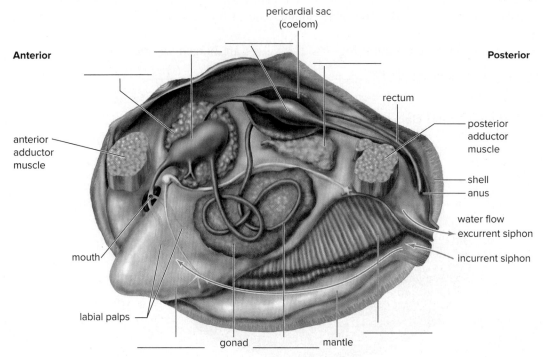

Anterior

pericardial sac
(coelom)

Posterior

rectum

posterior
adductor
muscle

anterior
adductor
muscle

shell

anus

water flow
excurrent siphon

incurrent siphon

mouth

labial palps

gonad          mantle

a.

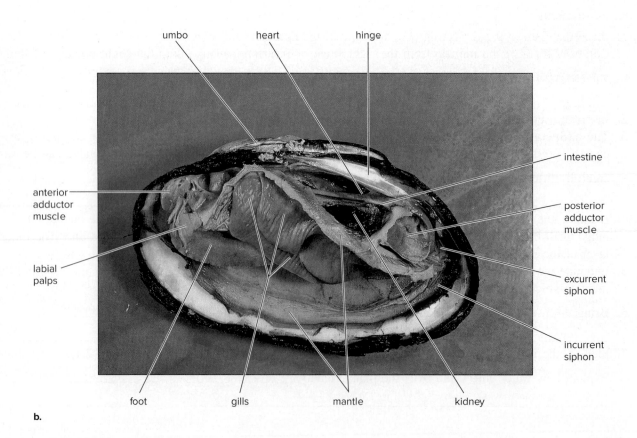

umbo          heart          hinge

intestine

anterior
adductor
muscle

posterior
adductor
muscle

labial
palps

excurrent
siphon

incurrent
siphon

foot          gills          mantle          kidney

b.

8. Cut away the free-hanging portion of the mantle to expose the **gills**. Does the clam have a respiratory organ? _____ If so, what type of respiratory organ? _____

9. A mucous layer on the gills entraps food particles brought into the mantle cavity, and the cilia on the gills convey these food particles to the mouth. Why is the clam called a filter feeder? _____

_____

10. The nervous system is composed of three pairs of ganglia (located anteriorly, posteriorly, and in the foot), all connected by nerves. The clam does not have a brain. A ganglion contains a limited number of neurons, whereas a brain is a large collection of neurons in a definite head region.

11. Identify the **foot,** a tough, muscular organ for locomotion, and the **visceral mass,** which lies above the foot and is soft and plump. The visceral mass contains the digestive and reproductive organs.

12. Identify the **labial palps** that channel food into the open mouth.

13. Identify the **anus,** which discharges into the excurrent siphon.

14. Find the **intestine** by its dark contents. Trace the intestine forward until it passes into a sac, the clam's only evidence of a coelom.

15. Locate the **pericardial sac** (**pericardium**) that contains the heart. The intestine passes through the heart. The heart pumps blood into the aortas, which deliver it to blood sinuses (open spaces) in the tissues.

A clam has an **open circulatory system.** Explain your answer. _____

_____

16. Cut the visceral mass and the foot into exact left and right halves, and examine the cut surfaces. Identify the digestive glands, greenish-brown; the stomach, embedded in the digestive glands; and the intestine, which winds about in the visceral mass. Reproductive organs (gonads) are also present.

## Anatomy of a Squid

Squids are cephalopods that have a well-defined head; the foot became the funnel surrounded by two arms and the many tentacles about the head. The head contains a brain and sensory organs. The squid moves quickly by jet propulsion of water, which enters the mantle cavity by way of a space that encircles the head. When the cavity is closed off, water exits by means of the funnel. Then the squid moves rapidly in the opposite direction.

The squid seizes fish with its tentacles; the mouth has a pair of powerful, beaklike jaws and a **radula,** a beltlike organ containing rows of teeth. The squid has a **closed circulatory system** composed of vessels and three hearts, one of which pumps blood to all the internal organs, while the other two pump blood to the gills located in the mantle cavity.

### Observation: Anatomy of a Squid

1. Examine a preserved squid.

2. Refer to Figure 17.5 for help in identifying the mouth (defined by beaklike jaws and containing a radula) and the tentacles and arms, which encircle the mouth.

3. Locate the head with its sense organs, notably the large, well-developed eye.

4. Find the funnel, where water exits from the mantle cavity, causing the squid to move backward.

5. If the squid has been dissected, note the heart, gills, and blood vessels.

## Figure 17.5 Anatomy of a squid.

The squid is an active predator and lacks the external shell of a clam. It captures fish with its tentacles and bites off pieces with its jaws. A strong contraction of the mantle forces water out the funnel, resulting in "jet propulsion." **a.** Drawing. *Fill in the missing labels with the help of your textbook.* **b.** Dissected specimen. (b) Ken Taylor/Wildlife Images

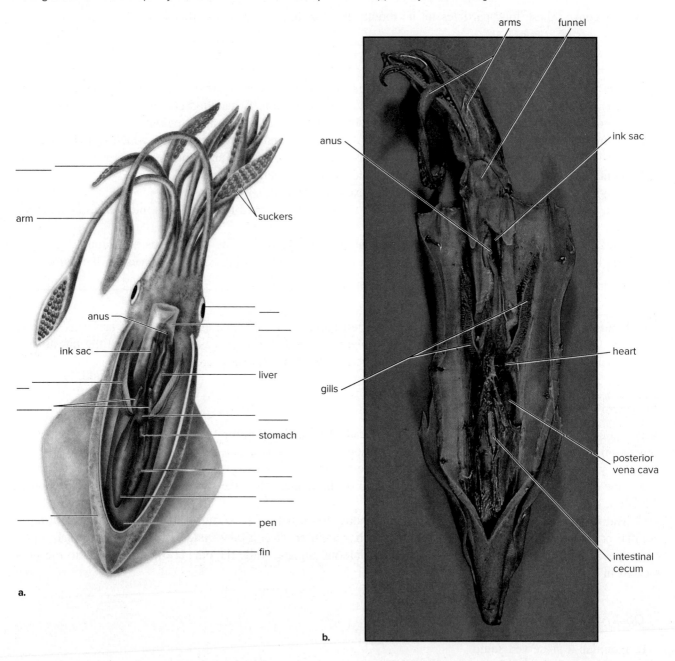

a.

b.

## Conclusion: Comparison of Clam to Squid

- Compare clam anatomy to squid anatomy by completing Table 17.1.
- Explain how both clams and squids are adapted to their way of life. _____

_____

_____

_____

## Table 17.1  Comparison of Clam to Squid

|  | Clam | Squid |
|---|---|---|
| Feeding mode |  |  |
| Skeleton |  |  |
| Circulation |  |  |
| Cephalization |  |  |
| Locomotion |  |  |

## 17.3  Annelids

**Pre-Lab**

4. *With the help of your textbook, complete the labeling of Figure 17.8.*

**Annelids** (phylum Annelida) are the **segmented** worms, so called because the body is divided into a number of segments and has a ringed appearance. The circular and longitudinal muscles work against the fluid-filled coelom to produce changes in width and length (Fig. 17.6). Therefore, annelids are said to have a **hydrostatic skeleton.**

**Figure 17.6  Locomotion in the earthworm.**
Contraction of first circular muscles and then longitudinal muscles allows the earthworm to move forward.

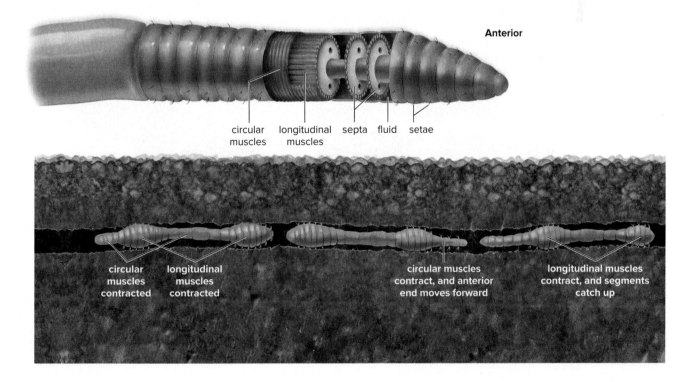

circular muscles    longitudinal muscles    septa    fluid    setae      Anterior

circular muscles contracted    longitudinal muscles contracted      circular muscles contract, and anterior end moves forward      longitudinal muscles contract, and segments catch up

Among annelids (Fig. 17.7), **polychaetes** have many slender bristles called **setae.** The polychaetes, almost all marine, are plentiful from the intertidal zone to the ocean depths. They are quite diverse, ranging from jawed forms that are carnivorous to fanworms that live in tubes and extend feathery filaments when filter feeding. Earthworms are called **oligochaetes** because they have few setae. Earthworms, have a worldwide distribution and can be found in almost any soil type. They are often present in large numbers, and some types may reach a length of as much as 3 meters. **Leeches,** annelids without setae, include the medicinal leech, which has been used in the practice of bloodletting for centuries.

**Figure 17.7  Annelid diversity.**

Aside from (**a**) earthworms, which are oligochaetes, there are (**b**) clam worms and (**c**) fan worms, both of which are polychaetes, and (**d**) leeches. Note the obvious segmentation.

(a) MikeLane45/iStockphoto/Getty Images; (b) blickwinkel/Hecker/Alamy Stock Photo; (c) Borut Furlan/Getty Images; (d) Bildagentur/age fotostock

head region

a. Earthworms, *Lumbricus*, mating

head
(sense organs and jaws)

b. Clam worm, *Nereis*

head region
(feathery arms)

c. European fan worm, *Sabella*

head region
(anterior sucker)

posterior
sucker

d. Leech, *Hirudo*

## Anatomy of the Earthworm

Earthworms are segmented in that the body has a series of ringlike segments. Earthworms have no head, and they burrow in the soil by alternately expanding and contracting segments along the length of the body.

Earthworms are scavengers that feed on decaying organic matter in the soil. They have a well-developed coelomic cavity, providing room for a well-developed digestive tract and both sets of reproductive organs. Earthworms are **hermaphroditic.**

### Observation: Anatomy of the Earthworm

#### External Anatomy

1. Examine a live or preserved specimen of an earthworm. Locate the small projection that sticks out over the mouth. Has cephalization occurred? _____ Explain your answer. _____

   _____

2. Count the total number of segments, beginning at the anterior end. The sperm duct openings are on segment 15 (somite XV) (Fig. 17.8). The enlarged section around a short length of the body is the **clitellum.** The clitellum secretes mucus that holds the worms together during mating. It also functions as a cocoon, in which fertilized eggs hatch and young worms develop. The anus is located on the worm's terminal segment.

3. Lightly pass your fingers over the earthworm's ventral and lateral sides. Do you feel the setae? _____

   _____

   Earthworms insert these slender bristles into the soil. Setae, along with circular and longitudinal muscles, enable the worm to locomote. Explain the action. _____

   _____

**Figure 17.8  External anatomy of an earthworm.**
In this drawing, the segments are numbered. *Fill in the missing labels with the help of your textbook.*

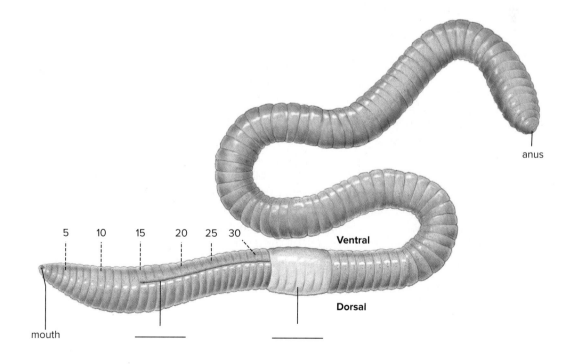

## Internal Anatomy

1. Place a preserved earthworm on its ventral side in the dissecting pan. With a scalpel or razor blade, make a shallow incision slightly to the side of the blackish median dorsal blood vessel (Fig. 17.9a). Start your incision about 10 segments after the clitellum, and proceed anteriorly to the mouth. If you see black ooze, you have accidentally cut the intestine.

2. Identify the thin partitions, the **septa,** between segments.

3. Lay out the body wall, and pin every 10th segment to the wax in your pan. Add water to prevent drying out. While alive, the earthworm body wall is always moist, and this facilitates gas exchange across the body wall. Notice the earthworm has no respiratory organ.

4. An earthworm feeds on **detritus** (organic matter) in the soil. Identify the digestive tract, which begins at the mouth and extends through body segments 1, 2, and 3 (Fig. 17.9a and b). It opens into the swollen, muscular, thick-walled **pharynx,** which extends from segment 3 to segment 6. The **esophagus** extends from the pharynx through segment 14 to the **crop.** Next is the **gizzard,** which lies in segments 17 through 17. The intestine extends from the gizzard to the anus. Does the digestive system show specialization

   of parts? _____ Explain your answer. _____

   _____

   _____

5. Identify the earthworm's circulatory system. The blood is always contained within vessels and never runs free. The **dorsal blood vessel** is readily seen along the dorsal side of the digestive tract. A series of aortic arches or "hearts" encircles the esophagus between segments 7 and 11, connecting the dorsal blood vessel with the ventral blood vessel. Does the earthworm have an open or closed circulatory

   system? _____ Explain your answer. _____

   _____

6. Locate the earthworm's nervous system. The two-lobed brain is located on the dorsal surface of the pharynx in segment 3. Two nerves, one on each side of the pharynx, connect the brain to a ganglion that lies below the pharynx in segment 4. The **ventral nerve cord** then extends along the floor of the body cavity to the last segment.

7. Find the earthworm's excretory system, which consists of a pair of minute, coiled, white tubules, the **nephridia,** located in every segment except the first three and the last. Each nephridium opens to the outside by means of an excretory pore. Does the excretory system show that the earthworm is segmented? _____

   Explain your answer. _____

   _____

   _____

8. Identify the earthworm's reproductive system, including **seminal vesicles,** light-colored bodies in segments 9 through 12, which house maturing sperm that have been formed in two pairs of testes within them; **sperm ducts** that pass to openings in segment 15; and **seminal receptacles** (four small, white, special bodies that lie in segments 9 and 10), which store sperm received from another worm. **Ovaries** are located in segment 13 but are too small to be seen.

9. During mating, earthworms are arranged so that the sperm duct openings of one worm are just about, but not quite, opposite the seminal receptacle openings of the other worm. After being released, the sperm pass down a pair of seminal grooves on the ventral surface (see Fig. 17.8) and then cross over at the level of the seminal receptacles of the opposite worm. Once the worms separate, eggs and sperm are released

   into a cocoon secreted by the clitellum. Is the earthworm hermaphroditic? _____ Explain your

   answer. _____

   _____

**10.** Does the earthworm have a respiratory organ or system (i.e., gills or lungs)? _____ How does the worm exchange gases? _____

_____

Why would you expect an earthworm to lack an exoskeleton? _____

_____

**Figure 17.9    Internal anatomy of an earthworm, dorsal view.**
**a.** Drawing shows internal organs and a cross section. The segments are numbered. *Fill in the missing labels with the help of your textbook.* **b.** Dissected specimen.

(b) Ken Taylor/Wildlife Images

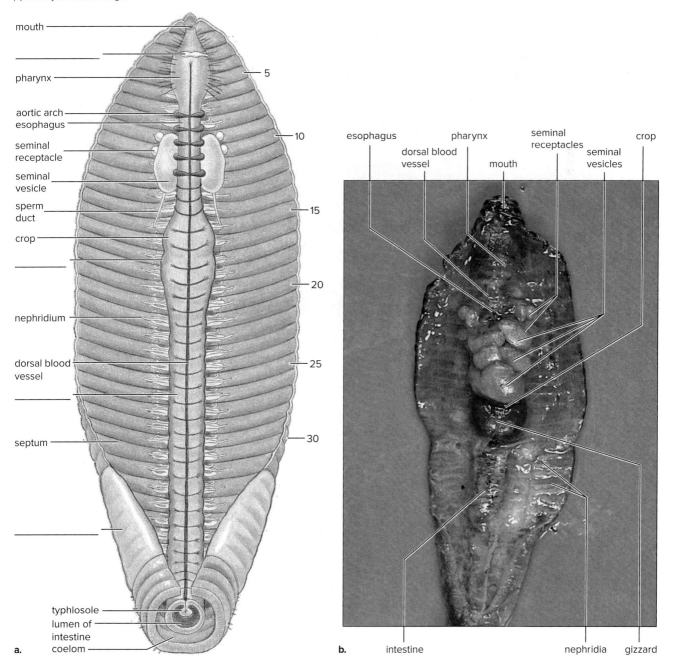

1. Obtain a prepared slide of a cross section of an earthworm (Fig. 17.10). Examine the slide under the dissecting microscope and under the light microscope.
2. Identify the following structures.
   a. **Body wall:** a thick outer circle of tissue, consisting of the **cuticle** and the **epidermis.**
   b. **Coelom:** a relatively clear space with scattered fragments of tissue.
   c. **Intestine:** an inner circle with a suspended fold.
   d. **Typhlosole:** a fold that increases the intestine's surface area.
   e. **Ventral nerve cord:** a white, threadlike structure.

3. Does the typhlosole help in nutrient absorption? _____ Explain your answer. _____

**Figure 17.10   Cross section of an earthworm.**
Cross-section slide as it would appear under the microscope.

blickwinkel/F. Fox/Alamy Stock Photo

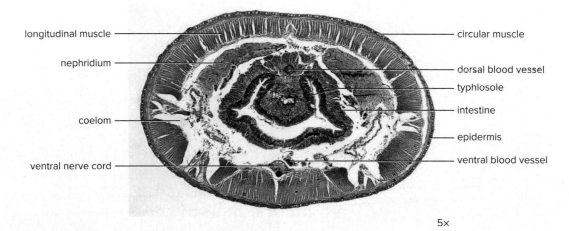

longitudinal muscle — circular muscle
nephridium — dorsal blood vessel
— typhlosole
— intestine
coelom — epidermis
ventral nerve cord — ventral blood vessel

5×

**Conclusion: Comparison of Clam to Earthworm**

- Complete Table 17.2 to compare the anatomy of a clam to that of an earthworm.

| Table 17.2   Comparison of Clam to Earthworm | | |
| --- | --- | --- |
| | **Clam** | **Earthworm** |
| Habitat (where they live) | | |
| Feeding mode | | |
| Skeleton | | |
| Segmentation | | |
| Circulation | | |
| Respiratory organs | | |
| Locomotion | | |
| Reproductive organs | | |

# 17.4 Arthropods

**Pre-Lab**

5. *With the help of your textbook, complete the labeling of Figures 17.12 and 17.13.*

**Arthropods** (phylum Arthropoda) have paired, jointed appendages and a hard exoskeleton that contains chitin. The chitinous exoskeleton consists of hardened plates separated by thin, membranous areas that allow movement of the body segments and appendages.

Figure 17.11*a* features insects and relatives. **Insects** with three pairs of legs, with or without wings, and three distinct body regions comprise 95% of all arthropods. **Millipedes** have two pairs of legs per segment, while **centipedes** have one pair of legs per segment. Figure 17.11*b* features spiders and relatives. Spiders and scorpions have four pairs of legs, no antennae, and a cephalothorax (head and thorax are fused). The horseshoe crab is called a living fossil because it has remained unchanged for thousands of years. The **crustaceans** (Fig. 17.11*c*), which include crabs, shrimp, and lobsters, have three to five pairs of legs, and two pairs of antennae. Barnacles are unusual, in that their legs are used to gather food.

**Figure 17.11   Arthropod diversity.**
**a.** Among arthropods, insects, millipedes, and centipedes are possibly related. **b.** Spiders, scorpions, and horseshoe crabs are related. **c.** Crabs, shrimp, and barnacles, among others, are crustaceans.

(a) (honeybee) Stephen Ausmus/USDA-ARS; (millipede) blickwinkel/Layer/Alamy Stock Photo; (centipede) Larry Miller/Science Source; (b) (spider) Horia Bogdan/iStock/Getty Images; (scorpion) Robert Pickett/Papilio/Alamy Stock Photo; (horseshoe crab) Robert Pos/U.S. Fish & Wildlife Service; (c) (crab) Tom McHugh/Science Source; (shrimp) Norbert Probst/imageBroker/Alamy Stock Photo; (barnacles) L. Newman & A. Flowers/Science Source

**a. Insects and relatives**

Honeybee, *Apis mellifera*

Millipede, *Rhapidostreptus virgator*

Centipede, *Scolopendra* sp.

**b. Spider and relatives**

Spider, *Argiope bruennichi*

Scorpion, *Hadrurus arizonensis*

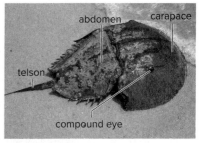
Horseshoe crab, *Limulus polyphemus*

**c. Crustaceans**

Crab, *Cancer productus*

Shrimp, *Lysmata grabhami*

Barnacles, *Lepas anatifera*

## Anatomy of a Crayfish

Crayfish also belong to the group of arthropods called crustaceans. Crayfish are adapted to an aquatic existence. They are known to be scavengers, but they also prey on other invertebrates. The mouth is surrounded by appendages modified for feeding, and there is a well-developed digestive tract. Dorsal, anterior, and posterior arteries carry **hemolymph** (blood plus lymph) to tissue spaces (hemocoel) and sinuses. Therefore, a crayfish has an open circulatory system.

### Observation: Anatomy of a Crayfish

#### External Anatomy

1. In a preserved crayfish, identify the chitinous **exoskeleton.** With the help of Figure 17.12, identify the head, thorax, and abdomen. Together, the head and thorax are called the **cephalothorax;** the

**Figure 17.12**
**Anatomy of a crayfish.**
**a.** Drawing shows external anatomy. *Fill in the missing labels with the help of your textbook.*
**b.** Internal anatomy of a female crayfish.

(b) Ken Taylor/Wildlife Images

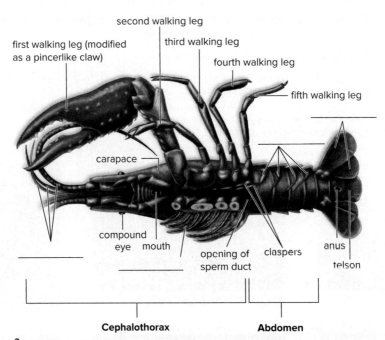

cephalothorax is covered by the **carapace.** Has specialization of segments occurred? _____

    Explain your answer. _____

2. Find the **antennae,** which project from the head. At the base of each antenna, locate a small, raised nipple containing an opening for the **green glands,** the organs of excretion. Crayfish excrete a liquid nitrogenous waste.

3. Locate the **compound eyes,** composed of many individual units for sight. Do crayfish demonstrate

    cephalization? _____ Explain your answer. _____

4. Identify the six pairs of appendages around the mouth for handling food.

5. Find the five pairs of walking legs attached to the cephalothorax. The most anterior pair is modified as pincerlike claws. In the female, identify the **seminal receptacles,** a swelling located between the bases of the third and fourth pairs of walking legs. Sperm from the male are deposited in the seminal receptacles. In the male, identify the opening of the sperm duct located at the base of the fifth walking leg.

6. Locate the five pairs of **swimmerets** on the abdomen. Crayfish use their swimmerets to swim and move water over their gills. In males, the anterior two pairs are stiffened and folded forward. They are claspers that aid in the transfer of sperm during mating. In females, clusters of eggs and larvae attach to the swimmerets.

7. Find the last abdominal segment, which bears a pair of broad, fan-shaped **uropods** that, together

    with a terminal extension of the body, form a tail. Has specialization of appendages occurred? _____

    Explain your answer. _____

## Internal Anatomy

1. Cut away the lateral surface of the carapace with scissors to expose the **gills** (Fig. 17.12*b*). Observe that

    the gills occur in distinct, longitudinal rows. How many rows of gills are there in your specimen? _____

    The outer row of gills is attached to the base of certain appendages. Which ones? _____

    _____

2. Remove a gill with your scissors by cutting it free near its point of attachment, and place it in a watch glass filled with water. Observe the numerous gill filaments arranged along a central axis.

3. Carefully cut away the dorsal surface of the carapace with scissors and a scalpel. The epidermis that adheres to the exoskeleton secretes the exoskeleton. Remove any epidermis adhering to the internal organs.

4. Identify the diamond-shaped heart lying in the middorsal region. A crayfish has an open circulatory system. Carefully remove the heart.

5. Locate the **gonads** anterior to the heart in both the male and female. The gonads are tubular structures bilaterally arranged in front of the heart and continuing behind it as a single mass. In the male, the testes are highly coiled, white tubes.

6. Find the **mouth;** the short, tubular **esophagus;** and the two-part **stomach,** with the attached **digestive gland,** that precedes the intestine.

7. Identify the **green glands,** two excretory structures just anterior to the stomach, on the ventral segment wall.

8. Remove the thoracic contents previously identified.

9. Identify the **brain** in front of the esophagus. The brain is connected to the ventral nerve cord by a pair of nerves that pass around the esophagus.

10. Remove the animal's entire digestive tract, and float it in water. Observe the various parts, especially the connections of the digestive gland to the stomach.

11. Cut through the stomach, and notice in the anterior region of the stomach wall the heavy, toothlike projections, called the **gastric mill,** which grind up food. Do you see any grinding stones ingested by the

    crayfish? _____

    If possible, identify what your specimen had been eating. _____

## Anatomy of a Grasshopper

The grasshopper is an **insect.** All insects have a head, a thorax, and an abdomen. Their appendages always include (1) three pairs of jointed legs and usually (2) two antennae as sensory organs. Grasshoppers are adapted to live on land. Wings and jumping legs are suitable for locomotion on land; **Malpighian tubules** save water by secreting a solid nitrogenous waste; the **tracheae** are tiny tubules that deliver air directly to the muscles; and the male has a penis with attached claspers to deliver sperm to the seminal receptacles of a female so they do not dry out.

### Observation: Anatomy of a Grasshopper

**External Anatomy**

1. Obtain a preserved grasshopper (*Romalea*), and study its external anatomy with the help of Figure 17.13*a*. Identify the head, thorax, and abdomen.
2. Use a hand lens or dissecting microscope to examine the grasshopper's special sense organs of the **head.** Identify the **antennae** (a pair of long, jointed feelers), the **compound eyes,** and the three dotlike **simple eyes.** The labial palps, labeled in Figure 17.13*a*, have sense organs for tasting food.
3. Note the sturdy **mouthparts,** which are used for chewing plant material. A grasshopper's mouthparts are quite different from those of a piercing and sucking insect.
4. Locate the leathery **forewings** and the inner, membranous **hindwings** attached to the **thorax.** Which pair of legs is used for jumping? _____ How many segments does each leg have? _____
5. Is locomotion in the grasshopper adapted to land? _____

   Explain your answer. _____

   _____

6. In the **abdomen,** identify the **tympana** (sing., **tympanum**), one on each side of the first abdominal segment (Fig. 17.13*a*). The grasshopper detects sound vibrations with these membranes.
7. Locate the **spiracles,** along the sides of the abdominal segments. These openings allow air to enter the tracheae, which constitute the respiratory system.
8. Find the **ovipositors** (Figs. 17.13*a* and 17.14*a*), four curved and pointed processes projecting from the abdomen of the female. These are used to dig a hole in which eggs are laid. The male has a **penis** with **claspers** used during copulation (Fig. 17.14*b*).

### Figure 17.13 Female grasshopper.

**a.** External anatomy. *Fill in the missing labels with the help of your textbook.* **b.** Internal anatomy.

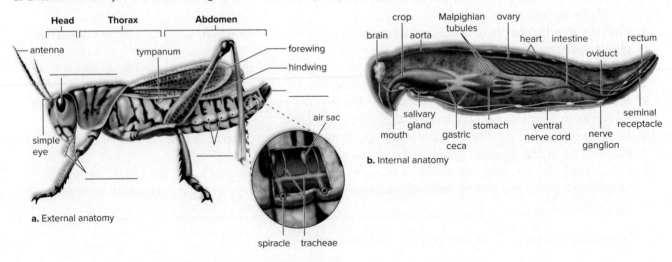

a. External anatomy

b. Internal anatomy

### Figure 17.14 Grasshopper genitalia.
**a.** Females have an ovipositor, and (**b**) males have claspers at the distal end of the penis.

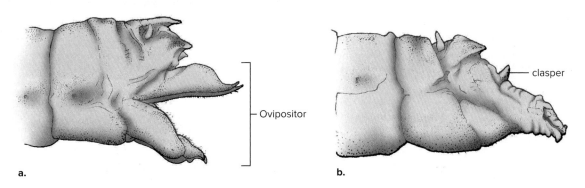

a.

Ovipositor

clasper

b.

*Internal Anatomy*

Observe a longitudinal section of a grasshopper if available on demonstration. Try to locate the structures shown in Figure 17.13*b*. What is the function of Malpighian tubules? _____

*Conclusion: Comparison of Crayfish to Grasshopper*

Compare the adaptations of a crayfish to those of a grasshopper by completing Table 17.3. Put a star beside each item that indicates an adaptation to life in the water (crayfish) and to life on land (grasshopper). Check with your instructor to see if you identified the maximum number of adaptations.

| Table 17.3 Comparison of Crayfish to Grasshopper | | |
|---|---|---|
| | **Crayfish** | **Grasshopper** |
| Locomotion | | |
| Respiration | | |
| Sense organs | | |
| Nervous system | | |
| External reproductive features<br>    Male<br>    Female | | |

## Grasshopper Metamorphosis

**Metamorphosis** is a change in form and shape over the course of an organism's life. Grasshoppers undergo *incomplete metamorphosis,* a gradual change in form rather than a drastic change. The immature stages of the grasshopper are called **nymphs,** and they are recognizable as grasshoppers even though they differ in shape and form (Fig. 17.15*a*). Some insects undergo what is called *complete metamorphosis,* in which case they have three stages of development: **larvae, pupa,** and **adult** (Fig. 17.15*b*). Metamorphosis occurs during the pupa stage when the animal is enclosed within a hard covering. The animals best known for complete metamorphosis are the butterfly and the moth, whose larval stage is called a caterpillar and whose pupa stage is the cocoon; the adult is the butterfly or moth.

## Figure 17.15 Metamorphosis.
**a.** During incomplete metamorphosis of a grasshopper, a series of **nymphs** lead to a full-grown grasshopper. **b.** During complete metamorphosis of a moth, a series of larvae lead to pupation. The **adult** hatches out of the pupa.

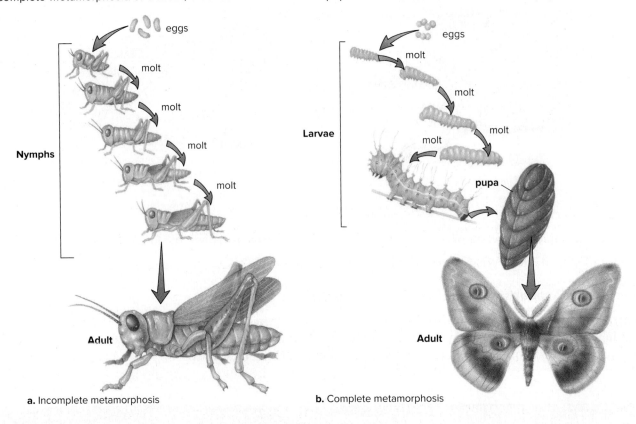

a. Incomplete metamorphosis

b. Complete metamorphosis

## Observation: Grasshopper Metamorphosis

1. Use Figure 17.15 to add the grasshopper and the moth to Table 17.4.
2. Examine any specific life cycle displays or plastomounts that illustrate complete and incomplete metamorphosis and add these examples to Table 17.4.

| Table 17.4 | Insect Metamorphosis |
|---|---|
| **Specimen** | **Complete or Incomplete Metamorphosis** |
| | |
| | |
| | |
| | |

## Conclusions: Insect Metamorphosis

- In insects with incomplete metamorphosis, do the nymphs or the adult have better developed wings? _____ What is the benefit of wings to an insect? _____

_____

- What stage is missing when an insect has incomplete metamorphosis? _____ What happens during this stage? _____

- What form, the larvae or the adult, disperses new individuals in flying insects that exhibit complete metamorphosis? _____

How is this a benefit? _____

_____

- In insects that undergo complete metamorphosis, the larvae and the adults utilize different food sources and habitats. Why might this be a benefit? _____

_____

## 17.5 Echinoderms

**Pre-Lab**

**6.** *With the help of your textbook, complete the labeling of Figure 17.17.*

The echinoderms (phylum Echinodermata) are the only invertebrate group that shares deuterostome development with the vertebrates. Unlike the vertebrates, all **echinoderms** are marine, and they dwell on the seabed, either attached to it, like sea lilies, or creeping slowly over it. The name *echinoderm* means "spiny-skinned," and most members of the group have defensive spines on the outside of their bodies. The spines arise from an **endoskeleton** composed of calcium carbonate plates. The endoskeleton supports the body wall and is covered by living tissue that may be soft (as in sea cucumbers) or hard (as in sea urchins).

Especially note that (1) adult echinoderms have radial symmetry, with generally five points of symmetry arranged around the axis of the mouth while the larval stage has bilateral symmetry; and (2) the echinoderms' most unique feature is their **water vascular system.** In those echinoderms in which the arms make contact with the substratum, the **tube feet** associated with the water vascular system are used for locomotion. In other echinoderms, the tube feet are used for gas exchange and food gathering.

Echinoderms belong to one of five groups: sea lilies and feather stars; sea stars; brittle stars; sea urchins and sand dollars; and sea cucumbers (Fig. 17.16).

## Figure 17.16  Echinoderm diversity.

**a.** Feather star,
*Oxymetra erinacea*

**b.** Sea star,
*Odontaster validus*

**c.** Brittle star,
*Ophiopholis aculeata*

**d.** Sea urchin,
*Stronglocentrotus pranciscanus*

**e.** Sand dollar,
*Echinarachnius parma*

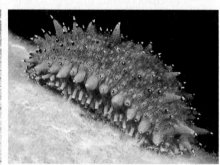

**f.** Sea cucumber,
*Apostichopus japonicus*

## Anatomy of a Sea Star

Sea stars (starfish) usually have five arms that radiate from a central disk. The mouth is normally oriented downward, and when sea stars feed on clams, they use the suction of their tube feet to force the shells open a crack. Then they evert the cardiac portion of the stomach, which releases digestive juices into the mantle cavity. Partially digested tissues are taken up by the pyloric portion of the stomach; digestion continues in this portion of the stomach and in the digestive glands found in the arms.

### Observation: Anatomy of a Sea Star

#### External Anatomy

1. Place a preserved sea star in a dissecting pan so that the aboral side is uppermost.
2. With the help of Figure 17.17, identify the **central disk** and five arms. What type of symmetry does an adult sea star have? _____

3. With the oral side uppermost, find the mouth, located in the center and protected by spines. Why is this side of the sea star called the oral side? _____

4. Locate the groove that runs along the middle of each arm and the **tube feet** (suctionlike disks) in rows on either side of the groove. Pluck away the tube feet from one area. How many rows of feet are there? _____

5. Turn the sea star over to its aboral side.

6. Locate the anal opening (Fig. 17.17). Why is this side of the sea star called the aboral side? _____

## Figure 17.17 Anatomy of a sea star.

**a.** Diagram *(fill in the missing labels with the help of your textbook)* and **(b)** image of dissected sea star. Both show the aboral side. **c.** Image of cut arm. **d.** Canals and tube feet of water vascular system. Seen from aboral side.

(b–c) Darrell S. Vodopich/BiologyImaging.com/Biology Imaging

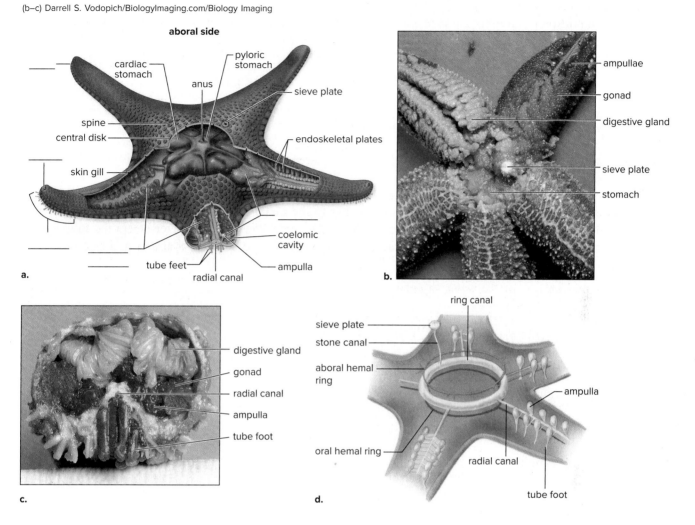

**a.**

aboral side

cardiac stomach
pyloric stomach
anus
sieve plate
spine
central disk
endoskeletal plates
skin gill
coelomic cavity
tube feet
radial canal
ampulla

**b.**

ampullae
gonad
digestive gland
sieve plate
stomach

**c.**

digestive gland
gonad
radial canal
ampulla
tube foot

**d.**

ring canal
sieve plate
stone canal
aboral hemal ring
ampulla
oral hemal ring
radial canal
tube foot

7. Lightly run your fingers over the spines extending from calcium carbonate plates that lie buried in the body wall beneath the surface. The plates form an endoskeleton of the animal.
8. Identify the **sieve plate** (**madreporite**), a brownish, circular spot between two arms where water enters the water vascular system.

### Internal Anatomy

1. Place the sea star so that the aboral side is uppermost. Refer to Figure 17.17*a* and *b* as you dissect the sea star following these instructions:
2. Cut the tip of one of the arms and, with scissors, carefully cut through the body wall along each side of this arm.
3. Carefully lift up the upper body wall. Separate any internal organs that may be adhering so that all internal organs are left intact.
4. Cut off the body wall near the central disk, but leave the sieve plate (madreporite) in place.
5. Remove the body wall of the central disk, being careful not to injure the internal organs.
6. Identify the digestive system. The mouth leads into a short **esophagus,** which is connected to the saclike **cardiac stomach.** When a sea star eats, the cardiac stomach sticks out through the sea star's mouth and

starts digesting the contents of a clam or oyster. Above the cardiac stomach is the **pyloric stomach,** which leads to a short intestine. Each arm contains one pair of **digestive glands.**

To which stomach do the digestive glands attach? _____

7. Cut off a portion of an arm, and examine the cut edge (Fig. 17.17c). Identify the digestive glands, ambulacral groove, radial canal, ampullae, and tube feet.

8. Remove the digestive glands of one arm.

9. Identify the **gonads** extending into the arm. What is the function of gonads? _____

_____

It is not possible to distinguish male sea stars from females by this observation.

10. Remove both stomachs.

11. In the **water vascular system** (Fig. 17.17d), you have already located the sieve plate and tube feet. Now try to identify the following components:

   a. **Stone canal:** takes water from the sieve plate to the ring canal.

   b. **Ring canal:** surrounds the mouth and takes water to the radial canals.

   c. **Radial canals:** send water into the ampulla. When the ampullae contract, water enters the tube feet. Each tube foot has an inner muscular sac called an ampulla. The ampulla contracts and forces water into the tube foot.

What is the function of the water vascular system? _____

_____

_____   **1.** What tissue layer lines the body cavity of an animal that contains a coelom?

_____   **2.** Is an animal a protostome or a deuterostome if the first opening becomes the mouth?

_____   **3.** What three components make up the body plan of a mollusk?

_____   **4.** What characteristic does an animal with a definite head region have?

_____   **5.** What organ is found inside the pericardial sac (pericardium)?

_____   **6.** What kind of skeleton do annelids have?

_____   **7.** To what system do the crop and gizzard of an earthworm belong?

_____   **8.** What is the location of the nerve cord in an earthworm?

_____   **9.** What phylum of animals have chitinous exoskeletons?

_____   **10.** What are the excretory structures in a crayfish called?

_____   **11.** How is air taken directly to the muscles in a grasshopper?

_____   **12.** What system of the echinoderms enables them to move about?

_____   **13.** What kind of symmetry does an adult starfish have?

_____   **14.** Where does water enter the water vascular system in a starfish?

### Thought Questions

**15.** Describe several of the specialized digestive organs in the animals studied in this laboratory given that they all have a coelom and a complete digestive system with a separate mouth and anus.

**16.** Explain why an earthworm will die if it dries out based on the type of skeleton annelids have and the absence of a respiratory organ or system in the earthworm.

**17.** Contrast an arthropod, the crayfish or the grasshopper, with the representative echinoderm, the starfish, with regard to signs of cephalization and how the type of symmetry each has is related.

# 18

# The Vertebrates

## Learning Outcomes

**18.1 Evolution of Chordates**
- Discuss the evolution of chordates in terms of seven derived characteristics.

**18.2 Invertebrate Chordates**
- Use the observation of lancelet anatomy to point out the common characteristics of all chordates.

**18.3 Vertebrates**
- Name the types of vertebrates, and give an example of an animal in each group.
- Identify and locate external and internal structures of a frog.

**18.4 Comparative Vertebrate Anatomy**
- Compare the organ systems (except musculoskeletal) of a frog, perch, pigeon, and rat.
- Relate aspects of an animal's respiratory system to the animal's environment and to features of the animal's circulatory system.

## Introduction

### Pre-Lab

1. Which group of animals will be studied in this lab? _____

Vertebrates are **chordates.** All chordates share the following characteristics (1) a dorsal tubular nerve cord; (2) a dorsal supporting rod, called a notochord, at some time in their life history; (3) a postanal tail (e.g., tailbone or coccyx); and (4) pharyngeal pouches. In chordates that breathe by means of gills, these pouches become gill slits. In terrestrial chordates, the pharyngeal pouches are modified for other purposes.

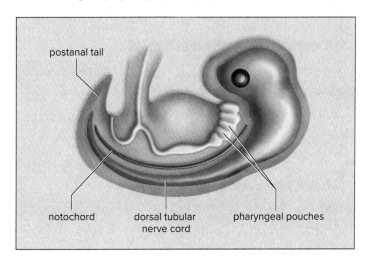

postanal tail

notochord          dorsal tubular          pharyngeal pouches
                   nerve cord

# 18.1 Evolution of Chordates

The evolutionary tree of the chordates (Fig. 18.1) shows that some chordates, notably the **tunicates** and **lancelets,** are not vertebrates. These chordates retain the notochord and are called the **invertebrate chordates.**

The other animal groups in Figure 18.1 are **vertebrates** in which the notochord develops into the vertebral column. Fishes include three groups: the **jawless fishes** were the first to evolve, followed by the **cartilaginous fishes** and then the **bony fishes.** The bony fishes include the **ray-finned fishes** (the largest group of vertebrates) and the **lobe-finned fishes.** The first lobe-finned fishes had a bony skeleton, fleshy appendages, and a lung. These lobe-finned fishes lived in shallow bodies of water and gave rise to the amphibians.

The terrestrial vertebrates are all **tetrapods** because they have four limbs. The limbs of tetrapods are jointed appendages just like those of arthropods. Amphibians still return to the water to reproduce, but **reptiles** are fully adapted to life on land because among other features they produce an **amniotic egg.**

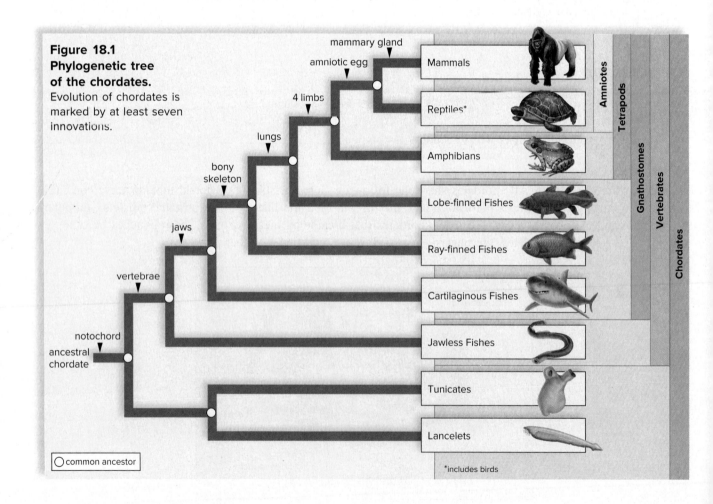

**Figure 18.1**
**Phylogenetic tree of the chordates.**
Evolution of chordates is marked by at least seven innovations.

The amniotic egg is so named because the embryo is surrounded by an amniotic membrane that encloses amniotic fluid. Therefore, amniotes develop in an aquatic environment of their own making. Do all animals develop in a water environment? _____ Explain. _____

_____

In placental mammals, such as humans, the fertilized egg develops inside the female, where the unborn mammal receives nutrients via the placenta. Reptiles (including birds) and mammals have many other adaptations that are suitable to living on land, as we will stress in later sections.

## 18.2 Invertebrate Chordates

Among the chordates (phylum Chordata), two groups contain invertebrates, and the others contain the vertebrates.

### Invertebrate Chordates

The two types of invertebrate chordates are urochordates and cephalochordates.

1. **Urochordates.** The tunicates, or sea squirts (Fig. 18.2), come in varying sizes and shapes, but all have incurrent and excurrent siphons. **Gill slits** are the only remaining chordate characteristic in adult tunicates. Examine any examples of tunicates on display.
2. **Cephalochordates.** Lancelets, also known as amphioxus (*Branchiostoma*), are small, fishlike animals that occur

 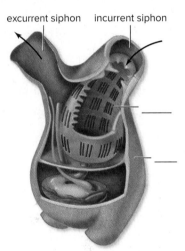

excurrent siphon    incurrent siphon

**Figure 18.2  Urochordates.**
The gill slits of a tunicate are the only chordate characteristic remaining in the adult. *Fill in the missing labels with the help of your textbook.*
Andrew J. Martinez/Science Source

in shallow marine waters in most parts of the world. They spend most of their time buried in the sandy bottom, with only the anterior end projecting.

### Observation: Lancelet Anatomy

**Preserved Specimen**
1. Examine a preserved lancelet (Fig. 18.3).
2. Identify the **caudal fin** (enlarged tail) used in locomotion, the **dorsal fin,** and the short **ventral fin.**
3. Examine the lancelet's V-shaped muscles.

**Figure 18.3 Anatomy of the lancelet, *Branchiostoma*.**

Lancelets feed on microscopic particles filtered out of the constant stream of water that enters the mouth and exits through the gill slits into a protective atrium formed by body folds. The water exits at the atriopore. *Fill in the missing labels with the help of your textbook.*

Heather Angel/Natural Visions

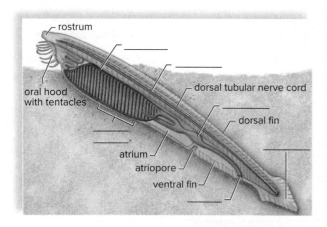

4. Find the tentacled **oral hood,** located anterior to the mouth and covering a vestibule. Water entering the mouth is channeled into the **pharynx,** where food particles are trapped before the water exits at the **atriopore.** Lancelets are filter feeders. Has cephalization occurred? _____

Explain your answer. _____

_____

## 18.3 Vertebrates

### Pre-Lab

5. List the different groups of vertebrates. _____

_____

6. List several unique characteristics of vertebrates. _____

_____

In vertebrates (subphylum Vertebrata) the embryonic notochord is replaced by a vertebral column composed of individual vertebrae that encase and protect the nerve cord. The internal jointed skeleton consists not only of a vertebral column, but also of a skull that encloses and protects the well-developed brain. There is an extreme degree of cephalization with complex sense organs. The eyes develop as outgrowths of the brain. The ears are primarily equilibrium devices in aquatic vertebrates; in land vertebrates, they also function as sound wave receivers.

The vertebrates are extremely motile and have well-developed muscles and usually paired appendages. They have bilateral symmetry and are segmented, as witnessed by the vertebral column. There is a large body cavity, a complete gut with both a mouth and an anus (or instead, a cloacal opening), and the circulatory system consists of a well-developed heart and many blood vessels. They have an efficient means of extracting oxygen from water (gills) or air (lungs) as appropriate. The kidneys are important excretory and water-regulating organs that conserve or rid the body of water as necessary. The sexes are generally separate, and reproduction is usually sexual.

## Figure 18.4 Vertebrate groups.

**Cartilaginous fishes**
Lack operculum and swim bladder; tail fin usually asymmetrical (sharks, skates, and rays)

Tiger shark

**Bony fishes**
Operculum; swim bladder or lungs; tail fin usually symmetrical: lung-fishes, lobe-finned fishes, and ray-finned fishes (herring, salmon, sturgeon, eels, and sea horse)

Banded butterflyfish

**Amphibians**
Tetrapods with nonamniotic egg; nonscaly skin; some show metamorphosis; three-chambered heart (salamanders, frogs, and toads)

Northern leopard frog

**Reptiles**
Tetrapods with amniotic egg; scaly skin (snakes, lizards, turtles, and tortoises)

Pearl River redbelly turtle

**Birds**
Now grouped with reptiles; tetrapods with feathers; bipedal with wings; double circulation (sparrows, penguins, and ostriches)

Scissor-tailed flycatcher

**Mammals**
Tetrapods with hair, mammary glands; double circulation; teeth differentiated: monotremes (spiny anteater and duckbill platypus), marsupials (opossum and kangaroo), and placental mammals (whales, rodents, dogs, cats, elephants, horses, bats, and humans)

Gray fox

## Anatomy of the Frog

In this laboratory, the anatomy of the frog will be used to represent the vertebrates. Frogs are amphibians, a group of animals in which metamorphosis occurs. Metamorphosis includes a change in structure, as when an aquatic tadpole becomes a frog with lungs and limbs (Fig. 18.5). Amphibians were the first vertebrates that evolved to live on land; however, many of them still require water for respiration and reproduction. Underline every structure mentioned in the following Observation that represents an adaptation to a land environment.

**Figure 18.5 External frog anatomy.**

Rod Planck/Science Source

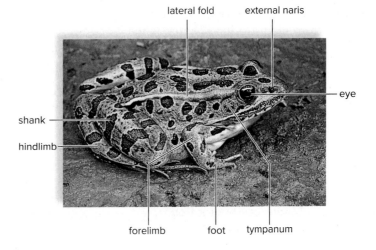

lateral fold   external naris

eye

shank

hindlimb

forelimb   foot   tympanum

## Observation: External Anatomy of the Frog

1. Place a preserved frog (*Rana pipiens*) in a dissecting tray.
2. Identify the bulging eyes, which have a nonmovable upper and lower lid but can be covered by a **nictitating membrane** that serves to moisten the eye.
3. Locate the **tympanum** behind each eye (Fig. 18.5). What is the function of a tympanum? _____ _____
4. Examine the external **nares** (sing., **naris,** or **nostril**). Insert a probe into an external naris, and observe that it protrudes from one of the paired small openings, the internal nares (Fig. 18.6), inside the mouth cavity. What is the function of the nares? _____
5. Identify the paired limbs. The bones of the fore- and hindlimbs are the same as in all tetrapods, in that the first bone articulates with a girdle and the limb ends in phalanges. The hind feet have five phalanges, and the forefeet have only four phalanges. Which pair of limbs is longest? _____

    How does a frog locomote on land? _____ _____

    What is a frog's means of locomotion in the water? _____ _____

## Observation: Internal Anatomy of the Frog

### Mouth

1. Open your frog's mouth very wide (Fig. 18.6), cutting the angles of the jaws if necessary.
2. Identify the tongue attached to the lower jaw's anterior end.
3. Find the **auditory (eustachian) tube** opening in the angle of the jaws. These tubes lead to the ears. Auditory tubes equalize air pressure in the ears.
4. Examine the **maxillary teeth** located along the rim of the upper jaw. Another set of teeth—**vomerine teeth**—is present just behind the midportion of the upper jaw.
5. Locate the **glottis,** a slit through which air passes into and out of the **trachea,** the short tube from glottis to lungs. What is the

    function of a glottis? _____
    _____

6. Identify the **esophagus,** which lies dorsal and posterior to the glottis and leads to the stomach.

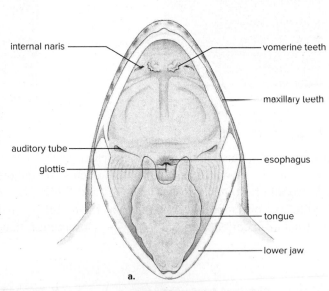

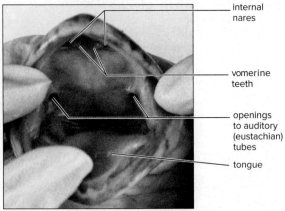

**Figure 18.6 Mouth cavity of a frog.**
**a.** Drawing. **b.** Dissected specimen.

(b) Sinhyu/iStock/Getty Images

### Opening the Frog

1. Place the frog ventral side up in the dissecting pan. Lift the skin with forceps, and use scissors to make a large, circular cut to remove the skin from the abdominal region as close to the limbs as possible. Cut only skin, not muscle.
2. Now, remove the muscles by cutting through them in the same circular fashion. At the same time, cut through any bones you encounter. A vein, called the abdominal vein, will be slightly attached to the internal side of the muscles.
3. Identify the **coelom,** or body cavity. Recall that vertebrates are deuterostomes in which the first embryonic opening becomes the anus and the second opening becomes the mouth.
4. If your frog is female, the abdominal cavity is likely to be filled by a pair of large, transparent **ovaries,** each containing hundreds of black and white eggs. Gently lift the left ovary with forceps, and find its place of attachment. Cut through the attachment, and remove the ovary in one piece.

### Respiratory System and Liver

1. Insert a probe into the glottis, and observe its passage into the trachea. Enlarge the glottis by making short cuts above and below it. When the glottis is spread open, you will see a fold on either side; these are the vocal cords used in croaking.
2. Identify the **lungs,** two small sacs on either side of the midline and partially hidden under the liver (Fig. 18.7). What path does air take through the respiratory tract? List all structures it passes from the external nares to the lungs. _____

_____

3. Locate the **liver,** the large, prominent, dark-brown organ in the midventral portion of the trunk (Fig. 18.7). Between the right half and left half of the liver, find the **gallbladder.**

### Circulatory System

1. Lift the liver gently. Identify the **heart,** covered by a membranous covering (the **pericardium**). With forceps, lift the covering, and gently slit it open. The heart consists of a single, thick-walled **ventricle** and two (right and left) anterior, thin-walled **atria.**
2. Locate the three large veins that join together beneath the heart to form the **sinus venosus.** (To lift the heart, you may have to snip the slender strand of tissue that connects the atria to the pericardium.) Blood from the sinus venosus enters the right atrium. The left atrium receives blood from the lungs.
3. Find the **conus arteriosus,** a single, wide arterial vessel leaving the ventricle and passing ventrally over the right atrium. Follow the conus arteriosus forward to where it divides into three branches on each side. The middle artery on each side is the **systemic artery,** which fuses behind the heart to become the **dorsal aorta.** The dorsal aorta transports blood through the body cavity and gives off many branches. The **posterior vena cava** begins between the two kidneys and returns blood to the sinus venosus. Which vessel lies above (ventral to) the other? _____

_____

### Digestive Tract

1. Identify the **esophagus,** a very short connection between the mouth and the stomach. Lift the left liver lobe, and identify the stomach, whitish and J-shaped. The **stomach** connects with the esophagus anteriorly and with the small intestine posteriorly.
2. Find the **small intestine** and the **large intestine,** which enters the **cloaca.** The cloaca lies beneath the pubic bone and is a general receptacle for the intestine, the reproductive system, and the urinary system. It opens to the outside by way of the anus. What path does food take through the digestive tract? List the organs in order from mouth to cloaca. _____

_____

## Accessory Glands

1. You identified the liver and gallbladder previously. Now try to find the **pancreas,** a yellowish tissue near the stomach and intestine.
2. Lift the stomach to see the **spleen,** a small, pea-shaped body.

**Figure 18.7 Internal organs of a female frog, ventral view.**

Ken Taylor/Wildlife Images

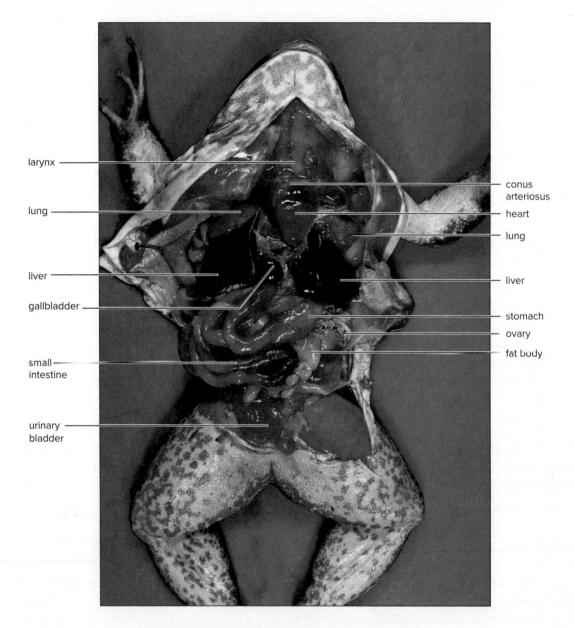

## Urogenital System

1. Identify the **kidneys,** long, narrow organs lying against the dorsal body wall (Fig. 18.8).
2. Locate the **testes** in a male frog (Fig. 18.8). Testes are yellow, oval organs attached to the anterior portions of the kidneys. Several small ducts, the **vasa efferentia,** carry sperm into kidney ducts that also carry urine from the kidneys. **Fat bodies,** which store fat, are attached to the testes.

---

3. Locate the ovaries in a female frog. The ovaries are attached to the dorsal body wall (Fig. 18.9). Fat bodies are also attached to the ovaries. Highly coiled **oviducts** lead to the cloaca. The ostium (opening) of the oviduct is dorsal to the liver.
4. Find the **mesonephric ducts**—thin, white tubes that carry urine from the kidney to the cloaca. In female frogs, you will have to remove the left ovary to see the mesonephric ducts.
5. Locate the **cloaca.** You will need to split through the bones of the pelvic girdle in the midventral line and carefully separate the bones and muscles to find the cloaca.
6. Identify the urinary bladder attached to the ventral wall of the cloaca. In frogs, urine backs up into the bladder from the cloaca.

7. Explain the term *urogenital system.* _____

_____  _____

8. The cloaca receives material from (1) _____,

(2) _____, and (3) _____.

9. Compare the frog's urogenital system to the human urinary system, which in females has no connection to the genital system.

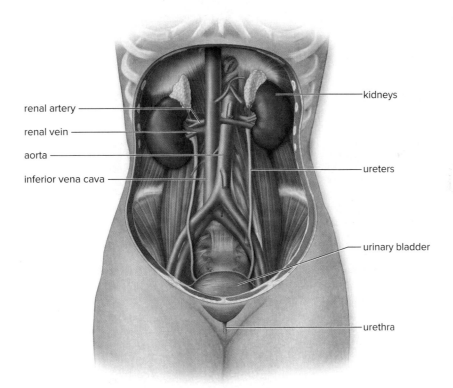

renal artery

renal vein

aorta

inferior vena cava

kidneys

ureters

urinary bladder

urethra

**Figure 18.8   Urogenital system of a male frog.**
**a.** Drawing. **b.** Dissected specimen.

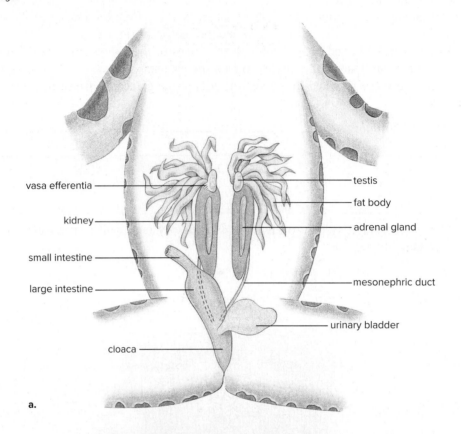

vasa efferentia

kidney

small intestine

large intestine

cloaca

testis

fat body

adrenal gland

mesonephric duct

urinary bladder

**a.**

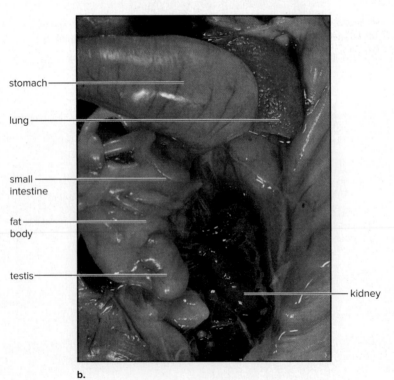

stomach

lung

small
intestine

fat
body

testis

kidney

**b.**

## Figure 18.9 Urogenital system of a female frog.
a. Drawing. b. Dissected specimen.

(b) Ken Taylor/Wildlife Images

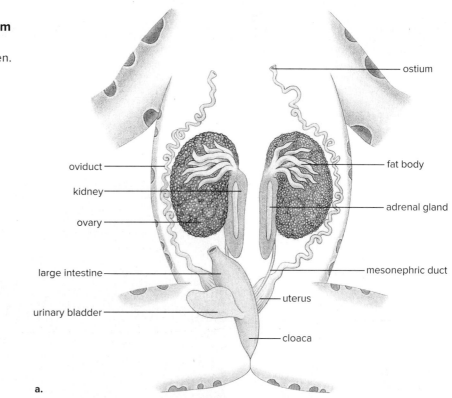

ostium

oviduct

kidney

ovary

large intestine

urinary bladder

fat body

adrenal gland

mesonephric duct

uterus

cloaca

a.

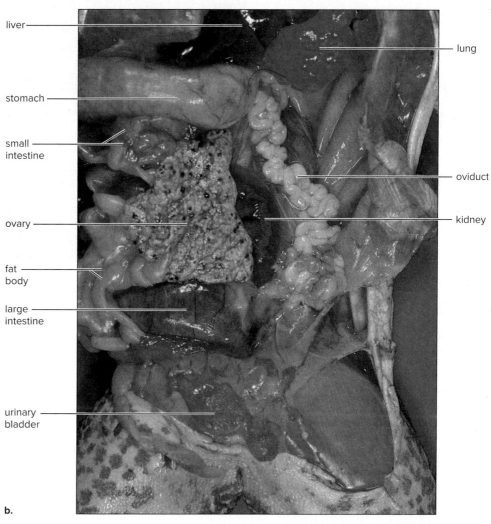

liver

stomach

small intestine

ovary

fat body

large intestine

urinary bladder

lung

oviduct

kidney

b.

## Nervous System

In the frog demonstration dissection, identify the **brain,** lying exposed within the skull. With the help of Figure 18.10, find the major parts of the brain.

**Figure 18.10   Frog brain, dorsal view.**
**a.** Drawing. **b.** Dissected specimen.

(b) Ken Taylor/Wildlife Images

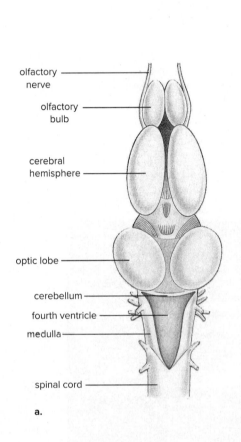

olfactory nerve

olfactory bulb

cerebral hemisphere

optic lobe

cerebellum

fourth ventricle

medulla

spinal cord

a.

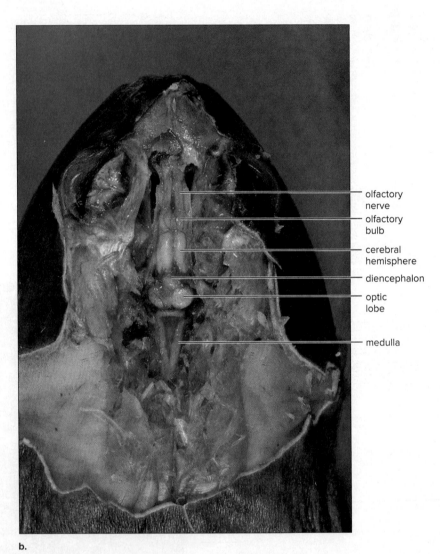

olfactory nerve

olfactory bulb

cerebral hemisphere

diencephalon

optic lobe

medulla

b.

# 18.4   Comparative Vertebrate Anatomy

## Pre-Lab

**7.** *Fill in the missing labels in Figures 18.11, 18.12, and 18.13.*

In addition to the frog (amphibian), examine the perch (fish), pigeon (reptile), and rat (mammal) on display.

1. Compare the external features of the perch, frog, pigeon, and rat by examining specimens in the laboratory. Answer the following questions and record your observations in Table 18.1.
   a. Is the skin smooth, scaly, hairy, or feathery?
   b. Is there any external evidence of segmentation?
   c. Are all forms bilaterally symmetrical?
   d. Is the body differentiated into regions?
   e. Is there a well-defined neck?
   f. Is there a postanal tail?
   g. Are there nares (nostrils)?
   h. Is there a cloaca opening, or are there urogenital and anal openings?
   i. Are eyelids present? How many?
   j. How many appendages are there? (Fins are considered appendages.) (Fig. 18.11)

| Table 18.1 Comparison of External Features | | | |
|---|---|---|---|
| | **Perch** | **Frog** | **Pigeon** | **Rat** |
| a. Skin | | | | |
| b. Segmentation | | | | |
| c. Symmetry | | | | |
| d. Regions | | | | |
| e. Neck | | | | |
| f. Postanal tail | | | | |
| g. Nares | | | | |
| h. Cloaca | | | | |
| i. Eyelids | | | | |
| j. Appendages | | | | |

**2.** Evidence that birds are reptiles: Birds
   **a.** have feathers, which are modified scales.
   **b.** have scales on their feet.
   **c.** and reptiles both lay eggs.
   **d.** and reptiles have similar internal organs.
   **e.** and reptiles also show some skeletal (skull) similarities.

Which of these can you substantiate by external examination? _____

_____

**3.** The perch, which lives in fresh water, and the pigeon and rat, which live on land, have a nearly

impenetrable covering. Why is this an advantage in each case? _____

_____

**4.** A frog uses its skin for breathing. Is the skin of a frog thick and dry or thin and moist? Explain. _____

_____

**Figure 18.11  Perch anatomy.**
All fins are shown except the pectoral fin. *Fill in the missing labels.*

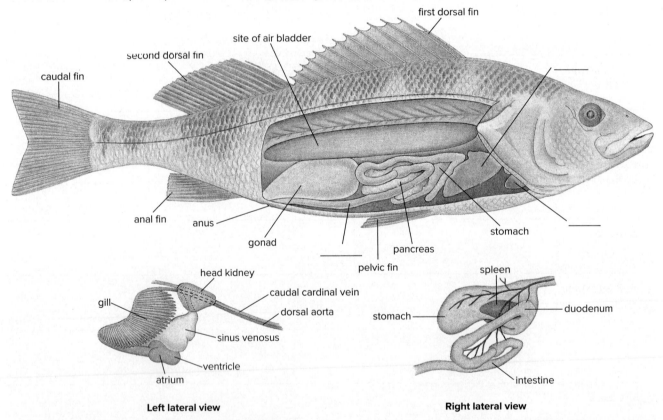

Left lateral view                    Right lateral view

---

### *Observation: Internal Anatomy of Vertebrates*

**1.** Examine the internal organs of the perch, frog, pigeon (Fig. 18.12) and rat (Fig. 18.13).
**2.** If necessary, make a median longitudinal incision in the ventral body wall, from the jaws to the anus.
   The body cavity is called a coelom because it is completely lined by mesoderm.
**3.** Which of these animals has a **diaphragm** dividing the body cavity into **thorax** and **abdomen**? _____

_____

# Figure 18.12  Pigeon anatomy.
*Fill in the missing labels.*

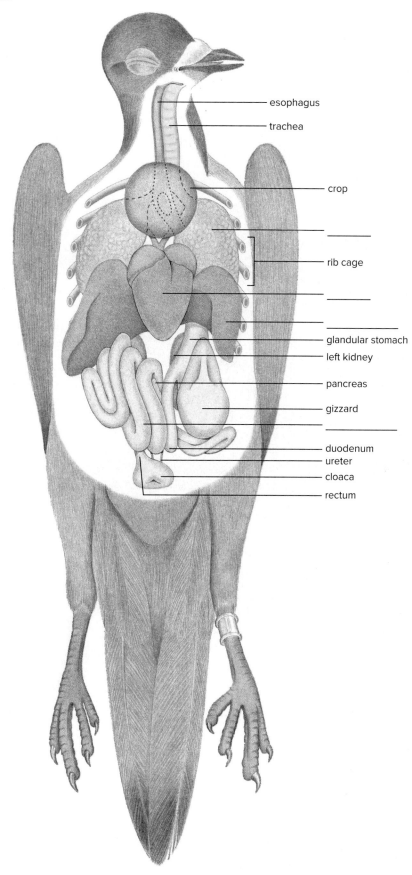

esophagus

trachea

crop

rib cage

glandular stomach

left kidney

pancreas

gizzard

duodenum

ureter

cloaca

rectum

**Figure 18.13  Rat anatomy.**

*Fill in the missing labels.*

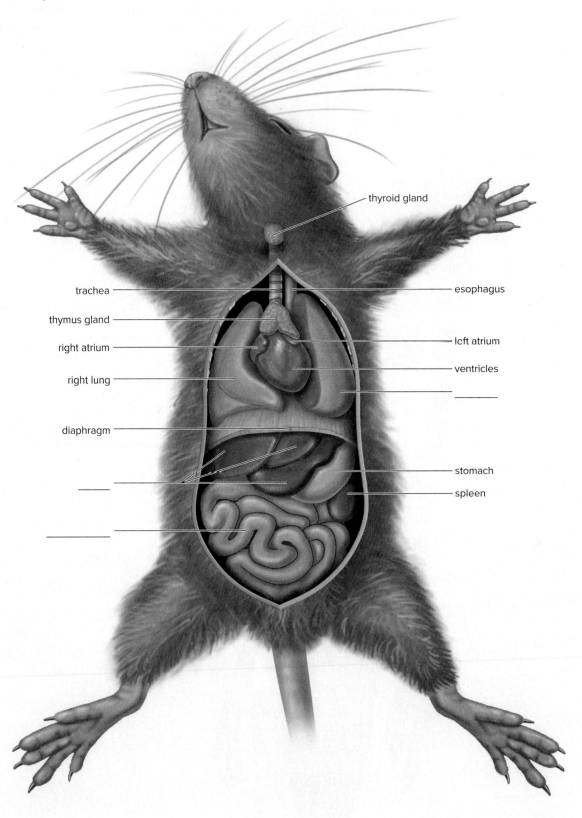

thyroid gland

trachea

thymus gland

right atrium

right lung

diaphragm

esophagus

left atrium

ventricles

stomach

spleen

### Digestive Systems and Urogenital Systems

1. All the vertebrates have a stomach, small intestine, and large intestine where food is processed. They also have a liver and pancreas. Which is the larger, more prominent organ (liver or pancreas) in the pigeon and rat? _____

2. All vertebrates have an anus. As noted in Table 18.1, which vertebrates studied have a cloaca (receptacle for the urogenital and digestive systems)? _____

   In the other vertebrates studied, the urogenital and digestive systems are separate.

3. Urogenital systems. All the animals have gonads and kidneys. Which of these organs is involved in urine production? _____

   Which of these organs is involved in reproduction? _____

   The sexes are separate in vertebrates: Females have ovaries and males have testes. In reptiles and mammals, males usually have a penis to pass sperm to the female. What is the chief biological benefit of the penis in terrestrial animals? _____

### Circulatory Systems

Study heart models for a fish, amphibian, bird, and mammal. Trace the path of the vessel that leaves the ventricle(s), and determine whether the animals have a **pulmonary system** (Fig. 18.14). The word *pulmonary* comes from the Latin *pulmonarius,* meaning "of the lungs."

**Figure 18.14  Cardiovascular systems in vertebrates.**
a. In a fish, the blood moves in a single loop. The heart has a single atrium and ventricle, which pumps the blood into the gill region, where gas exchange takes place. b. Amphibians have a double-loop system in which the heart pumps blood to both the lungs and the body. c. In birds and mammals, the right side pumps blood to the lungs, and the left side pumps blood to the body.

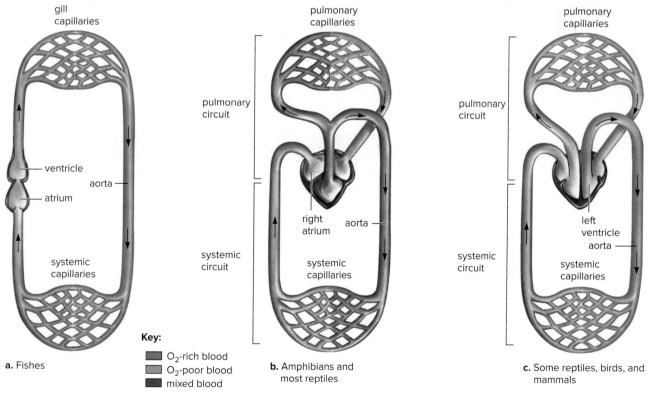

a. Fishes

**Key:**
- $O_2$-rich blood
- $O_2$-poor blood
- mixed blood

b. Amphibians and most reptiles

c. Some reptiles, birds, and mammals

Complete Table 18.2.

### Table 18.2 Comparative Circulatory Systems

| Animal | Number of Heart Chambers | Pulmonary Circuit (Yes or No) |
|--------|--------------------------|-------------------------------|
| Perch | | |
| Frog | | |
| Pigeon | | |
| Rat | | |

1. Do fish have a blood vessel that returns blood from the gills to the heart? _____ Would you expect blood pressure to be high or low after blood has moved through the gills? _____

2. What animals studied have pulmonary vessels that take blood from the heart to the respiratory organ and back to the heart? _____ What is the advantage of a pulmonary circuit? _____
   _____

3. Which of these animals has a four-chambered heart? _____
   What is the advantage of having separate ventricles? _____
   _____

4. The circulatory system distributes the heat of muscle contraction in birds and mammals. Is the anatomy of birds and mammals conducive to maintaining a warm internal temperature? _____
   Explain your answer. _____
   _____

### Respiratory Systems

Compare the respiratory systems of the perch, frog, pigeon, and rat, and complete Table 18.3 by checking the anatomical features that appear in each animal.

### Table 18.3 Respiratory Systems

| | Gills | Trachea | Lungs | Rib Cage* | Diaphragm | Air Sacs |
|--------|-------|---------|-------|-----------|-----------|----------|
| Perch | | | | | | |
| Frog | | | | | | |
| Pigeon | | | | | | |
| Rat | | | | | | |

*A rib cage consists of ribs plus a sternum. Some ribs are connected to the sternum, which lies at the midline in the anterior portion of the rib cage.

1. Among the animals studied, only a perch breathes by ＿＿＿＿＿＿. Can the particular respiratory organ be related to the environment of the animals? ＿＿＿＿＿ Explain your answer. ＿＿＿＿＿＿＿

＿＿＿＿＿＿＿＿＿＿＿＿＿＿＿＿＿＿＿＿＿＿＿＿＿＿＿＿＿＿＿＿＿＿＿＿＿

2. Knowing that gills are attached to the pharynx (throat in humans), explain why fish have no trachea. ＿＿

＿＿＿＿＿＿＿＿＿＿＿＿＿＿＿＿＿＿＿＿＿＿＿＿＿＿＿＿＿＿＿＿＿＿＿＿＿

＿＿＿＿＿＿＿＿＿＿＿＿＿＿＿＿＿＿＿＿＿＿＿＿＿＿＿＿＿＿＿＿＿＿＿＿＿

3. A rib cage is present in the rat and pigeon but missing in the frog. Can this difference be related to the fact that frogs breathe by positive pressure, while birds and mammals breathe by negative pressure? ＿＿＿＿＿ (A frog swallows air and then pushes the air into its lungs; in birds and mammals, the thorax expands first, and then the air is drawn in.) Explain. ＿＿＿＿＿＿＿＿＿＿＿＿＿＿＿

＿＿＿＿＿＿＿＿＿＿＿＿＿＿＿＿＿＿＿＿＿＿＿＿＿＿＿＿＿＿＿＿＿＿＿＿＿

4. A diaphragm is present only in mammals (e.g., rat). Of what benefit is this feature to the expansion of lungs in mammals? ＿＿＿＿＿＿＿＿＿＿＿＿＿＿＿＿＿＿＿＿＿＿＿＿＿＿＿＿＿＿＿

＿＿＿＿＿＿＿＿＿＿＿＿＿＿＿＿＿＿＿＿＿＿＿＿＿＿＿＿＿＿＿＿＿＿＿＿＿

＿＿＿＿＿＿＿＿＿＿＿＿＿＿＿＿＿＿＿＿＿＿＿＿＿＿＿＿＿＿＿＿＿＿＿＿＿

5. Air sacs (not shown in Fig. 18.12) are present only in birds. This feature allows air to pass one way through the lungs of a bird and greatly increases the bird's ability to extract oxygen from the air.

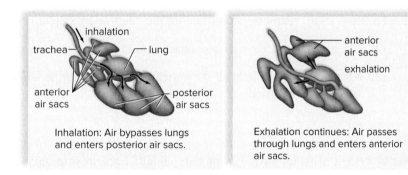

Inhalation: Air bypasses lungs and enters posterior air sacs.

Exhalation continues: Air passes through lungs and enters anterior air sacs.

_____ **1.** Are chordates and echinoderms classified as protostomes or deuterostomes?

_____ **2.** Which chordate features are present in the tunicates and lancelets?

_____ **3.** What structure develops into the vertebral column in vertebrate chordates?

_____ **4.** Do cartilaginous fishes or ray-finned fishes have a bony skeleton?

_____ **5.** What term refers to terrestrial vertebrates that have four limbs?

_____ **6.** What do reptiles produce that ensures their embryos won't dry out on land?

_____ **7.** What term refers to the changes that occur when an aquatic tadpole becomes a frog with limbs and lungs?

_____ **8.** What is the name of the membranous covering associated with the heart?

_____ **9.** What structure in the frog receives materials from the large intestine, mesonephritic ducts, and the reproductive organs?

_____ **10.** Which animals have modified scales covering their bodies, have scales on their feet, lay shelled eggs, and have internal organs and skulls that provide evidence to classify them as reptiles?

_____ **11.** What organs are present if a vertebrate has a pulmonary system of circulation with pulmonary capillaries?

_____ **12.** How many ventricles are present in the heart of an amphibian and most reptiles?

_____ **13.** Which vertebrates have air sacs attached to their lungs that make the lungs more efficient?

_____ **14.** What organs, present in all the vertebrates, produce urine and maintain the proper balance of fluids and salts in the blood?

## Thought Questions

**15. a.** Does the circulatory system of fishes have a pulmonary circuit? Explain your answer.

**b.** Given the attachment of gills to the pharynx, explain the absence of a trachea in fishes and the need for a trachea in mammals.

**16.** What is the role of the four-chambered hearts of birds and mammals in maintaining a warm body temperature?

**17. a.** Which of the animals studied in this laboratory exercise have a rib cage? Explain the importance of a rib cage to respiration.

**b.** Which of the animals studied in this laboratory exercise have a diaphragm? Explain the importance of a diaphragm to respiration.

# 19

# Basic Mammalian Anatomy I

## Learning Outcomes

### 19.1 External Anatomy
- Compare the limbs of a pig to the limbs of a human.
- Identify the sex of a fetal pig.

### 19.2 Oral Cavity and Pharynx
- Find and identify the teeth, tongue, epiglottis, glottis, esophagus, and palates.

### 19.3 Thoracic and Abdominal Incisions
- Identify the thoracic cavity, abdominal cavity, and diaphragm.

### 19.4 Neck Region
- Find, identify, and state a function for the thymus gland, the larynx, and the thyroid gland.

### 19.5 Thoracic Cavity
- Identify the three compartments and the organs of the thoracic cavity.

### 19.6 Abdominal Cavity
- Find, identify, and state a function for the liver, stomach, spleen, small intestine, gallbladder, pancreas, and large intestine. Describe where these organs are positioned in relation to one another.

## Introduction

### Pre-Lab

1. _____ deliver blood away from the heart, and _____ deliver blood toward the heart.

In this laboratory, you will dissect a fetal pig. Alternately, your instructor may choose to have you observe a pig that has already been dissected. Both pigs and humans are mammals, which means they will share a large degree of similarity in their anatomy. The piglets used in class will usually be within 1 to 2 weeks of birth.

The pigs may have a slash in the right neck region, indicating the site of blood drainage. A red latex solution may have been injected into the **arterial system,** and a blue latex solution may have been injected into the **venous system** of the pigs. If so, when a vessel appears red, it is an artery, and when a vessel appears blue, it is a vein.

As a result of this laboratory, you should gain an appreciation of which organs work together. For example, the liver and the pancreas help to digest fat in the small intestine.

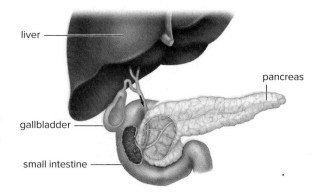

# 19.1 External Anatomy

## Pre-Lab

2. How can you tell the difference between a male and female piglet? _____

_____

Mammals are characterized by the presence of mammary glands and hair. Mammals also occur in two distinct sexes, males and females, often distinguishable by their external **genitals,** the reproductive organs.

Both pigs and humans are placental mammals, which means that development occurs within the uterus of the mother. An **umbilical cord** stretches externally between the fetal animal and the **placenta,** where carbon dioxide and organic wastes are exchanged for oxygen and organic nutrients.

Pigs and humans are tetrapods, meaning they have four limbs. Pigs walk on all four of their limbs; in fact, they walk on their toes, and their toenails have evolved into hooves. In contrast, humans walk only on the feet of their legs.

## Observation: External Anatomy

### Body Regions and Limbs

1. Place your animal in a dissecting pan, and observe the following body regions: the rather large head; the short, thick neck; the cylindrical trunk with two pairs of appendages (forelimbs and hindlimbs); and the short tail (Fig. 19.1a). The tail is an extension of the vertebral column.

> ⚠️ **Latex gloves:** Wear protective latex gloves when handling preserved animal organs. Use protective eyewear and exercise caution when using sharp instruments during this experiment. Wash hands thoroughly upon completion of this experiment.

2. Examine the four limbs, and feel for the joints of the digits, wrist, elbow, shoulder, hip, knee, and ankle.
3. Determine which parts of the forelimb correspond to your arm, elbow, forearm, wrist, and hand.
4. Do the same for the hindlimb, comparing it with your leg.
5. The pig walks on its toenails, which would be like a ballet dancer on "tiptoe." Notice how your heel touches the ground when you walk. Where is the heel of the pig? _____

### Umbilical Cord

1. Locate the umbilical cord arising from the ventral (toward the belly) portion of the abdomen.
2. What is the function of the umbilical cord? _____

_____

### Nipples and Hair

1. Locate the small **nipples,** the external openings of the **mammary glands.** The nipples are *not* an indication of sex, since both males and females possess them. How many nipples does a pig have? _____

When is it advantageous for a pig to have so many nipples? _____

_____

---

**Directional Terms for Dissecting Fetal Pig**

| | |
|---|---|
| Anterior: toward the head end | Ventral: toward the belly |
| Posterior: toward the hind end | Dorsal: toward the back |

---

## Figure 19.1 External anatomy of the fetal pig.

**a.** Body regions and limbs. **b, c.** The sexes can be distinguished by the external genitals.

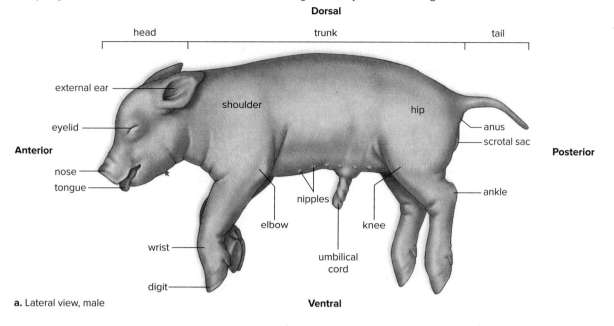

**a.** Lateral view, male

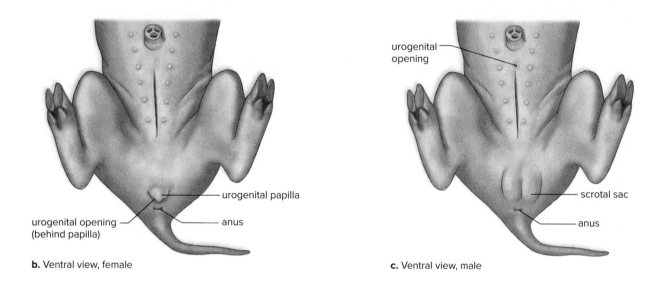

**b.** Ventral view, female

**c.** Ventral view, male

### Anus and External Genitals

1. Locate the **anus** under the tail. Name the organ system that ends in the opening called the anus. _____
2. In females, locate the **urogenital opening,** just anterior to the anus, and a small, fleshy **urogenital papilla** projecting from the urogenital opening (Fig. 19.1*b*).
3. In males, locate the urogenital opening just posterior to the umbilical cord (Fig. 19.1*c*). The duct leading to it runs forward from between the legs in a long, thick tube, the **penis,** which can be felt under the skin. In males, the urinary system and the genital system are always joined.
4. You are responsible for identifying pigs of both sexes. What sex is your pig? _____
   Be sure to look at a pig of the opposite sex that another group of students is dissecting.

## 19.2 Oral Cavity and Pharynx

The **oral cavity** is the space in the mouth that contains the tongue and the teeth. The **pharynx** is dorsal to the oral cavity and has three openings: The **glottis** is an opening through which air passes on its way to the **trachea** (the windpipe) and lungs. The **esophagus** is a portion of the digestive tract that leads through the neck and thorax to the stomach. The **nasopharynx** leads to the nasal passages.

### Observation: Oral Cavity and Pharynx

#### Oral Cavity

1. Insert a sturdy pair of scissors into one corner of the specimen's mouth, and cut posteriorly (toward the hind end) for approximately 4 cm. Repeat on the opposite side until the mouth is open as in Figure 19.2.
2. Place your thumb on the tongue at the front of the mouth, and gently push downward on the lower jaw. This will tear some of the tissue in the angles of the jaws so that the mouth will remain partly open (Fig. 19.2).
3. Note small, underdeveloped teeth in both the upper and lower jaws. Care should be taken because teeth can be very sharp. Other embryonic, nonerupted teeth may also be found within the gums. The teeth are used to chew food.
4. Examine the tongue, which is partly attached to the lower jaw region but extends posteriorly and is attached to a bony structure at the back of the oral cavity (Fig. 19.2). The tongue manipulates food for swallowing.
5. Locate the hard and soft palates (Fig. 19.2). The **hard palate** is the ridged roof of the mouth that separates the oral cavity from the nasal passages. The **soft palate** is a smooth region posterior to the hard palate. An extension of the soft palate—the **uvula**—hangs down into the throat in humans. (A pig does not have a uvula.)

**Figure 19.2   Oral cavity of the fetal pig.**
The roof of the oral cavity contains the hard and soft palates, and the tongue lies above the floor of the oral cavity.

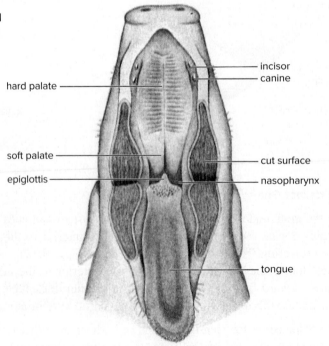

## Pharynx

1. Push down on the tongue until you open the jaws far enough to see a slightly pointed flap of tissue pointing dorsally (toward the back) (Fig. 19.2). This flap is the **epiglottis,** which covers the glottis. The **glottis** leads to the trachea (Fig. 19.3*a*).

2. Posterior and dorsal to the glottis, find the opening into the **esophagus,** a tube that takes food to the stomach. Note the proximity of the glottis and the opening to the esophagus. Each time the pig—or a human—swallows, the epiglottis instantly closes to keep food and fluids from going into the lungs via the trachea.

3. Insert a blunt probe into the glottis, and note that it enters the trachea. Remove the probe, insert it into the esophagus, and note the position of the esophagus beneath (dorsal to) the trachea.

4. Make a midline cut in the soft palate from the epiglottis to the hard palate. Then make two lateral cuts at the edge of the hard palate.

5. Posterior to the soft palate, locate the openings to the nasal passages.

6. Explain why it is correct to say that the air and food passages cross in the pharynx. _____

_____

_____

_____

**Figure 19.3   Air and food passages in the fetal pig.**
The air and food passages cross in the pharynx. **a.** Drawing. **b.** Dissection of specimen.
(b) Ken Taylor/Wildlife Images

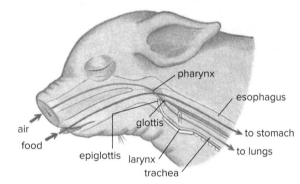

a.

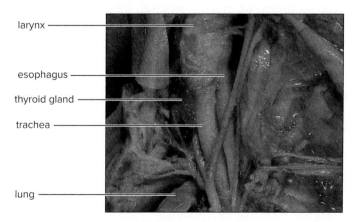

b.

## 19.3 Thoracic and Abdominal Incisions

### Pre-Lab

4. Which organs do you expect to find in the thoracic cavity? _____

_____

First, prepare your pig according to the following directions, and then make thoracic and abdominal incisions so that you will be able to study the internal anatomy of your pig.

### Preparation of Pig for Dissection

1. Place the fetal pig on its back in the dissecting pan.
2. Tie a cord around one forelimb, and then bring the cord around underneath the pan to fasten back the other forelimb.
3. Spread the hindlimbs in the same way.
4. With scissors always pointing up (never down), make the following incisions to expose the thoracic and abdominal cavities. The incisions are numbered on Figure 19.4 to correspond with the following steps.

### Thoracic Incisions

1. Starting at the diaphragm, a structure that separates the thoracic cavity from the abdominal cavity, cut anteriorly until you reach the hairs in the throat region.
2. Make two lateral cuts, one on each side of the midline incision anterior to the forelimbs, taking extra care not to damage the blood vessels around the heart.
3. Make two lateral cuts, one on each side of the midline just posterior to the forelimbs and anterior to the diaphragm, following the ends of the ribs. Pull back the flaps created by these cuts (do not remove

   them) to expose the **thoracic cavity.** List the organs you find in the thoracic cavity. _____

_____

### Abdominal Incisions

4. With scissors pointing up, cut posteriorly from the diaphragm to the umbilical cord.
5. Make a flap containing the umbilical cord by cutting a semicircle around the cord and by cutting posteriorly to the left and right of the cord.
6. Make two cuts, one on each side of the midline incision posterior to the diaphragm. Examine the diaphragm, attached to the chest wall by radially arranged muscles. The central region of the diaphragm, called the **central tendon,** is a membranous area.
7. Make two more cuts, one on each side of the flap containing the umbilical cord and just anterior to the hindlimbs. Pull back the side flaps created by these cuts to expose the **abdominal cavity.**
8. Lifting the flap with the umbilical cord requires cutting the **umbilical vein.** Before cutting the umbilical vein, tie a thread on each side of where you will cut to mark the vein for future reference.
9. Rinse out your pig as soon as you have opened the abdominal cavity. If you have a problem with excess fluid, obtain a disposable plastic pipet to suction off the liquid.

10. Name the two cavities separated by the diaphragm. _____

11. List the organs located in the abdominal cavity. _____

_____

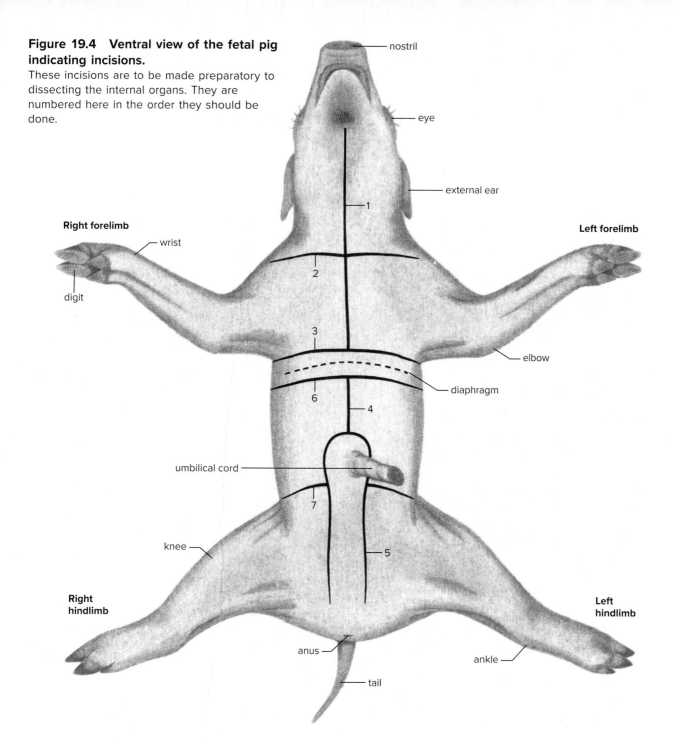

**Figure 19.4  Ventral view of the fetal pig indicating incisions.**
These incisions are to be made preparatory to dissecting the internal organs. They are numbered here in the order they should be done.

nostril

eye

external ear

**Right forelimb**

wrist

digit

**Left forelimb**

elbow

diaphragm

umbilical cord

knee

**Right hindlimb**

**Left hindlimb**

anus

ankle

tail

## 19.4  Neck Region

### Pre-Lab

5. Which structures do you expect to find while dissecting the neck? _____

_____

You will locate several organs in the neck region. Use Figures 19.3b and 19.5 as a guide, but *keep all the flaps* in order to close the thoracic and abdominal cavities at the end of the laboratory session.

The **thymus gland** is a part of the lymphatic system. Certain white blood cells called T (for thymus) lymphocytes mature in the thymus gland and help us fight disease. The **larynx,** or voice box, sits atop (anterior to) the **trachea,** or windpipe. The esophagus is a portion of the digestive tract that leads to the stomach. The **thyroid gland** secretes hormones that travel in the blood and act upon other body cells. These hormones (e.g., thyroxine) regulate the rate at which metabolism occurs in cells.

## Observation: Neck Region

### Thymus Gland

1. Move the skin apart in the neck region just below the hairs mentioned earlier. If necessary, cut the body wall laterally to make flaps. You will most likely be viewing exposed muscles.
2. *Cut through and clear away muscle* to expose the thymus gland, a diffuse gland that lies among the muscles. Later you will notice that the glandular thymus flanks the thyroid and overlies the heart (Fig. 19.5). The thymus is particularly large in fetal pigs, since their immune systems are still developing.

### Larynx, Trachea, and Esophagus

1. Probe down into the deeper layers of the neck. Medially (toward the center), beneath several strips of muscle, find the hard-walled larynx and the trachea, which are parts of the respiratory passage to be examined later. Dorsal to the trachea, find the esophagus.
2. Open the mouth and insert a probe into the glottis and esophagus from the pharynx to better understand the orientation of these two organs.

### Thyroid Gland

Locate the thyroid gland just posterior to the larynx, lying ventral to (on top of) the trachea.


# 19.5   Thoracic Cavity

## Pre-Lab

6. Why does the left lung have fewer lobes? _____

_____

As previously mentioned, the body cavity of mammals, including humans, is divided by the diaphragm into the thoracic cavity and the abdominal cavity. The heart and lungs are in the thoracic cavity (Figs. 19.5 and 19.6). The **heart** is a pump for the cardiovascular system, and the **lungs** are organs of the respiratory system where gas exchange occurs.

## Observation: Thoracic Cavity

### Heart and Lungs

1. If you have not yet done so, fold back the chest wall flaps. To do this, you will need to tear the thin membranes that divide the thoracic cavity into three compartments: the **left pleural cavity** containing the left lung, the **right pleural cavity** containing the right lung, and the **pericardial cavity** containing the heart.
2. Examine the lungs. Locate the four lobes of the right lung and the three lobes of the left lung. The trachea, dorsal to the heart, divides into the **bronchi,** which enter the lungs. Later, when the heart is removed, you will be able to see the trachea and bronchi.
3. List the structures air passes as it moves through the respiratory tract from the nasal passages to the lungs.

_____

_____

# Figure 19.5   Internal anatomy of the fetal pig.

The major organs are featured in this drawing. In the fetal pig, a red color tells you a vessel is an artery, and a blue color tells you it is a vein. (It does not tell you whether this vessel carries $O_2$-rich or $O_2$-poor blood.) **Contrary to this drawing, *keep all the flaps on your pig*** so you can close the thoracic and abdominal cavities at the end of the laboratory session.

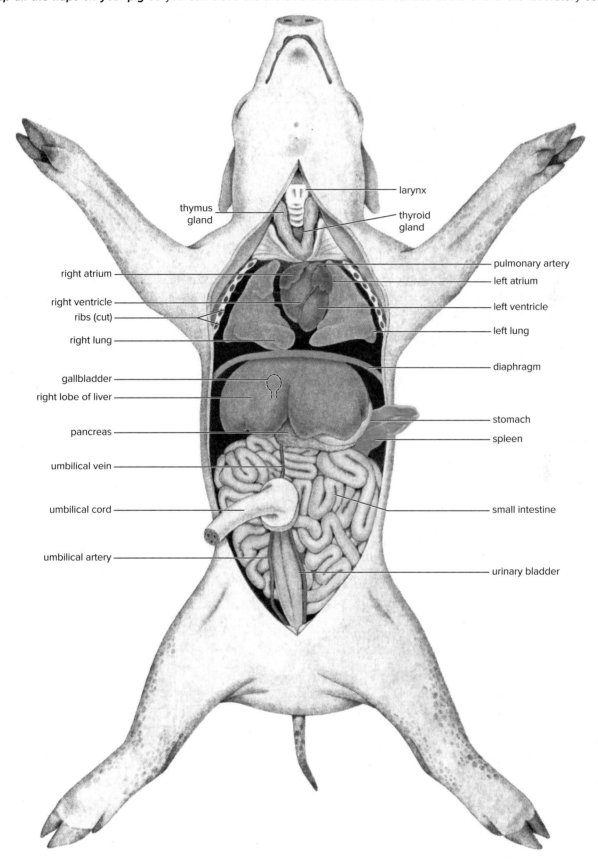

## 19.6    Abdominal Cavity

**Pre-Lab**

7. What is the largest structure within the abdominal cavity? _____

8. Which organ in the abdominal cavity appears as a thin, winding tube? _____

The abdominal wall and organs are lined by a membrane called **peritoneum,** consisting of epithelium supported by connective tissue. Double-layered sheets of peritoneum, called **mesenteries,** project from the body wall and support the organs.

### Observation: Abdominal Cavity

If your pig is partially filled with dark, brownish material, take your animal to the sink and rinse it out. This material is clotted blood. Consult your instructor before removing any red or blue latex masses, since they may enclose organs you will need to study.

#### Liver

The **liver,** the largest organ in the abdomen (Fig. 19.6), performs numerous vital functions, including (1) disposing of worn-out red blood cells, (2) producing bile, (3) storing glycogen, (4) maintaining the blood glucose level, and (5) producing blood proteins.

1. Locate the liver, a large, brown organ. Its anterior surface is smoothly convex and fits snugly into the concavity of the diaphragm.

2. Name several functions of the liver. _____

_____

#### Stomach and Spleen

The organs of the digestive tract include the stomach, small intestine, and large intestine. The **stomach** (see Fig. 19.5) stores food and has numerous gastric glands. These glands secrete a juice that digests protein. The **spleen** (see Fig. 19.5) is a lymphoid organ in the lymphatic system that contains both white and red blood cells. It purifies blood and disposes of worn-out red blood cells.

1. Push aside and identify the stomach, a large sac dorsal to the liver on the left side.
2. Locate the point near the midline of the body where the **esophagus** penetrates the diaphragm and joins the stomach.
3. Find the spleen, a long, flat, reddish organ attached to the stomach by mesentery.
4. What is the function of the stomach? _____
5. What is the function of the spleen? _____

#### Small Intestine

The **small intestine** is the part of the digestive tract that receives secretions from the pancreas and gallbladder. Besides being an area for the digestion of all components of food, carbohydrate, protein, and fat, the small intestine absorbs the products of digestion: glucose, amino acids, glycerol, and fatty acids.

1. Look posteriorly where the stomach makes a curve to the right and narrows to join the anterior end of the small intestine, called the **duodenum.**
2. From the duodenum, the small intestine runs posteriorly for a short distance and is then thrown into an irregular mass of bends and coils held together by a common mesentery.
3. The small intestine is a part of what system? _____

   What is its function? _____

# Figure 19.6  Internal anatomy of the fetal pig.

Most of the major organs are shown in this photograph. The stomach has been removed. The spleen, gallbladder, and pancreas are not visible. *Do not* remove any organs or flaps from your pig.

Ken Taylor/Wildlife Images

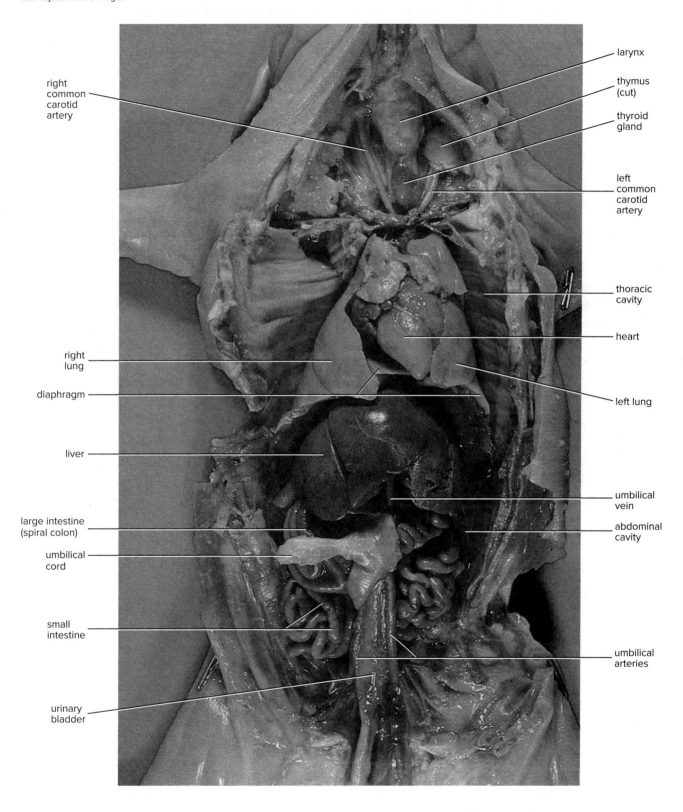

larynx

thymus (cut)

thyroid gland

left common carotid artery

thoracic cavity

heart

left lung

umbilical vein

abdominal cavity

umbilical arteries

right common carotid artery

right lung

diaphragm

liver

large intestine (spiral colon)

umbilical cord

small intestine

urinary bladder

**Laboratory 19**  Basic Mammalian Anatomy I

### Gallbladder and Pancreas

The **gallbladder** stores and releases bile, which aids in the digestion of fat. The **pancreas** (see Fig. 19.5) is both an exocrine and an endocrine gland. As an exocrine gland, it produces and secretes pancreatic juice, which digests all the components of food in the small intestine. Both bile and pancreatic juice enter the duodenum by way of ducts. As an endocrine gland, the pancreas secretes the hormones insulin and glucagon into the bloodstream. Insulin and glucagon regulate blood glucose levels.

1. Locate the **bile duct,** which runs in the mesentery stretching between the liver and the duodenum. Find the gallbladder, embedded in the liver on the underside of the right lobe. It is a small, greenish sac.
2. Lift the stomach and locate the pancreas, the light-colored, diffuse gland lying in the mesentery between the stomach and the small intestine. The pancreas has a duct that empties into the duod enum of the small intestine.
3. What is the function of the gallbladder? _____
   _____
4. What is the function of the pancreas? _____
   _____

### Large Intestine

The **large intestine** is the part of the digestive tract that absorbs water and prepares feces for defecation at the anus. The first part of the large intestine, called the **cecum,** has a projection called the vermiform (meaning wormlike) appendix.

1. Locate the distal (far) end of the small intestine, which joins the large intestine posteriorly, in the left side of the abdominal cavity (right side in humans). At this junction, note the cecum, a blind pouch.
2. Follow the main portion of the large intestine, known as the **colon,** as it runs from the point of juncture with the small intestine into a tight coil (spiral colon), then out of the coil anteriorly, then posteriorly again along the midline of the dorsal wall of the abdominal cavity. In the pelvic region, the **rectum** is the last portion of the large intestine. The rectum leads to the **anus.**
3. The large intestine is a part of what system? _____
4. What is the function of the large intestine? _____
5. List the structures food passes as it moves through the digestive tract from the mouth to the anus. _____
   _____

## Storage of Pigs

1. Before leaving the laboratory, place your pig in the plastic bag provided.
2. Expel excess air from the bag, and tie it shut.
3. Write your name and section on the tag provided, and attach it to the bag. Your instructor will indicate where the bags are to be stored until the next laboratory period.
4. Clean the dissecting tray and tools, and return them to their proper location.
5. Wipe off your goggles.
6. Wash your hands.

_____ **1.** What directional term refers to the head end of the fetal pig?

_____ **2.** If the urogenital opening is just anterior to the anus, is the fetal pig male or female?

_____ **3.** What space is dorsal to the oral cavity and has three openings?

_____ **4.** What is the name of the flap that closes during swallowing to prevent food and fluids from entering the trachea?

_____ **5.** What is the structure that separates the thoracic cavity from the abdominal cavity?

_____ **6.** What glandular organ is especially large in fetal pigs since their immune systems are still developing?

_____ **7.** Is the trachea ventral or dorsal to the esophagus?

_____ **8.** In what thoracic cavities are the lungs located?

_____ **9.** What is the largest organ in the abdominal cavity?

_____ **10.** What organ do secretions from the gallbladder and pancreas enter?

_____ **11.** What is the name of the tissue that holds the small intestine together?

_____ **12.** What is the name of the first part of the large intestine?

_____ **13.** What organ produces digestive enzymes and hormones that regulate blood glucose?

_____ **14.** To what organ system do the kidneys belong?

## Thought Questions

**15.** Why would someone born without a thymus gland be likely to suffer from a greater number of infections?

**16.** Treatment for a person's laryngeal cancer might require surgical removal of the epiglottis. What difficulties would this cause the person?

**17.** Based on the location of the spleen, explain why it is damaged so often in car accidents, especially when the driver's side is struck.

# 20

# Chemical Aspects of Digestion

## Learning Outcomes

**Introduction**
- Sequence the organs of the digestive tract from the mouth to the anus.

**20.1 Protein Digestion by Pepsin**
- Associate the enzyme pepsin with the ability of the stomach to digest protein.
- Explain why stomach contents are acidic and how a warm body temperature aids digestion.

**20.2 Fat Digestion by Pancreatic Lipase**
- Explain why the emulsification process assists the action of lipase.
- Explain why a change in pH indicates that fat digestion has occurred.
- Explain the relationship between time and enzyme activity.

**20.3 Starch Digestion by Pancreatic Amylase**
- Associate the enzyme pancreatic amylase with the ability of the small intestine to digest starch.

**20.4 Requirements for Digestion**
- List four factors that can affect the activity of all enzymes.

## Introduction

### Pre-Lab

1. *Label the organs in Figure 20.1 with the help of your textbook.*

In this lab, we will examine the process of digestion by learning the organs associated with digestion and studying the action of digestive enzymes.

Enzymes are molecules (typically proteins) that catalyze chemical reactions. They are very specific and usually participate in only one type of reaction. The active site of an enzyme has a shape that accommodates its substrate, and if an environmental factor such as a change in temperature or pH that pushes the enzyme out of its functional range. This causes the enzyme to change it's shape and prevents it from binding to the substrate causing the reaction to stop. We will have an opportunity to make these observations with controlled experiments.

> **Planning Ahead** Be advised that protein digestion requires 1½ hours and fat digestion requires 1 hour. Also, a boiling water bath is required for starch digestion.

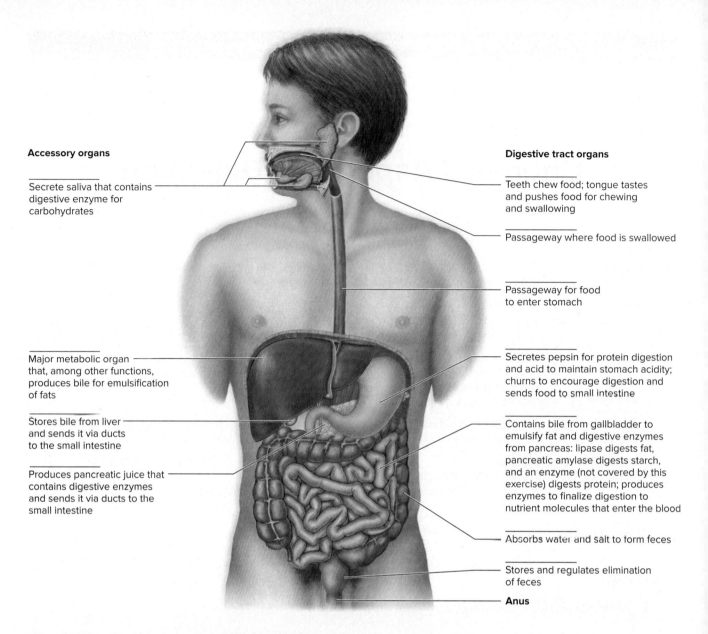

**Accessory organs**

Secrete saliva that contains
digestive enzyme for
carbohydrates
_____

Major metabolic organ
that, among other functions,
produces bile for emulsification
of fats
_____

Stores bile from liver
and sends it via ducts
to the small intestine
_____

Produces pancreatic juice that
contains digestive enzymes
and sends it via ducts to the
small intestine
_____

**Digestive tract organs**

Teeth chew food; tongue tastes
and pushes food for chewing
and swallowing
_____

Passageway where food is swallowed
_____

Passageway for food
to enter stomach
_____

Secretes pepsin for protein digestion
and acid to maintain stomach acidity;
churns to encourage digestion and
sends food to small intestine
_____

Contains bile from gallbladder to
emulsify fat and digestive enzymes
from pancreas: lipase digests fat,
pancreatic amylase digests starch,
and an enzyme (not covered by this
exercise) digests protein; produces
enzymes to finalize digestion to
nutrient molecules that enter the blood
_____

Absorbs water and salt to form feces
_____

Stores and regulates elimination
of feces
_____

**Anus**

**Figure 20.1   Organs of the digestive tract** *(right)* **and accessory organs** *(left).*
Label *each structure with the help of your textbook.*

## 20.1   Protein Digestion by Pepsin

**Pre-Lab**

2. Do you expect a higher temperature to increase or decrease the rate of pepsin activity? _____

_____

3. Do you expect a lower pH to increase or decrease the rate of pepsin activity? _____

_____

Certain foods, such as meat and egg whites, are rich in protein. Egg whites contain albumin, which is the protein used in this Experimental Procedure. Protein is digested by **pepsin** in the stomach (Fig. 20.2), a process described by the following reaction:

$$\text{protein} + \text{water} \xrightarrow{\text{pepsin (enzyme)}} \text{peptides}$$

The stomach has a very low pH. Does this indicate that pepsin works effectively in an acidic or a basic environment? _____ This is the pH that allows the enzyme to maintain its normal shape so that it will combine with the substrate. A warm temperature causes molecules to move about more rapidly and increases the encounters between enzyme and substrate.

## Test for Protein Digestion

Biuret reagent is used to test for protein digestion. If digestion has not occurred, biuret reagent turns purple, indicating that protein is present. If digestion has occurred, biuret reagent turns pinkish-purple, indicating that peptides are present.

> ⚠ **Biuret reagent** is highly corrosive. Exercise care in using this chemical. If any should spill on your skin, wash the area with mild soap and water. Follow your instructor's directions for its disposal.

### Experimental Procedure: Protein Digestion

1. *Label four clean test tubes (1 to 4)*. Using the designated graduated pipet, add 2 ml of the albumin solution to all tubes. Albumin is a protein.
2. Add 2 ml of the pepsin solution to tubes 1 to 3, as listed in Table 20.1.
3. Add 2 ml of 0.2% HCl to tubes 1 and 2. HCl simulates the acidic conditions of the stomach.
4. Add 2 ml of water to tube 3 and 4 ml of water to tube 4, as listed in Table 20.1.
5. Swirl to mix the tubes. Tube 2 remains at room temperature, but the other three are incubated for 1½ hours. Record the temperature for each tube in Table 20.1.
6. Remove the tubes from the incubator and place all four tubes in a tube rack. Add 2 ml of biuret reagent to all tubes and observe. Record your results in Table 20.1 as + or − to indicate digestion or no digestion.

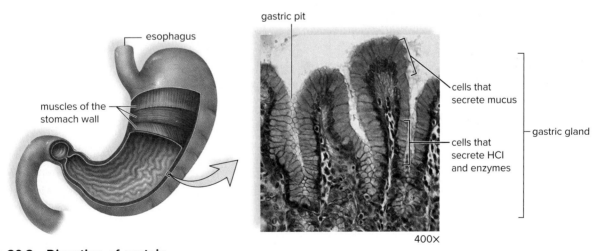

**Figure 20.2  Digestion of protein.**
Pepsin, produced by the gastric glands of the stomach, helps digest protein.
(b) Ed Reschke

## Table 20.1 Protein Digestion by Pepsin

| Tube | Contents | Temperature | Digestion (+ or −) | Explanation |
|------|----------|-------------|--------------------|-------------|
| 1 | Albumin<br>Pepsin<br>HCl<br>Biuret reagent | | | |
| 2 | Albumin<br>Pepsin<br>HCl<br>Biuret reagent | | | |
| 3 | Albumin<br>Pepsin<br>Water<br>Biuret reagent | | | |
| 4 | Albumin<br>Water<br>Biuret reagent | | | |

### Conclusions: Protein Digestion

- Explain your results in Table 20.1 by giving an explanation why digestion did or did not occur. To be complete, consider all the requirements for an enzymatic reaction as listed in Table 20.4. Now show here that tube 1 met all the requirements for digestion:

  Pepsin is the correct _____.

  Albumin is the correct _____.

  37°C is the optimum _____.

  HCL provides the optimum _____.

  1½ hours provides _____ for the reaction to occur.

- Which tube was the control? _____

  Explain. _____

  _____

- If this control tube had given a positive result for protein digestion, what could you conclude about this experiment? _____

  _____

## 20.2 Fat Digestion by Pancreatic Lipase

Lipids include fats (e.g., butterfat) and oils (e.g., sunflower, corn, olive, and canola). Lipids are digested by **pancreatic lipase** in the small intestine (Fig. 20.3).

**Figure 20.3  Emulsification and digestion of fat.**
Bile from the liver (stored in the gallbladder) enters the small intestine, where lipase in pancreatic juice from the pancreas digests fat.

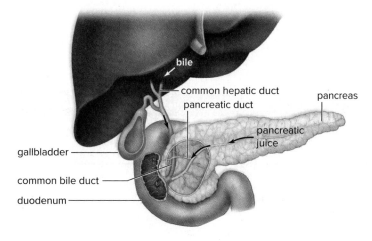

The following two reactions describe fat digestion:

1.     fat $\xrightarrow{\text{bile (emulsifier)}}$ fat droplets

2.     fat droplets + water $\xrightarrow{\text{lipase (enzyme)}}$ glycerol + fatty acids

With regard to the first step, consider that fat is not soluble in water; yet, lipase makes use of water when it digests fat. Therefore, bile is needed to emulsify fat—cause it to break up into fat droplets that disperse in water. The reason for dispersal is that bile contains molecules with two ends. One end (the nonpolar end) is soluble in fat, and the other end (the polar end) is soluble in water. Bile can emulsify fat because of this.

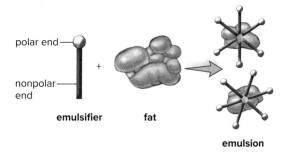

With regard to the second step, would the pH of the solution be lower before or after the enzymatic reaction? (*Hint:* Remember that an acid decreases pH and a base increases pH.) _____

## Test for Fat Digestion

In the test for fat digestion, you will be using a pH indicator, which changes color as the solution in the test tube goes from basic conditions to acidic conditions. Phenol red is a pH indicator that is red in basic solutions and yellow in acidic solutions.

### Experimental Procedure: Fat Digestion

1. *Label three clean test tubes (1 to 3).* Using the designated graduated pipet, add 1 ml of vegetable oil to all tubes.
2. Add 2 ml of phenol red solution to each tube. What role does phenol red play? _____
3. Add 2 ml of pancreatic lipase (pancreatin) to tubes 1 and 2 and 2 ml of water to tube 3, as listed in Table 20.2. What role does lipase play? _____
4. Add a pinch of bile salts to tube 1.
5. Record the initial color of all tubes in Table 20.2.
6. Incubate all three tubes at 37°C and check every 20 minutes.
7. Record any color change and how long it took to see this color change in Table 20.2.

### Table 20.2  Fat Digestion by Pancreatic Lipase

| Tube | Contents | Color | | Time Taken | Explanation |
| --- | --- | --- | --- | --- | --- |
| | | *Initial* | *Final* | | |
| 1 | Vegetable oil<br>Phenol red<br>Pancreatin<br>Bile salts | | | | |
| 2 | Vegetable oil<br>Phenol red<br>Pancreatin | | | | |
| 3 | Vegetable oil<br>Phenol red<br>Water | | | | |

### Conclusions: Fat Digestion

- Explain your results in Table 20.2 by giving an explanation why digestion did or did not occur.
- What role did bile salts play in this experiment? _____

  _____

- What role did phenol red play in this experiment? _____
- Which test tube in this experiment could be considered a control? _____

  _____

## 20.3 Starch Digestion by Pancreatic Amylase

**Pre-Lab**

5. Why does a positive Benedict's test for sugar indicate starch digestion has taken place? _____

_____

Starch is present in bakery products as well as foods like potatoes, rice, and corn. Starch is digested by **pancreatic amylase** in the small intestine, a process described by the following reaction:

$$\text{starch} + \text{water} \xrightarrow{\text{amylase (enzyme)}} \text{maltose}$$

1. If digestion *does not* occur, which will be present—starch or maltose? _____

2. If digestion *does* occur, which will be present—starch or maltose? _____

### Tests for Starch Digestion

You will be using two tests for starch digestion:

1. If digestion has not taken place, the iodine test for starch will be positive (+) and a blue-black color will be observed. If digestion has occurred, the iodine test for starch will be negative (−) and the iodine will remain yellowish-brown.

2. If digestion has taken place, the Benedict's test for sugar (maltose) will be positive (+) and a color change ranging from green to red will be observed. If digestion has not taken place, the Benedict's test for sugar will be negative (−) and the solution will remain blue.

> ⚠ **Benedict's reagent** is highly corrosive. Use protective eyewear when performing this experiment. Exercise care in using this chemical. If any should spill on your skin, wash the area with mild soap and water. Follow your instructor's directions for disposal of this chemical.

To test for sugar, add five drops of Benedict's reagent to each test tube. Place the tube in a boiling water bath for a few minutes, and note any color changes. Boiling the test tube is necessary for the Benedict's reagent to react.

### Experimental Procedure: Starch Digestion

1. *Label six clean test tubes (1 to 6).*
2. Using the designated graduated transfer pipet, add 1 ml of pancreatic amylase solution to tubes 1 to 4 and 1 ml of water to tubes 5 and 6.
3. Test tubes 1 and 2 immediately.

   **Tube 1**    Shake the starch solution and add 1 ml of starch solution. Immediately add five drops of iodine to test for starch. Put this tube in a test tube rack and record your results in Table 20.3.

   **Tube 2**    Shake the starch solution and add 1 ml of starch solution. Immediately test for sugar with the Benedict's test following the preceding directions. Put this tube in a test tube rack and record your results in Table 20.3.

4. Shake the starch solution and add 1 ml of starch solution to tubes 3 to 6. Allow the tubes to stand for 30 minutes.

   **Tubes 3 and 5**    After the 30 minutes have passed, test for starch using the iodine test. Place these tubes in the test tube rack and record your results in Table 20.3.

   **Tubes 4 and 6**    After the 30 minutes have passed, test for sugar with the Benedict's test following the preceding directions. Place these tubes in the test tube rack and record your results in Table 20.3.

5. Examine all your tubes in the test tube rack and decide whether digestion occurred (+) or did not occur (−). Complete Table 20.3.

## Table 20.3  Starch Digestion by Amylase

| Tube | Contents | Time* | Type of Test | Results | Explanation |
|------|----------|-------|--------------|---------|-------------|
| 1 | Pancreatic amylase<br>Starch | 0 | Iodine | + | |
| 2 | Pancreatic amylase<br>Starch | | | | |
| 3 | Pancreatic amylase<br>Starch | | | | |
| 4 | Pancreatic amylase<br>Starch | | | | |
| 5 | Water<br>Starch | | | | |
| 6 | Water<br>Starch | | | | |

*Enter either 0 for immediately or T for after 30 minutes.

### Conclusions: Starch Digestion

- Considering tubes 1 and 2, this Experimental Procedure showed that _____ must pass for digestion to occur.

- Considering tubes 5 and 6, this Experimental Procedure showed that an active _____ must be present for digestion to occur.

- Why would you not recommend doing the test for starch and the test for sugar on the same tube? _____
  _____

  _____

- Which test tubes served as a control in this experiment? _____

## Absorption of Sugars and Other Nutrients

Figure 20.4 shows that the folded lining of the small intestine has many fingerlike projections called villi. The small intestine not only digests food, it also absorbs the products of digestion, such as sugars from carbohydrate digestion, amino acids from protein digestion, and glycerol and fatty acids from fat digestion at the villi.

**Figure 20.4   Anatomy of the small intestine.**
Nutrients enter the bloodstream across the much-convoluted walls of the small intestine.

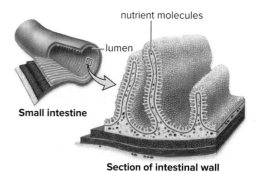

# 20.4   Requirements for Digestion

**Pre-Lab**

6. What happens to enzymes if they are not in their ideal temperature or pH? _____

_____

7. Do all enzymes have the same ideal temperature and pH? _____

Explain in Table 20.4 how each of the requirements listed influences effective digestion.

| Table 20.4   Requirements for Digestion | |
| --- | --- |
| **Requirement** | **Explanation** |
| Specific enzyme | |
| Specific substrate | |
| Warm temperature | |
| Specific pH | |
| Time | |
| Fat emulsifier | |

_____ 1. What part of an enzyme allows it to bind to a specific substrate?

_____ 2. What two environmental conditions affect the ability of an enzyme to function properly?

_____ 3. Where are carbohydrates first digested by an enzyme?

_____ 4. What organ produces the lipase and amylase that perform digestion in the small intestine?

_____ 5. What is the name given to a sample that contains the factor being tested and that goes through all the steps of the experiment?

_____ 6. Describe the environment in which pepsin functions.

_____ 7. What are the end products of protein digestion?

_____ 8. What are the end products of fat digestion?

_____ 9. What molecule emulsifies fats in the small intestine?

_____ 10. What organ stores bile?

_____ 11. What component of foods is digested by pancreatic amylase?

_____ 12. What molecule is present after pancreatic amylase digests starch?

_____ 13. What two tests can be done to test for starch digestion?

_____ 14. What structures absorb the products of digestion in the small intestine?

## Thought Questions

15. Salivary amylase digests starch to maltose in the mouth while chewing takes place. Why is it beneficial to have amylase present in two different locations in the body?

16. How would the digestion of fat be impacted by the removal of the gallbladder? Explain.

17. Explain why consuming excess antacids might affect protein digestion. How could you investigate the impact of consuming antacids on protein digestion?

# 21

# Energy Requirements and Ideal Weight

## Learning Outcomes

**21.1 Average Daily Energy Intake**
- Keep a food diary and use it to calculate your average daily energy intake.

**21.2 Average Daily Energy Requirement**
- Keep a physical activity diary and use it to calculate average daily energy required for physical activity.
- Calculate the daily energy required for your basal metabolism.
- Calculate the daily energy required for your specific dynamic action.
- Calculate your total average daily energy requirement.

**21.3 Comparison of Average Daily Energy Intake and Average Daily Energy Requirement**
- Predict whether weight gain or weight loss will occur after energy intake is compared to energy required.

**21.4 Ideal Weight**
- Consult height and weight tables to determine ideal weight.
- Calculate ideal weight based on body mass index (BMI).
- Calculate ideal weight based on body composition.
- Make a recommendation regarding daily energy intake and daily energy required for achieving or maintaining ideal weight.

## Pre-Lab

**1.** What term describes the amount of energy used by your body at rest? _____

_____

## Introduction

This laboratory helps you determine your average daily energy (kcal) requirements, your average daily energy (kcal) intake, and your ideal weight. In addition to the energy content of food, other nutritional aspects are also important. Minerals and vitamins, along with specific types of biological molecules, contribute to a healthy diet. So, while the focus of this lab is on energy in our food, there is more to diet than just calories.

Before you come to the laboratory, it is necessary for you to keep two daily diaries for 3 days. The physical activity diary is a record of the energy you expend for physical activity. Physical activity is only a portion of your daily energy requirement. You also need energy for **basal metabolism** (energy needed when the body is resting) and **specific dynamic action** (energy needed to process food), also known as the thermal effect of food.

The food diary is a record of the energy you take in as food. If the energy you expend and the energy you take in are in balance, you should maintain a stable weight. If your ideal weight is different from your present weight, you will be able to determine whether to increase or decrease your physical activity and the amount of food you eat in order to attain your ideal weight.

## 21.1 Average Daily Energy Intake

**Pre-Lab**

2. What does *daily energy intake* mean? _____

_____

_____

To complete this laboratory, you must calculate an average daily energy intake. The first method given is preferred, because it makes use of your actual intake. However, when appropriate, a hypothetical method is also provided.

### Personal Diet

Follow these directions.

1. Keep a written food diary of your own design for 3 days. Your diary should list all food and beverages consumed throughout an entire 3-day period, including snacks and alcoholic beverages.
2. Make sure your food diary contains a column for recording the energy in kcal (kilocalories)* of the food you eat. Table 21.1 may be helpful in this regard. Also, food packages often contain kcal information, as do various calorie books and computer programs. For example, after eating two slices of bread, your diary might look like this:

| Day of Week _____ | | | |
|---|---|---|---|
| **Type of Food** | **Portion Size** | **kcal** | **Total kcal** |
| Bread | Two slices | 70/one slice | 140 |

Continue in this manner until you have listed all the foods you eat in 1 day. Do the same for 2 more days.

3. Add up your total kcal intake for each day, and record the information here:

   Day 1 energy intake _____ kcal

   Day 2 energy intake _____ kcal

   Day 3 energy intake _____ kcal

   Total _____ kcal

4. Divide the total by 3 to calculate the average daily energy intake.

   Av. energy intake/day = _____ kcal

*The amount of heat required to raise 1 kg (kilogram) of water 1°C.

## Table 21.1  Nutrient and Energy Content of Foods

**Breads**

1 slice bread or any of the following:

¾ cup ready-to-eat cereal

⅓ cup corn

1 small potato

(1 bread = 15 g carbohydrate, 2 g protein, and 70 kcal)

**Milk**

1 cup skim milk or:                  (2% milk—add 5 g fat)

1 cup skim-milk yogurt, plain     (whole milk—add 8 g fat)

1 cup buttermilk

½ cup evaporated skim milk or milk dessert

(1 milk = 12 g carbohydrate, 8 g protein, and 80 kcal)

**Vegetables**

½ cup greens

½ cup carrots

½ cup beets

(1 vegetable = 5 g carbohydrate, 1 g protein, and 25 kcal)

**Fruits**

½ small banana or

1 small apple

½ cup orange juice or ½ grapefruit

(1 fruit = 10 g carbohydrate and 40 kcal)

**Meats (lean)**

1 oz. lean meat or:

1 oz. chicken meat without the skin

1 oz. any fish

¼ cup canned tuna or 1 oz. low-fat cheese

(1 oz. low-fat meat = 7 g protein, 3 g fat, and 55 kcal)

**Meats (medium fat)**

1 oz. pork loin

1 egg

¼ cup creamed cottage cheese

(1 medium-fat meat = 7 g protein,

5½ g fat, and about 80 kcal)

**Meats (high fat)**

1 oz. high-fat meat is like:

1 oz. country-style ham

1 oz. cheddar cheese

small hot dog (frankfurter)

(1 high-fat meat = 7 g protein, 8 g fat, and
100 kcal)

**Peanut butter**

Peanut butter is like a meat in terms of its
protein content but is very high in fat. It is
estimated as (2 tbsp peanut butter = 7 g protein,
15½ g fat, and about 170 kcal).

**Fats**

1 tsp butter or margarine

1 tsp any oil

1 tbsp salad dressing

1 strip crisp bacon

5 small olives

10 whole Virginia peanuts

(1 fat = 5 g fat and 45 kcal)

**Legumes (beans and peas)**

Legumes are like meats because they are rich in
protein and iron but are lower in fat than meat.
They contain much starch. They can be treated as
(½ cup legumes = 15 g carbohydrate, 9 g protein,
3 g fat, and 125 kcal).

### Miscellaneous Foods

| | g (grams) | | | kcal | | g (grams) | | | kcal |
|---|---|---|---|---|---|---|---|---|---|
| | Protein | Fat | Carbohydrate | | | Protein | Fat | Carbohydrate | |
| Ice cream (1 cup) | 5 | 14 | 32 | 274 | Beer (1 can) | 1 | 0 | 14 | 60 |
| Cake (1 piece) | 3 | 1 | 32 | 149 | Soft drink | 0 | 0 | 37 | 148 |
| Doughnuts (1) | 3 | 11 | 22 | 199 | Soup (1 cup) | 2 | 3 | 15 | 95 |
| Pie | 3 | 15 | 51 | 351 | Coffee and tea | 0 | 0 | 0 | 0 |
| Caramel candy (1 oz.) | 1 | 3 | 22 | 119 | | | | | |

## Alternate Option: Hypothetical Diet

Suppose you have decided to eat your meals at the fast-food restaurants listed in the table.

1. Circle the restaurant and the items you have chosen for breakfast, lunch, and dinner. Choose a different restaurant for each meal.
2. Record the number of kcal for the items chosen.
3. Add the number of kcal.
4. This will be your average energy intake/day.

### Fast-Food Menus and kcal

| Restaurant/Menu | kcal | Restaurant/Menu | kcal |
|---|---|---|---|
| **Taco Bell (www.tacobell.com)** | | **McDonald's (www.mcdonalds.com)** | |
| Steak Baja chalupa | 390 | Egg McMuffin | 300 |
| Soft beef taco supreme | 240 | Sausage McMuffin w/egg | 450 |
| Chicken Baja gordita | 320 | Ham and egg cheese bagel | 550 |
| Spicy chicken soft taco | 170 | Hash browns | 130 |
| Nachos Bellgrande | 760 | Hotcakes (w/margarine & syrup) | 350 |
| Soft drink (small) | 150 | Breakfast burrito | 290 |
| **Subway (www.subway.com)** | | 6 pc. chicken nugget | 280 |
| Meatball marinara | 560 | Barbeque sauce | 50 |
| 6" roasted chicken breast | 310 | Hamburger | 250 |
| 6" roast beef | 290 | Quarter pounder w/cheese | 510 |
| Soft drink (small) | 150 | Big Mac | 540 |
| **Burger King (www.burgerking.com)** | | Filet-O-Fish | 380 |
| Croissan'wich w/sausage, egg & cheese | 520 | Chicken McGrill w/mayo | 420 |
| Croissan'wich w/egg & cheese | 320 | Med. french fries | 380 |
| Egg'wich w/Canadian bacon & egg | 380 | Large french fries | 500 |
| French toast sticks | 390 | Hot fudge sundae | 340 |
| Lg. hash brown rounds | 390 | Vanilla milk shake | 570 |
| Double Whopper w/cheese and mayo | 1,150 | Soft drink (small) | 150 |
| Whopper w/cheese and mayo | 760 | **Papa John's (www.papajohns.com)** | |
| BK big fish | 710 | Cheese pizza | 283 |
| King-sized fries w/salt | 600 | All-meat pizza | 390 |
| Med.-sized fries w/salt | 360 | Garden special pizza | 280 |
| King-sized onion rings | 550 | The works pizza | 342 |
| Med.-sized onion rings | 320 | Cheese sticks | 180 |
| Dutch apple pie | 340 | Breadsticks | 140 |
| Hershey's sundae pie | 310 | Garlic sauce | 235 |
| Soft drink (small) | 150 | Pizza sauce | 25 |
| Vanilla milk shake (medium) | 720 | Nacho cheese sauce | 60 |
| | | Soft drink (small) | 150 |

Av. energy intake/day = _____ kcal

## 21.2 Average Daily Energy Requirement

As stated at the beginning of this laboratory, the body needs energy for three purposes: (1) energy for physical activity, (2) energy to support basal metabolism, and (3) energy for the specific dynamic action of processing food. We will calculate the daily energy requirement for each of these.

### Calculating Average Daily Energy Required for Physical Activity

You will need to know your average daily energy requirement for physical activity before coming to this laboratory.

1. Keep a physical activity diary for 3 days. Make three copies of Table 21.2, one for each day.
2. Record your body weight in kg (kilograms) in your physical activity diary (Table 21.2).
3. Consult Table 21.3 and fill in column 4 of your physical activity diary.
4. Fill in column 5 of your physical activity diary by using a calculator and this formula:

$$\text{time spent} \times \text{energy cost} \times \text{body weight} = \text{total energy expended}$$
$$\text{(min)} \qquad\qquad\qquad\qquad \text{(kg)} \qquad\qquad \text{(kcal)}$$

Therefore, if 100 minutes are spent doing word processing, and the body weight is 77.3 kg, the kcal expended for typing is

$$100 \text{ min} \times 0.01 \text{ kcal/kg min} \times 77.3 \text{ kg} = 77.3 \text{ kcal}$$

5. Total the number of kcal expended for each of the 3 days, and divide this total by 3 to get the average (av.) daily energy required for physical activity:

Energy for physical activity (day 1)  _____ kcal

Energy for physical activity (day 2)  _____ kcal

Energy for physical activity (day 3)  _____ kcal

Total energy for physical activity/3 days  _____ kcal

Av. energy required for physical activity/day  _____ kcal

## Table 21.2  Physical Activity Diary for One Day

| Time of Day | Activity | Time Spent (min) | Factor (kcal/kg min) | Weight (kg) | Total Energy Expended (kcal) |
|---|---|---|---|---|---|
| | | | | | |
| | | | | | |
| | | | | | |
| | | | | | |
| | | | | | |
| | | | | | |
| | | | | | |

Total energy expended _____

## Table 21.3  Energy Cost for Activities (Exclusive of Basal Metabolism and Specific Dynamic Action)

| Type of Activity | Energy Cost (kcal/kg min) | Type of Activity | Energy Cost (kcal/kg min) |
|---|---|---|---|
| Sitting or standing still | 0.010 | Heavy exercise | 0.070 |
| Studying | | Fast dancing | |
| Writing | | Walking uphill | |
| Word processing | | Jogging | |
| TV watching | | Fast swimming | |
| Eating | | Severe exercise | 0.110 |
| Very light activity | 0.020 | Tennis | |
| Driving car | | Racquetball | |
| Walking slowly | | Running | |
| Light exercise | 0.025 | Aerobic dancing | |
| Light housework | | Soccer | |
| Walking at moderate speed | | Very severe exercise | 0.140 |
| Carrying books or packages | | Wrestling | |
| Moderate exercise | 0.040 | Boxing | |
| Fast walking | | Racing | |
| Slow dancing | | Rowing | |
| Slow bicycling | | Full-court basketball | |
| Golf | | | |

## Calculating BMR/Day

The **basal metabolic rate** (**BMR**) is the rate at which kcal are spent for basal metabolism. Basal metabolism is the minimum amount of energy the body needs at rest in the fasting state. The beating of the heart, breathing, maintaining body temperature, and sending nerve impulses are some of the activities that maintain life. BMR varies according to a person's body surface area, age, and sex. You will multiply your body surface area by a basal metabolic rate constant to arrive at your BMR/hour. This hourly rate multiplied by 24 hr/day will give you the BMR/day.

**Figure 21.1 Method to estimate body surface area from height and weight.**
A straight line is drawn from the subject's height (Scale 1) to the subject's weight (Scale 3). The point at which the line intersects Scale 2 is the subject's body surface area in $m^2$ (meters squared).

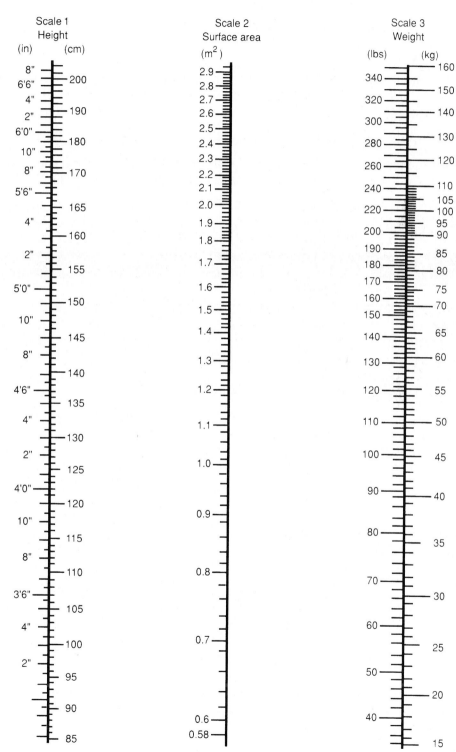

## Body Surface Area

1. Use a scale to determine your weight and a measuring device (e.g., measuring tape) to determine your height. It is assumed that you are fully clothed and wearing shoes with a 1-inch sole.
2. Consult Figure 21.1 and with a ruler draw a straight line from your height to your weight. The point where that line crosses the middle column shows your surface area in $m^2$ (squared meters). For example, a person who is 6 ft tall and weighs 170 lb has a body surface area of 1.99 $m^2$.

What is your body surface area? _____

## BMR/Day

1. Consult Table 21.4 to find the BMR constant for your age and sex. Multiply your surface area by this factor to calculate your BMR/hr. For example, a 17-year-old male has a BMR constant of 41.5 kcal/$m^2$ hr. If his surface area is 1.99 $m^2$, his BMR is 1.99 $m^2$ × 41.5 kcal/$m^2$ = 82.6 kcal/hr.

What is your BMR/hr? _____

2. Multiply your BMR/hr by 24 to obtain the total number of kcal you need for BMR/day. For example, if the BMR is 82.6 kcal/hr, then the daily BMR rate is 24 hr × 82.6 kcal/hr = 1,982 kcal/day.

What is your BMR/day? _____

### Table 21.4  Basal Metabolic Rate Constants

| Age | BMR (kcal/$m^2$ hr) Males | Females | Age | BMR (kcal/$m^2$ hr) Males | Females |
|-----|-------|---------|-----|-------|---------|
| 10 | 47.7 | 44.9 | 29 | 37.7 | 35.0 |
| 11 | 46.5 | 43.5 | 30 | 37.6 | 35.0 |
| 12 | 45.3 | 42.0 | 31 | 37.4 | 35.0 |
| 13 | 44.5 | 40.5 | 32 | 37.2 | 34.9 |
| 14 | 43.8 | 39.2 | 33 | 37.1 | 34.9 |
| 15 | 42.9 | 38.3 | 34 | 37.0 | 34.9 |
| 16 | 42.0 | 37.2 | 35 | 36.9 | 34.8 |
| 17 | 41.5 | 36.4 | 36 | 36.8 | 34.7 |
| 18 | 40.8 | 35.8 | 37 | 36.7 | 34.6 |
| 19 | 40.5 | 35.4 | 38 | 36.7 | 34.5 |
| 20 | 39.9 | 35.3 | 39 | 36.6 | 34.4 |
| 21 | 39.5 | 35.2 | 40–44 | 36.4 | 34.1 |
| 22 | 39.2 | 35.2 | 45–49 | 36.2 | 33.8 |
| 23 | 39.0 | 35.2 | 50–54 | 35.8 | 33.1 |
| 24 | 38.7 | 35.1 | 55–59 | 35.1 | 32.8 |
| 25 | 38.4 | 35.1 | 60–64 | 34.5 | 32.0 |
| 26 | 38.2 | 35.0 | 65–69 | 33.5 | 31.6 |
| 27 | 38.0 | 35.0 | 70–74 | 32.7 | 31.1 |
| 28 | 37.8 | 35.0 | 75+ | 31.8 | |

## Calculating Energy Required for SDA

The **specific dynamic action** (**SDA**) is the amount of energy needed to process food. For example, muscles that move food along the digestive tract and glands that make digestive juices use up energy. To calculate the amount of average energy you require for daily SDA, (1) add the average energy required for daily physical activity and the energy for daily BMR, and (2) multiply the total by 10% to obtain an estimated average daily SDA. For example, if 1,044 kcal/day is required for physical activity and 1,882 kcal/day is required for BMR, then an average SDA = 293 kcal/day.

What is your **average SDA/day**? _____ kcal

## Calculating Average Daily Energy Requirement

Total the amounts you have calculated for average daily physical activity, daily BMR, and average daily SDA. This is your average (av.) energy requirement/day.

Av. physical activity/day _____ kcal

BMR/day _____ kcal

Av. SDA/day _____ kcal

Av. energy requirement/day = _____ kcal

# 21.3  Comparison of Average Daily Energy Intake and Average Daily Energy Requirement

### Pre-Lab

4. Which component of a person's average daily energy requirement can be changed to affect weight loss or gain? _____
_____

Figure 21.2 illustrates that, if the average daily energy intake is the same as the average daily energy requirement, weight is likely to remain the same.

1. Compare your average daily energy intake to your average daily energy requirement:

_____

2. Our methods of calculation are very generalized. If your two figures are within 20% of each other, you will most likely neither lose nor gain weight. If the two figures are not within 20% of each other,

are you apt to lose weight or gain weight? _____

Explain. _____
_____

**Figure 21.2  Comparison of average daily energy intake and average daily energy requirement.**
If average daily energy intake is the same as the average daily energy requirement, the person's weight stays the same.

energy content of food

SDA (specific dynamic action)

BMR (basal metabolic rate)

physical activity

## 21.4  Ideal Weight

**Pre-Lab**

5. What is the ideal range for BMI? _____

_____

6. What procedure can be used to determine lean body weight? _____

_____

Three methods of determining your ideal weight are explained next. Your instructor will choose which one you are to use.

### Ideal Weight Based on Height and Weight Tables

Tables 21.5 and 21.6 are constructed by life insurance companies from data on thousands of people, at an age when bone and muscle growth are complete. Such tables are not highly accurate. Determine your ideal weight

| Height | | Small Frame | Medium Frame | Large Frame |
|---|---|---|---|---|
| **Table 21.5  Weight and Height Table for Males** | | | | |
| (ft) | (in) | (lb) | (lb) | (lb) |
| 5 | 2 | 128–134 | 131–141 | 138–150 |
| 5 | 3 | 130–136 | 133–143 | 140–153 |
| 5 | 4 | 132–138 | 135–145 | 142–156 |
| 5 | 5 | 134–140 | 137–148 | 144–160 |
| 5 | 6 | 136–142 | 139–151 | 146–164 |
| 5 | 7 | 138–145 | 142–154 | 149–168 |
| 5 | 8 | 140–148 | 145–157 | 152–172 |
| 5 | 9 | 142–151 | 148–160 | 155–176 |
| 5 | 10 | 144–154 | 151–163 | 158–180 |
| 5 | 11 | 146–157 | 154–166 | 161–184 |
| 6 | 0 | 149–160 | 157–170 | 164–188 |
| 6 | 1 | 152–164 | 160–174 | 168–192 |
| 6 | 2 | 155–168 | 164–178 | 172–197 |
| 6 | 3 | 158–172 | 167–182 | 176–202 |
| 6 | 4 | 162–176 | 171–187 | 181–207 |

## Table 21.6  Weight and Height Table for Females

| Height | | Small Frame | Medium Frame | Large Frame |
|---|---|---|---|---|
| (ft) | (in) | (lb) | (lb) | (lb) |
| 4 | 10 | 102–111 | 109–121 | 118–131 |
| 4 | 11 | 103–113 | 111–123 | 120–134 |
| 5 | 0 | 104–115 | 113–126 | 122–137 |
| 5 | 1 | 106–118 | 115–129 | 125–140 |
| 5 | 2 | 108–121 | 118–132 | 128–143 |
| 5 | 3 | 111–124 | 121–135 | 131–147 |
| 5 | 4 | 114–127 | 124–138 | 134–151 |
| 5 | 5 | 117–130 | 127–141 | 137–155 |
| 5 | 6 | 120–133 | 130–144 | 140–159 |
| 5 | 7 | 123–136 | 133–147 | 143–163 |
| 5 | 8 | 126–139 | 136–150 | 146–167 |
| 5 | 9 | 129–142 | 139–153 | 149–170 |
| 5 | 10 | 132–145 | 142–156 | 152–173 |
| 5 | 11 | 135–148 | 145–159 | 155–176 |
| 6 | 0 | 138–151 | 148–162 | 158–179 |

Weight at ages 25–59 based on lowest mortality. Weight according to frame (indoor clothing weighing 3 lb is allowed for females; 5 lb for males), shoes with 1-inch heels.

range. A person more than 10% over the weight in the tables is overweight; a person 20% over is obese. Similarly, a person 10% below the weight in the table is underweight. What is your ideal weight based on these height and weight tables?

Ideal weight _____ lb

## Ideal Weight Based on Body Mass Index (BMI)

Your **body mass index (BMI)** is a measure of your weight relative to your height (Table 21.7). A BMI of 19–25 is considered ideal. Below 19 a person is considered underweight, and above 25 a person is considered overweight. An obese person has a BMI of 30 or greater. Keep in mind, BMI is not an absolute indicator of health. However, it is a good starting point for discussions of health concerns associated with overweight and obesity.

## Table 21.7 Body Mass Index

| Height (inches) | 19 | 20 | 21 | 22 | 23 | 24 | 25 | 26 | 27 | 28 | 29 | 30 | 31 | 32 | 33 | 34 | 35 |
|---|---|---|---|---|---|---|---|---|---|---|---|---|---|---|---|---|---|
| | | | | Ideal | | | | Body Weight (pounds) | | | | Not Ideal | | | | | |
| 58 | 91 | 96 | 100 | 105 | 110 | 115 | 119 | 124 | 129 | 134 | 138 | 143 | 148 | 153 | 158 | 162 | 167 |
| 59 | 94 | 99 | 104 | 109 | 114 | 119 | 124 | 128 | 133 | 138 | 143 | 148 | 153 | 158 | 163 | 168 | 173 |
| 60 | 97 | 102 | 107 | 112 | 118 | 123 | 128 | 133 | 138 | 143 | 148 | 153 | 158 | 163 | 168 | 174 | 179 |
| 61 | 100 | 106 | 111 | 116 | 122 | 127 | 132 | 137 | 143 | 148 | 153 | 158 | 164 | 169 | 174 | 180 | 185 |
| 62 | 104 | 109 | 115 | 120 | 126 | 131 | 136 | 142 | 147 | 153 | 158 | 164 | 169 | 175 | 180 | 186 | 191 |
| 63 | 107 | 113 | 118 | 124 | 130 | 135 | 141 | 146 | 152 | 158 | 163 | 169 | 175 | 180 | 186 | 191 | 197 |
| 64 | 110 | 116 | 122 | 128 | 134 | 140 | 145 | 151 | 157 | 163 | 169 | 174 | 180 | 186 | 192 | 197 | 204 |
| 65 | 114 | 120 | 126 | 132 | 138 | 144 | 150 | 156 | 162 | 168 | 174 | 180 | 186 | 192 | 198 | 204 | 210 |
| 66 | 118 | 124 | 130 | 136 | 142 | 148 | 155 | 161 | 167 | 173 | 179 | 186 | 192 | 198 | 204 | 210 | 216 |
| 67 | 121 | 127 | 134 | 140 | 146 | 153 | 159 | 166 | 172 | 178 | 185 | 191 | 198 | 204 | 211 | 217 | 223 |
| 68 | 125 | 131 | 138 | 144 | 151 | 158 | 164 | 171 | 177 | 184 | 190 | 197 | 203 | 210 | 216 | 223 | 230 |
| 69 | 128 | 135 | 142 | 149 | 155 | 162 | 169 | 176 | 182 | 189 | 196 | 203 | 209 | 216 | 223 | 230 | 236 |
| 70 | 132 | 139 | 146 | 153 | 160 | 167 | 174 | 181 | 188 | 195 | 202 | 209 | 216 | 222 | 229 | 236 | 243 |
| 71 | 136 | 143 | 150 | 157 | 165 | 172 | 179 | 186 | 193 | 200 | 208 | 215 | 222 | 229 | 236 | 243 | 250 |
| 72 | 140 | 147 | 154 | 162 | 169 | 177 | 184 | 191 | 199 | 206 | 213 | 221 | 228 | 235 | 242 | 250 | 258 |
| 73 | 144 | 151 | 159 | 166 | 174 | 182 | 189 | 197 | 204 | 212 | 219 | 227 | 235 | 242 | 250 | 257 | 265 |
| 74 | 148 | 155 | 163 | 171 | 179 | 186 | 194 | 202 | 210 | 218 | 225 | 233 | 241 | 249 | 256 | 264 | 272 |
| 75 | 152 | 160 | 168 | 176 | 184 | 192 | 200 | 208 | 216 | 224 | 232 | 240 | 248 | 256 | 264 | 272 | 279 |
| 76 | 156 | 164 | 172 | 180 | 189 | 197 | 205 | 213 | 221 | 230 | 238 | 246 | 254 | 263 | 271 | 279 | 287 |

### Procedure to Determine Ideal Weight Based on Body Mass Index (BMI)

1. Consult Table 21.7 and find your height in inches.
2. Go across that row to your weight.
3. Read the number at the top of that column. What is your BMI? _____
4. Is your BMI within the normal range (19–25)? _____ What is your ideal weight range based on BMI? _____

### Comparison

How does this weight range compare to your desirable weight range from Table 21.5 or Table 21.6?

_____

## Ideal Weight Based on Body Composition

*Body composition* refers to the lean body weight plus the body fat. Ideal weight based on body composition is preferred, because then you know how much of your weight is due to fat. In adult males, lean body weight should equal about 84% of weight, and body fat should be only 16% of weight. In adult females, lean body weight should equal about 77% of weight, and body fat should be only 23% of weight. Well-conditioned male athletes, such as marathon runners and swimmers, usually have 10–12% body fat, whereas football players can have as high as 19–20% body fat. It can be unhealthy for the percentage of body fat to drop below 5% in males and 10% in females for extended periods of time.

To calculate **lean body weight (LBW),** it is necessary to subtract the amount of your body fat from your present weight. The amount of body fat is to be determined by taking skin-fold measurements, because the fat attached to the skin is roughly proportional to total body fat. The more areas of the body measured, the more accurate the estimate of body fat will be. In this laboratory, we will take only those measurements that permit you to remain fully clothed.

## Procedure to Determine Lean Body Weight

1. Work in pairs. Obtain calipers designed to measure skin folds. All measurements should be done on the right side in mm (millimeters). Firmly grasp the fold of skin between the left thumb and the other four fingers and lift. Pinch and lift the fold several times to make sure no musculature is grasped. Hold the skin fold firmly, and place the contact side of the calipers $1/2$ inch below the thumb and fingers; do not let go of the fold. Close the calipers on the skin fold, and do the measurements noted in items 2–6 following. Take the reading to the nearest $1/2$ mm. Release the grip on the caliper and release the fold. To make sure the reading is accurate, repeat the measurement a few times. If the second measurement is within 1–2 mm of the first, it is reliable. Record two readings.

2. **Triceps (females only).** The triceps skin fold is measured on the back of the upper arm, halfway between the elbow and the tip of the shoulder, while the arm is hanging loosely at the subject's side. Grasp the skin fold parallel to the long axis of the arm and measure as described in step 1. Record in 7 below. Record two readings.

3. **Ilium—hip (males and females).** Fold the skin diagonally just above the top of the hip bone on an imaginary line that would divide the body into front and back halves. Measure and record as before. Record two readings.

4. **Abdomen (males and females).** Fold the skin vertically 1 inch to the right of the navel. Measure and record as before. Record two readings.

5. **Chest (males only).** Fold the skin diagonally, midway between the nipple and the armpit. Measure and record as before. Record two readings.

6. **Axilla—side (males only).** Fold the skin vertically at the level of nipple on an imaginary line that would divide the body into front and back halves. Measure and record as before. Record two readings.

7. Total any four of your skin-fold measurements.

|  | **Females** | | **Males** | |
|---|---|---|---|---|
| Triceps | _____ _____ mm | | | |
| Ilium | _____ _____ mm | | _____ _____ mm | |
| Abdomen | _____ _____ mm | | _____ _____ mm | |
| Chest | | | _____ _____ mm | |
| Axilla | | | _____ _____ mm | |
| Total of any four skin-fold measurements | _____ mm | | _____ _____ mm | |

8. Consult Table 21.8 (males) and Table 21.9 (females) to determine percentage (%) body fat from the sum of four skin-fold measurements and your age.

What is your % body fat? _____ %

## Table 21.8 Percentage Fat Estimates for Males

| Sum of Four Skin Folds (mm) | Age | | | | | | | | |
|---|---|---|---|---|---|---|---|---|---|
| | 18–22 | 23–27 | 28–32 | 33–37 | 38–42 | 43–47 | 48–52 | 53–57 | 58–older |
| 8–12 | 1.9% | 2.5% | 3.2% | 3.8% | 4.4% | 5.0% | 5.7% | 6.3% | 6.9% |
| 13–17 | 3.3% | 3.9% | 4.5% | 5.1% | 5.7% | 6.4% | 7.0% | 7.6% | 8.2% |
| 18–22 | 4.5% | 5.2% | 5.8% | 6.4% | 7.0% | 7.7% | 8.3% | 8.9% | 9.5% |
| 23–27 | 5.8% | 6.4% | 7.1% | 7.7% | 8.3% | 8.9% | 9.5% | 10.2% | 10.8% |
| 28–32 | 7.1% | 7.7% | 8.3% | 8.9% | 9.5% | 10.2% | 10.8% | 11.4% | 12.0% |
| 33–37 | 8.3% | 8.9% | 9.5% | 10.1% | 10.8% | 11.4% | 12.0% | 12.6% | 13.2% |
| 38–42 | 9.5% | 10.1% | 10.7% | 11.3% | 11.9% | 12.6% | 13.2% | 13.8% | 14.4% |
| 43–47 | 10.6% | 11.6% | 11.9% | 12.5% | 13.1% | 13.7% | 14.4% | 15.0% | 15.6% |
| 48–52 | 11.8% | 12.4% | 13.0% | 13.6% | 14.2% | 14.9% | 15.5% | 16.1% | 16.7% |
| 53–57 | 12.9% | 13.5% | 14.1% | 14.7% | 15.4% | 16.0% | 16.6% | 17.2% | 17.9% |
| 58–62 | 14.0% | 14.6% | 15.2% | 15.8% | 16.4% | 17.1% | 17.7% | 18.3% | 18.9% |
| 63–67 | 15.0% | 15.6% | 16.3% | 16.9% | 17.5% | 18.1% | 18.8% | 19.4% | 20.0% |
| 68–72 | 16.1% | 16.7% | 17.3% | 17.9% | 18.5% | 19.2% | 19.8% | 20.4% | 21.0% |
| 73–77 | 17.1% | 17.7% | 18.3% | 18.9% | 19.5% | 20.2% | 20.8% | 21.4% | 22.0% |
| 78–82 | 18.0% | 18.7% | 19.3% | 19.9% | 20.5% | 21.0% | 21.8% | 22.4% | 23.0% |
| 83–87 | 19.0% | 19.6% | 20.2% | 20.8% | 21.5% | 22.1% | 22.7% | 23.3% | 24.0% |
| 88–92 | 19.9% | 20.5% | 21.2% | 21.8% | 22.4% | 23.0% | 23.6% | 24.3% | 24.9% |
| 93–97 | 20.8% | 21.4% | 22.1% | 22.7% | 23.3% | 23.9% | 24.8% | 25.2% | 25.8% |
| 98–102 | 21.7% | 22.6% | 22.9% | 23.5% | 24.2% | 24.8% | 25.4% | 26.0% | 26.7% |
| 103–107 | 22.5% | 23.2% | 23.8% | 24.4% | 25.0% | 25.6% | 26.3% | 26.9% | 27.5% |
| 108–112 | 23.4% | 24.0% | 24.6% | 25.2% | 25.8% | 26.5% | 27.1% | 27.7% | 28.3% |
| 113–117 | 24.1% | 24.8% | 25.4% | 26.0% | 26.6% | 27.3% | 27.9% | 28.5% | 29.1% |
| 118–122 | 24.9% | 25.5% | 26.2% | 26.8% | 27.4% | 28.0% | 28.6% | 29.3% | 29.9% |
| 123–127 | 25.7% | 26.3% | 26.9% | 27.5% | 28.1% | 28.8% | 29.4% | 30.0% | 30.6% |
| 128–132 | 26.4% | 27.0% | 27.6% | 28.2% | 28.8% | 29.5% | 30.1% | 30.7% | 31.3% |
| 133–137 | 27.1% | 27.7% | 28.3% | 28.9% | 29.5% | 30.2% | 30.8% | 31.4% | 32.0% |
| 138–142 | 27.7% | 28.3% | 29.0% | 29.6% | 30.2% | 30.8% | 31.4% | 32.1% | 32.7% |
| 143–147 | 28.3% | 29.0% | 29.6% | 30.2% | 30.8% | 31.5% | 32.1% | 32.7% | 33.3% |
| 148–152 | 29.0% | 29.6% | 30.2% | 30.8% | 31.4% | 32.7% | 32.7% | 33.3% | 33.9% |
| 153–157 | 29.5% | 30.2% | 30.8% | 31.4% | 32.0% | 32.7% | 33.3% | 33.9% | 34.5% |
| 158–162 | 30.1% | 30.7% | 31.3% | 31.9% | 32.6% | 33.2% | 33.8% | 34.4% | 35.1% |
| 163–167 | 30.6% | 31.2% | 31.9% | 32.5% | 33.1% | 33.7% | 34.3% | 35.0% | 35.6% |
| 168–172 | 31.1% | 31.7% | 32.4% | 33.0% | 33.6% | 34.2% | 34.8% | 35.5% | 36.1% |
| 173–177 | 31.6% | 32.2% | 32.8% | 33.5% | 34.1% | 34.7% | 35.3% | 35.9% | 36.6% |
| 178–182 | 32.0% | 32.7% | 33.3% | 33.9% | 34.5% | 35.2% | 35.8% | 36.4% | 37.0% |
| 183–187 | 32.5% | 33.1% | 33.7% | 34.3% | 34.9% | 35.6% | 36.2% | 36.8% | 37.4% |
| 188–192 | 32.9% | 33.5% | 34.1% | 34.7% | 35.3% | 36.0% | 36.6% | 37.2% | 37.8% |
| 193–197 | 33.2% | 33.8% | 34.5% | 35.1% | 35.7% | 36.3% | 37.0% | 37.8% | 38.2% |
| 198–202 | 33.6% | 34.2% | 34.8% | 35.4% | 36.1% | 36.7% | 37.3% | 37.9% | 38.5% |
| 203–207 | 33.9% | 34.5% | 35.1% | 35.7% | 36.4% | 37.0% | 37.6% | 38.2% | 38.9% |

## Table 21.9 Percentage Fat Estimates for Females

| Sum of Four Skin Folds (mm) | Age | | | | | | | | |
|---|---|---|---|---|---|---|---|---|---|
| | 18–22 | 23–27 | 28–32 | 33–37 | 38–42 | 43–47 | 48–52 | 53–57 | 58–older |
| 8–12 | 8.8% | 9.0% | 9.2% | 9.4% | 9.5% | 9.7% | 9.9% | 10.1% | 10.3% |
| 13–17 | 10.8% | 10.9% | 11.1% | 11.3% | 11.5% | 11.7% | 11.8% | 12.0% | 12.2% |
| 18–22 | 12.6% | 12.8% | 13.0% | 13.2% | 13.4% | 13.5% | 13.7% | 13.9% | 14.1% |
| 23–27 | 14.5% | 14.6% | 14.8% | 15.0% | 15.2% | 15.4% | 15.6% | 15.7% | 15.9% |
| 28–32 | 16.2% | 16.4% | 16.6% | 16.8% | 17.0% | 17.1% | 17.3% | 17.5% | 17.7% |
| 33–37 | 17.9% | 18.1% | 18.3% | 18.5% | 18.7% | 18.9% | 19.0% | 19.2% | 19.4% |
| 38–42 | 19.6% | 19.8% | 20.0% | 20.2% | 20.3% | 20.5% | 20.7% | 20.9% | 21.1% |
| 43–47 | 21.2% | 21.4% | 21.6% | 21.8% | 21.9% | 22.1% | 22.3% | 22.5% | 22.7% |
| 48–52 | 22.8% | 22.9% | 23.1% | 23.3% | 23.5% | 23.7% | 23.8% | 24.0% | 24.2% |
| 53–57 | 24.2% | 24.4% | 24.6% | 24.8% | 25.0% | 25.2% | 25.3% | 25.5% | 25.7% |
| 58–62 | 25.7% | 25.9% | 26.0% | 26.2% | 26.4% | 26.6% | 26.8% | 27.0% | 27.1% |
| 63–67 | 27.1% | 27.2% | 27.4% | 27.6% | 27.8% | 28.0% | 28.2% | 28.3% | 28.5% |
| 68–72 | 28.4% | 28.6% | 28.7% | 28.9% | 29.1% | 29.3% | 29.5% | 29.7% | 29.8% |
| 73–77 | 29.6% | 29.8% | 30.0% | 30.2% | 30.4% | 30.6% | 30.7% | 30.9% | 31.1% |
| 78–82 | 30.9% | 31.0% | 31.2% | 31.4% | 31.6% | 31.8% | 31.9% | 32.1% | 32.3% |
| 83–87 | 32.0% | 32.2% | 32.4% | 32.6% | 32.7% | 32.9% | 33.1% | 33.3% | 33.5% |
| 88–92 | 33.1% | 33.3% | 33.5% | 33.7% | 33.8% | 34.0% | 34.2% | 34.4% | 34.6% |
| 93–97 | 34.1% | 34.3% | 34.5% | 34.7% | 34.9% | 35.1% | 35.2% | 35.4% | 35.6% |
| 98–102 | 35.1% | 35.3% | 35.5% | 35.7% | 35.9% | 36.0% | 36.2% | 36.4% | 36.6% |
| 103–107 | 36.1% | 36.2% | 36.4% | 36.6% | 36.8% | 37.0% | 37.2% | 37.3% | 37.5% |
| 108–112 | 36.9% | 37.1% | 37.3% | 37.5% | 37.7% | 37.9% | 38.0% | 38.2% | 38.4% |
| 113–117 | 37.8% | 37.9% | 38.1% | 38.3% | 39.2% | 39.4% | 39.6% | 39.8% | 40.0% |
| 118–122 | 38.5% | 38.7% | 38.9% | 39.1% | 39.4% | 39.6% | 39.8% | 40.0% | 40.5% |
| 123–127 | 39.2% | 39.4% | 39.6% | 39.8% | 40.0% | 40.1% | 40.3% | 40.5% | 40.7% |
| 128–132 | 39.9% | 40.1% | 40.2% | 40.4% | 40.6% | 40.8% | 41.0% | 41.2% | 41.3% |
| 133–137 | 40.5% | 40.7% | 40.8% | 41.0% | 41.2% | 41.4% | 41.6% | 41.7% | 41.9% |
| 138–142 | 41.0% | 41.2% | 41.4% | 41.6% | 41.7% | 41.9% | 42.1% | 42.3% | 42.5% |
| 143–147 | 41.5% | 41.7% | 41.9% | 42.0% | 42.2% | 42.4% | 42.6% | 42.8% | 43.0% |
| 148–152 | 41.9% | 42.1% | 42.3% | 42.8% | 42.6% | 42.8% | 43.0% | 43.2% | 43.4% |
| 153–157 | 42.3% | 42.5% | 42.6% | 42.8% | 43.0% | 43.2% | 43.4% | 43.6% | 43.7% |
| 158–162 | 42.6% | 42.8% | 43.0% | 43.1% | 43.3% | 43.5% | 43.7% | 43.9% | 44.1% |
| 163–167 | 42.9% | 43.0% | 43.2% | 43.4% | 43.6% | 43.8% | 44.0% | 44.1% | 44.3% |
| 168–172 | 43.1% | 43.2% | 43.4% | 43.6% | 43.8% | 44.0% | 44.2% | 44.3% | 44.5% |
| 173–177 | 43.2% | 43.4% | 43.6% | 43.8% | 43.9% | 44.1% | 44.3% | 44.5% | 44.7% |
| 178–182 | 43.3% | 43.5% | 43.7% | 43.8% | 44.0% | 44.2% | 44.4% | 44.6% | 44.8% |

9. Multiply your present body weight in pounds (lb) by your present % body fat to determine how much of your body weight is fat. For example, if a male weighs 180 lb and his percentage body fat is 20%, then $180 \times 0.20 = 36$ lb. His body fat is 36 lb.

   How much of your body weight is fat? _____ lb

10. Subtract the weight of fat from your present body weight. This is lean body weight (LBW). For example, if a male weighs 180 lb and his body fat is 36 lb, then $180 - 36 = 144$. His LBW is 144 lb.

    What is your LBW? _____ lb

11. Calculate ideal weight based on body composition.

    For males, 84% of ideal weight should be LBW. Therefore, ideal weight = LBW/0.84. What is your ideal weight based on body composition? For example, if a male has a LBW of 144 lb, his ideal weight is 144/0.84 = 171 lb.

    > Male ideal weight = _____ lb

    (At this weight, your body will contain 16% fat, the amount generally recommended for males.)

    For females, 77% of ideal weight should be LBW. Therefore, ideal weight = LBW/0.77. What is your ideal weight based on body composition?

    > Female ideal weight = _____ lb

    (At this weight, your body will contain 23% fat, the amount generally recommended for females.)

## Comparison

How does your ideal weight based on body composition compare to your ideal weight based on BMI?

_____

_____

## Conclusions: Ideal Body Weight

- What is your average daily energy requirement? _____ kcal
- What is your average daily energy intake? _____ kcal
- What is your ideal weight? (Take an average of any you have calculated.) _____ lb

    For your information, 1 lb of fat represents 3,500 kcal. Therefore, if you want to lose 1 lb of fat, you must either reduce your intake by 3,500 kcal or increase your activity by 3,500 kcal. Assuming the same amount of physical activity, you could lose 1 lb per week by reducing your kcal intake by 500 kcal per day or gain 1 lb per week by increasing your kcal intake by 500 kcal per day.

- What is your recommendation for maintaining or achieving your ideal weight? _____

    _____

_____ 1. Which units are used to describe energy intake?

_____ 2. Why does your body need energy?

_____ 3. How many kcal are expended during 45 minutes of jogging for a 68-kg person (150 lb)?

_____ 4. How many kcal are expended during 45 minutes of studying for a 68-kg person (150 lb)?

_____ 5. What term describes the amount of energy a body needs at rest for basic functions?

_____ 6. Calculate the BMR/day for a 19-year-old female with a surface area of 1.6 m$^2$.

_____ 7. Specific dynamic action is the amount of energy needed to _____ _____.

_____ 8. A person with a greater energy intake compared to their energy expenditure will _____ weight.

_____ 9. Which BMI range is considered ideal?

_____ 10. What is the BMI for an individual who is 70 inches tall and weighs 150 lb?

_____ 11. When comparing height/weight tables, BMI, and body composition, which is most accurate for determining ideal weight?

_____ 12. What tool is used to determine skin-fold measurements?

_____ 13. Based on skin-fold measurements, what is the percent body fat of a 24-year-old female who had skin folds totaling 35 mm?

_____ 14. If a 160-lb individual has 28% body fat, how much of their body weight is fat in lb?

## Thought Questions

15. Of the factors examined during this lab, which can be controlled by an individual attempting to lose, maintain, or gain weight?

16. Why do males typically have a higher basal metabolic rate compared to females?

17. Why is body composition more accurate for determining ideal weight compared to height/weight tables or BMI?

# 22

# Cardiovascular System

## Learning Outcomes

### 22.1 The Heart
- Name the four chambers of the heart and the blood vessels connected to these chambers.
- Name and locate the valves of the heart, and trace the path of blood through the heart.
- Distinguish the role of the pulmonary circuit from that of the systemic circuit.

### 22.2 Vessels of the Pulmonary Circuit
- Identify the pulmonary blood vessels, and state whether they carry $O_2$-rich or $O_2$-poor blood.

### 22.3 Vessels of the Systemic Circuit
- Identify the coronary arteries and cardiac veins.
- Identify the arteries and veins in the thoracic cavity and abdominal cavity.

### 22.4 Blood Vessel Comparison
- Describe each of the layers of a blood vessel wall.
- Contrast the wall of an artery with that of a vein.


## Introduction

### Pre-Lab

1. What are the differences between the pulmonary and systemic circuit? _____

_____


Blood must circulate in order to carry oxygen throughout the body. The **heart,** located in the thoracic cavity, pumps the blood, which moves away from the heart in **arteries** and **arterioles** and returns to the heart in **venules** and **veins.** When the heart contracts, blood leaves under high pressure. Therefore, arteries have thicker walls than veins, which return blood under low pressure to the heart. Skeletal muscle contractions in the limbs add pressure to venous blood by pushing on the thin walls of veins, which have valves to keep blood moving toward the heart. Capillaries, which connect arterioles to venules, have thin walls that allow exchange of molecules with tissue fluid.

   In today's laboratory, you will learn the anatomy of the heart and how to trace the path of blood through the

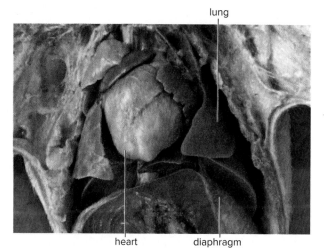

David F. Cox/McGraw Hill

heart in two major circuits. In the **pulmonary circuit,** blood moves from the heart to the lungs and back. In the **systemic circuit,** the blood moves from the heart to the rest of the body and then back to the heart. Work done today will give you an appreciation of your own heart and blood vessels and of how the blood moves through these vessels to service your organs.

## 22.1 The Heart

### Pre-Lab

2. *Label Figure 22.1a with the help of your textbook and Figure 22.1b.*
3. *Label Figure 22.2a with the help of your textbook and Figure 22.2b.*

The heart is divided by the **septum** creating a right side and a left side. To tell the right from the left side, position Figure 22.1 so it corresponds to your body. There are four **chambers:** two upper, thin-walled atria and two lower, thick-walled ventricles. The heart has valves that keep the blood flowing in one direction; special muscles and tendons secure the valves to prevent backflow from the ventricles to the atria. The right side of the heart sends blood through the lungs, and the left side sends blood into the body. Therefore, the heart is called a double pump.

### Observation: External Anatomy of the Heart

1. In a heart model and/or sheep heart, identify the **right atrium** and its attached blood vessels, the superior and inferior **venae cavae** (Fig. 22.1). The superior vena cava and the inferior vena cava return blood from the head and body, respectively, to the right atrium. Is the blood that enters the right atrium $O_2$-poor or

   $O_2$-rich? _____

   Explain. _____

   _____

2. Identify the **right ventricle** and its attached blood vessel, the **pulmonary trunk.** The pulmonary trunk leaves the ventral side of the heart from the top of the right ventricle and then passes diagonally forward before branching into the **right** and **left pulmonary arteries.**

3. Identify the **left atrium** and its attached blood vessels, the left and right **pulmonary veins.** The pulmonary veins return blood from the lungs to the left atrium. Is the blood that enters the left atrium $O_2$-poor or

   $O_2$-rich? _____ Explain. _____

   _____

4. Identify the **left ventricle** and its attached blood vessel, the **aorta,** which arises from the anterior end of the left ventricle, just dorsal to the origin of the pulmonary trunk. The aorta soon bends to the animal's left as the aortic arch. The aorta carries blood to the body proper.

5. Identify the **coronary arteries** and the **cardiac veins,** which service the needs of the heart wall. The coronary arteries branch off the aorta as soon as it leaves the heart and appear on the surface of the heart. The cardiac veins, also on the surface of the heart, join and then enter the right atrium through the coronary sinus on the dorsal side of the heart.

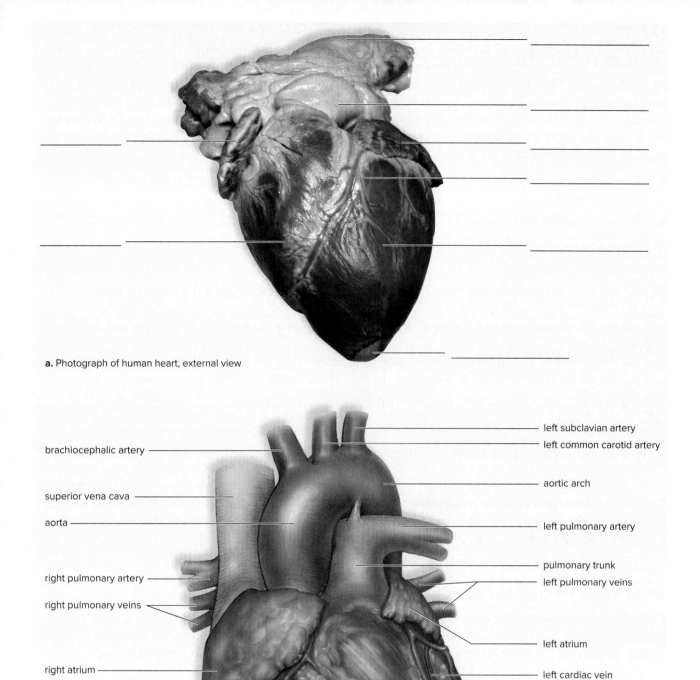

**a.** Photograph of human heart, external view

brachiocephalic artery

superior vena cava

aorta

right pulmonary artery

right pulmonary veins

right atrium

right coronary artery

right ventricle

inferior vena cava

left subclavian artery

left common carotid artery

aortic arch

left pulmonary artery

pulmonary trunk

left pulmonary veins

left atrium

left cardiac vein

left ventricle

apex

**b.** Drawing of human heart, external view

## Figure 22.1 External view of heart.
*Label Figure 22.1a with the help of your textbook and Figure 22.1b.* Externally, notice the coronary arteries and cardiac veins that serve the heart.

(a) Peter Arnold/A. & F. Michler/Getty Images

Remove the ventral half of the human heart model (Fig. 22.2) and/or sheep heart. To use a sheep heart that requires dissection, position the heart as it is shown in Figure 22.2*a*. Using a sharp scalpel, cut through the left atrium and the left ventricle to the apex of the heart. Be sure the cut is deep enough to reach the septum. Then position the heart so the apex is uppermost, and cut from the apex of the heart through the right atrium, being sure the cut is deep enough to reach the septum of the heart.

1. Identify the four chambers of the heart in longitudinal section: right atrium, right ventricle, left atrium, and left ventricle.

2. Which ventricle is more muscular? _____

   Why is this appropriate? _____

   _____

3. Find the **right atrioventricular** (tricuspid) valve, located between the right atrium and the right ventricle.

4. Find the **left atrioventricular** (bicuspid or mitral) valve, located between the left atrium and the left ventricle.

5. Find the **pulmonary semilunar** valve, located in the base of the pulmonary trunk.

6. Find the **aortic semilunar** valve, located in the base of the aorta.

   What is the function of the atrioventricular valves? _____

   _____

   What is the function of the semilunar valves? _____

   _____

7. Note the **chordae tendineae** ("heartstrings") that hold the atrioventricular valves in place while the heart contracts. These extend from the papillary muscles. The chordae tendineae prevent the atrioventricular valves from inverting into the atria when the ventricles contract.

## Path of Blood Through the Heart

To demonstrate that $O_2$-poor blood is kept separate from $O_2$-rich blood, identify the path blood takes from the right side of the heart to the aorta by filling in the following blanks. The blood passes through the lungs to go from the right to the left sides of the heart. Which side of the heart pumps $O_2$-poor

blood? _____ Which side pumps $O_2$-rich blood? _____

*Venae Cavae*                                        *Lungs*

_____                                  _____

_____ valve                          _____

_____                                  _____ valve

_____ valve                          _____

_____                                  _____ valve

Lungs                                               Aorta

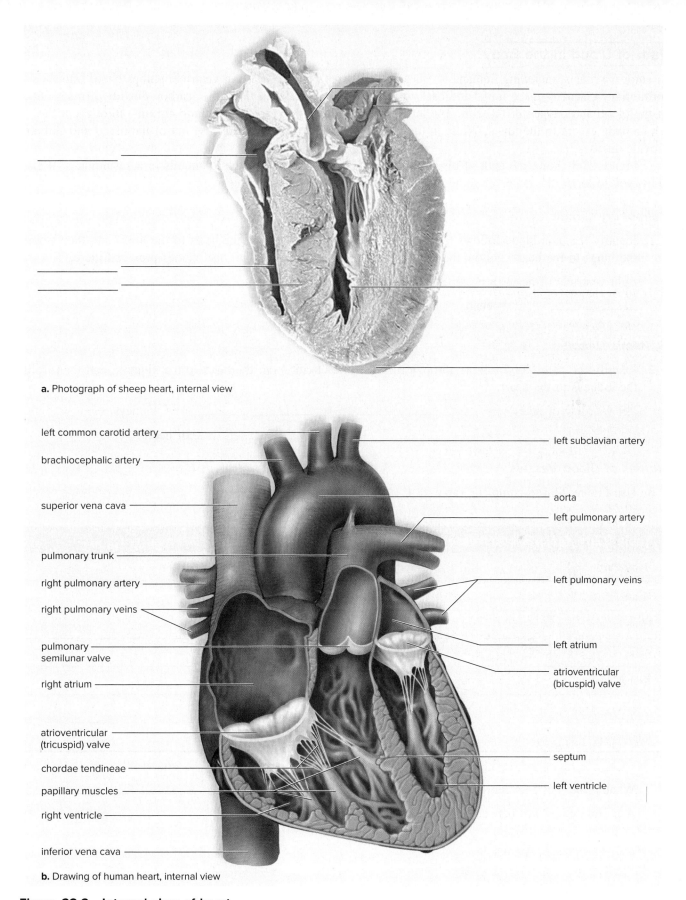

**a.** Photograph of sheep heart, internal view

left common carotid artery

brachiocephalic artery

superior vena cava

pulmonary trunk

right pulmonary artery

right pulmonary veins

pulmonary
semilunar valve

right atrium

atrioventricular
(tricuspid) valve

chordae tendineae

papillary muscles

right ventricle

inferior vena cava

left subclavian artery

aorta

left pulmonary artery

left pulmonary veins

left atrium

atrioventricular
(bicuspid) valve

septum

left ventricle

**b.** Drawing of human heart, internal view

## Figure 22.2   Internal view of heart.

*Label Figure 22.2a with the help of your textbook and Figure 22.2b.* Internally, the heart has four chambers, and there is a septum that separates the left side from the right side.

(a) Eric Wise

## Path of Blood in the Body

In adult mammals, including humans, the heart is a double pump. The right ventricle pumps blood into the **pulmonary circuit**—to the lungs and back to the heart (Fig. 22.3). In the lungs, carbon dioxide diffuses out of the blood and oxygen diffuses in. The left ventricle pumps blood to the **systemic circuit**—throughout the whole body except to the lungs. While in the systemic circuit, oxygen will diffuse out of the blood and carbon dioxide diffuses in.

Figure 22.3 shows the path of blood in both the pulmonary and systemic circuits in adult humans. It also helps you to learn the names of some of the major blood vessels.

### Pulmonary Circuit

1. Identify the path blood follows through the pulmonary circuit from the heart to the lungs and then from the lungs to the heart. Follow the arrows in Figure 22.3, and use the label names provided there.

right ventricle of heart $\longrightarrow$ _____ $\longrightarrow$ lungs $\longrightarrow$ _____ $\longrightarrow$

_____ of heart

### Systemic Circuit

2. Identify the path blood follows through the systemic circuit from the heart to the kidneys, and then from the kidneys to the heart.

left ventricle of heart $\longrightarrow$ _____ $\longrightarrow$ _____ $\longrightarrow$ kidneys $\longrightarrow$

_____ $\longrightarrow$ _____ $\longrightarrow$ _____ of heart

### Names of Blood Vessels

3. Use Figure 22.3 to complete Table 22.1.

| Table 22.1 Major Blood Vessels in the Systemic Circuit | | |
|---|---|---|
| **Body Part** | **Artery** | **Vein** |
| Heart | | |
| Head | | |
| Arms | | |
| Kidney | | |
| Legs | | |
| Intestines | | |

What vessel lies between the digestive tract and the liver? _____

A portal system lies between two sets of capillaries.

## Figure 22.3  Diagram of the human cardiovascular system.

In the pulmonary circuit, the pulmonary arteries take $O_2$-poor blood to the lungs (i.e., the capillaries in the lungs), and the pulmonary veins return $O_2$-rich blood to the heart. In the systemic circuit, the aorta branches into the various arteries that go to all other parts of the body. After blood passes through arterioles, capillaries, and venules, it enters various veins and then the superior and inferior venae cavae, which return it to the heart.

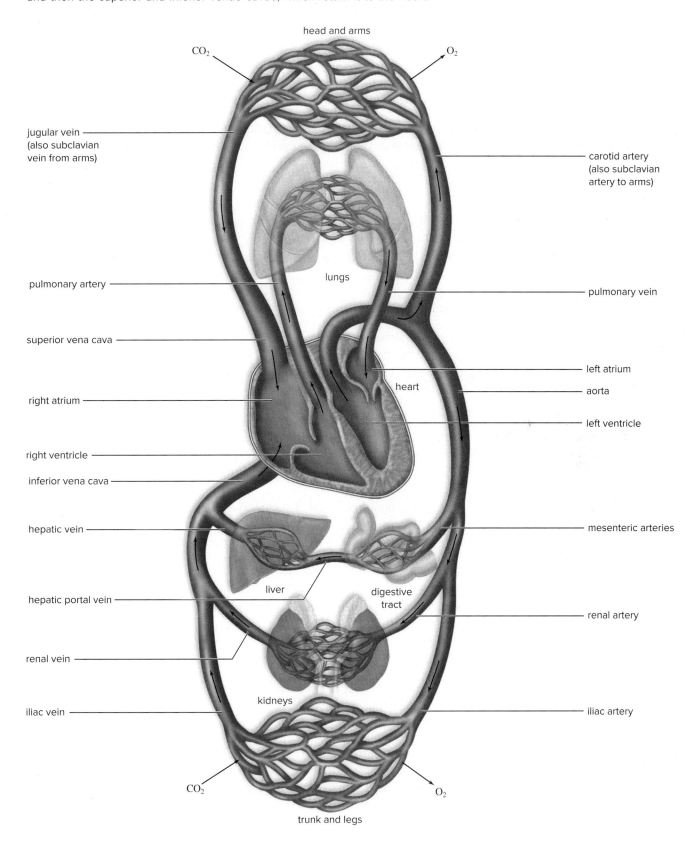

head and arms

$CO_2$

$O_2$

jugular vein
(also subclavian
vein from arms)

carotid artery
(also subclavian
artery to arms)

lungs

pulmonary artery

pulmonary vein

superior vena cava

left atrium

heart

aorta

right atrium

left ventricle

right ventricle

inferior vena cava

hepatic vein

mesenteric arteries

liver

digestive
tract

hepatic portal vein

renal artery

renal vein

kidneys

iliac vein

iliac artery

$CO_2$

$O_2$

trunk and legs

# 22.2 Vessels of the Pulmonary Circuit

## Pre-Lab

4. Which structure links fetal circulation to the placenta? _____

In this section, we will use the fetal pig to examine the pulmonary arteries and veins that occur in mammals, including humans. The fetal pig will have pulmonary arteries and veins, even though they are not functional until after the pig is born. Fetal circulation has two features, numbered 1 and 2 in Table 22.2, that keep blood from entering the pulmonary arteries.

### Table 22.2  Unique Features of Fetal Circulation

1. Oval opening (foramen ovale): an opening between the atria of the heart that allows blood to bypass the lungs
2. Arterial duct (ductus arteriosus): a short, stout vessel leading directly from the pulmonary trunk to the aorta; provides another way to bypass the lungs
3. Umbilical arteries: take blood from the iliac arteries to the placenta, the organ that nourishes the fetus
4. Umbilical vein: returns blood from the placenta to the liver
5. Venous duct (ductus venosus): a continuation of the umbilical vein that takes blood to the inferior vena cava

### Observation: Vessels of the Pulmonary Circuit

In the pulmonary circuit of adult mammals, pulmonary arteries take blood away from the heart to the lungs, and pulmonary veins take blood from the lungs to the heart. Remember that if your pig has been injected, arteries will be red and veins will be blue.

> ⚠ **Wear protective latex gloves and eyewear** when handling preserved animal organs. Exercise caution when using sharp instruments during this laboratory. Wash hands thoroughly upon completion of this laboratory.

#### Pulmonary Trunk and Pulmonary Arteries

1. Locate the pulmonary trunk, which arises from the ventral side of the heart. It may appear white because the thick wall prevents the color of the red latex from showing through.
2. If necessary, remove the pericardial sac from the heart to reveal the blood vessels entering the heart. Veins take blood to the heart.
3. Observe the path of the pulmonary trunk, and notice that it seems to connect directly with the aorta, the major artery. This is the **arterial duct,** a connection found only in a fetus (see Table 22.2).
4. In addition to this duct, look closely and you will find the pulmonary arteries, which leave the pulmonary trunk and go to the lungs.

#### Pulmonary Veins

1. The pulmonary veins are hard to find. To view them, clean away the membrane dorsal to the heart, and carefully note the vessels (pulmonary veins) that leave the lungs. Observe how they connect to the left atrium of the heart.
2. In the adult, which blood vessels—pulmonary arteries or pulmonary veins—carry $O_2$-rich blood?

# 22.3 Vessels of the Systemic Circuit

## Pre-Lab

5. What gas do arteries deliver to the body? _____

6. Which arteries and veins serve the kidneys? _____

In adult mammals, the systemic circuit serves all parts of the body except where gas exchange occurs in the lungs. Arteries take $O_2$-rich blood from the heart to the organs, and veins take $O_2$-poor blood from the organs to the heart. The aorta is the major artery, and the venae cavae are the major veins. It will be possible for you to identify the arteries branching from the aorta and the corresponding veins branching from the venae cavae. In the fetal pig, the venae cavae are called the *anterior* vena cava and the *posterior* vena cava because the normal position of the body is horizontal rather than vertical. Use the figures to identify the blood vessels, but **do not remove any organs.**

## Observation: Aorta and Venae Cavae

1. Follow the aorta as it extends through the thoracic cavity (Fig. 22.4). To do this, gently move the lungs and heart to the right side of the thoracic cavity. Notice how the thoracic or dorsal aorta is a large, whitish vessel extending through the thoracic cavity and then through the diaphragm to become the abdominal aorta. Also notice the esophagus, a smaller, flattened tube more toward the midline. The esophagus goes through the diaphragm to join with the stomach. Locate the anterior vena cava coming off the top of the heart and the posterior vena cava just to the right of the midline. The posterior vena cava also passes through the diaphragm.

2. Name three structures that pass through the diaphragm. 1. _____

   2. _____     3. _____

## Blood Vessels of the Upper Body

The **coronary arteries** and **cardiac veins** lie on the surface of the heart. The **carotid arteries** and **jugular veins** serve the neck and head regions of the body.

## Observation: Blood Vessels of the Upper Body

### Coronary Arteries and Cardiac Veins

1. Locate the coronary arteries and cardiac veins, which are easily visible on the heart's surface (see Fig. 22.1).
2. The coronary arteries arise from the aorta just as it leaves the heart, and the cardiac veins go directly into the right atrium.

### Carotid Arteries and Jugular Veins

1. Find the aorta as it leaves the left ventricle and curves downward. As it arches, it gives off branches. Its first branch, the **brachiocephalic arterial trunk,** divides almost immediately into the **right subclavian** (to the pig's right shoulder) **artery** and the **right** and **left common carotid arteries** (Figs. 22.4 and 22.5).
2. Find the second branch of the aorta, which is the **left subclavian artery.**
3. Both the right and left subclavian arteries have branches called the right and left brachial arteries, which serve the upper limbs.
4. Find the jugular veins alongside the carotid arteries (Figs. 22.4 and 22.6). Follow the carotid arteries and jugular veins as far as possible toward the head. What part of the body is serviced by the carotid arteries and the jugular veins? _____

### Subclavian Arteries and Veins

1. Locate the **subclavian arteries** (Figs. 22.4 and 22.5) and **subclavian veins** (Figs. 22.4 and 22.6), which serve the forelimbs and are easily identified.
2. Note that while the left subclavian artery branches from the aorta, the right subclavian artery branches from the brachiocephalic arterial trunk.
3. Note that the jugular veins and the subclavian veins join to form the brachiocephalic venous trunk, which enters the anterior vena cava.

# Figure 22.4 Ventral view of fetal pig arteries and veins.

Use this diagram to identify blood vessels, *but do not remove any organs.* (a. = artery; v. = vein.)

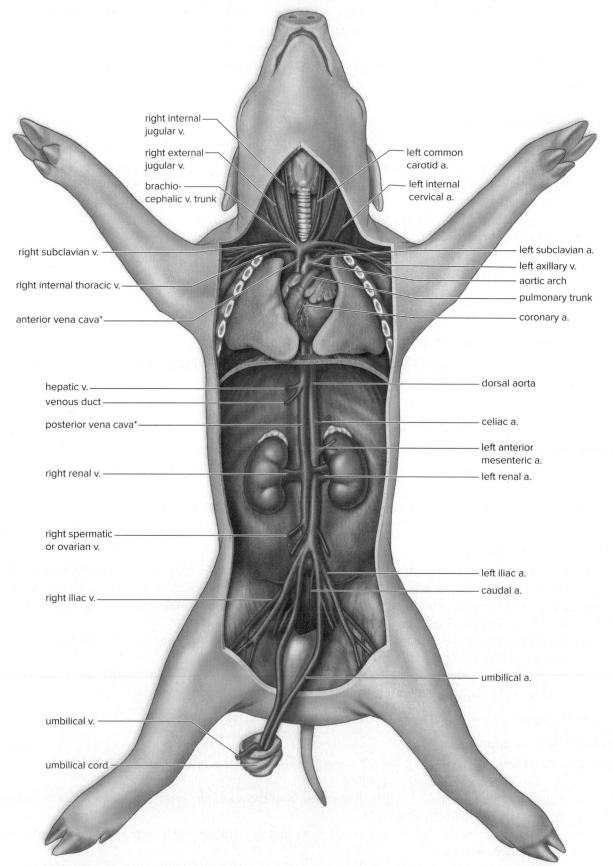

right internal jugular v.

right external jugular v.

brachio-cephalic v. trunk

left common carotid a.

left internal cervical a.

right subclavian v.

right internal thoracic v.

anterior vena cava*

left subclavian a.

left axillary v.

aortic arch

pulmonary trunk

coronary a.

hepatic v.

venous duct

posterior vena cava*

right renal v.

dorsal aorta

celiac a.

left anterior mesenteric a.

left renal a.

right spermatic or ovarian v.

right iliac v.

left iliac a.

caudal a.

umbilical a.

umbilical v.

umbilical cord

*Because the pig walks on all four limbs, the superior vena cava in humans is called the anterior vena cava in pigs, and the inferior vena cava in humans is called the posterior vena cava in pigs.

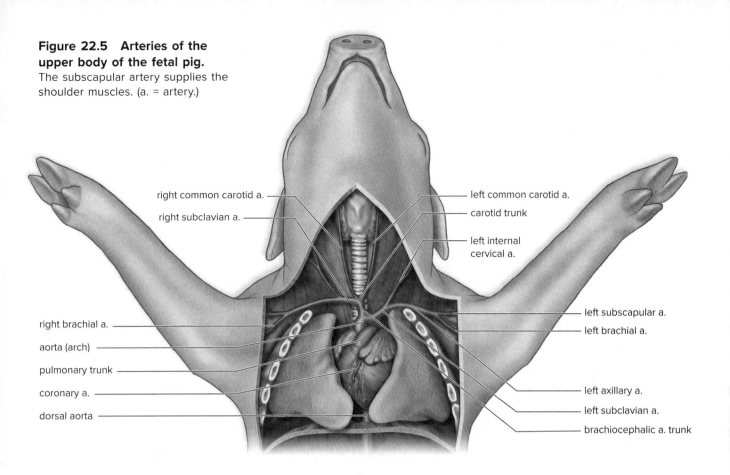

**Figure 22.5 Arteries of the upper body of the fetal pig.**
The subscapular artery supplies the shoulder muscles. (a. = artery.)

right common carotid a.
right subclavian a.
right brachial a.
aorta (arch)
pulmonary trunk
coronary a.
dorsal aorta

left common carotid a.
carotid trunk
left internal cervical a.
left subscapular a.
left brachial a.
left axillary a.
left subclavian a.
brachiocephalic a. trunk

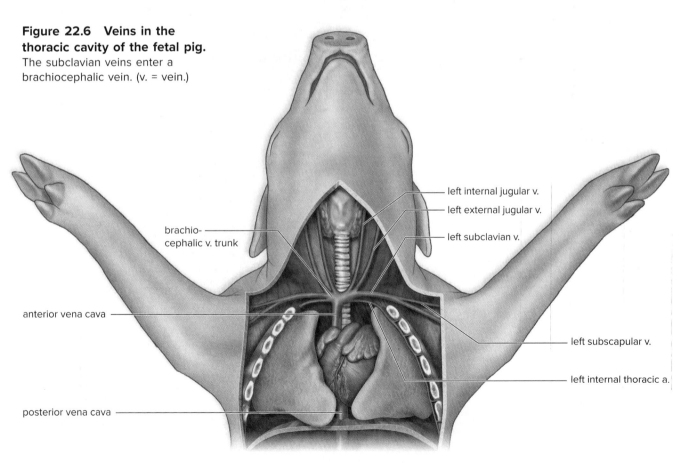

**Figure 22.6 Veins in the thoracic cavity of the fetal pig.**
The subclavian veins enter a brachiocephalic vein. (v. = vein.)

brachio-cephalic v. trunk
anterior vena cava
posterior vena cava

left internal jugular v.
left external jugular v.
left subclavian v.
left subscapular v.
left internal thoracic a.

# Vessels of the Abdominal Cavity

Several major blood vessels are in the abdominal cavity. Use Figures 22.7 and 22.8 to identify the blood vessels, but **do not remove any organs.**

## Observation: Vessels of the Abdominal Cavity

### Celiac and Mesenteric Arteries

1. Carefully lift up the liver and stomach, and put them aside to your left. Dissect the dorsal mesentery to see the **celiac artery** as it leaves the aorta. Tributaries from this vessel eventually reach the stomach, duodenum, liver, and spleen.
2. Branching from the aorta just posterior to the celiac artery is a long, unpaired trunk, called the **anterior mesenteric artery,** which has tributaries to the pancreas and small intestine (see Fig. 22.7).
3. The celiac and mesenteric arteries take blood to the intestines. Thereafter, the hepatic portal vein takes blood from the intestinal capillaries to capillaries in the liver. A portal system is defined as a circulatory unit that goes from one capillary bed to another without passing through the heart (see Fig. 22.3).

### Renal Arteries and Veins

1. Locate the **renal arteries** as they branch from the aorta, and identify these arteries as they go into the kidneys.
2. Locate the renal veins as they leave the kidneys (Fig. 22.8), and identify these veins as they join the posterior vena cava.

### Iliac Arteries and Veins

1. At its posterior end, the aorta branches into the paired **iliac arteries.** Locate the iliac arteries at the posterior end of the aorta, and follow these arteries into the hindlimbs (see Fig. 22.7).

**Figure 22.7 Arteries in the abdominal cavity of the fetal pig.**
The celiac and mesenteric artery serve the digestive and associated organs. The external iliac arteries proceed into the hindlimbs from the caudal end of the dorsal aorta. They give rise to the umbilical arteries and continue as much smaller vessels. (aa. = arteries.) Use this diagram to identify blood vessels, *but do not remove any organs.*

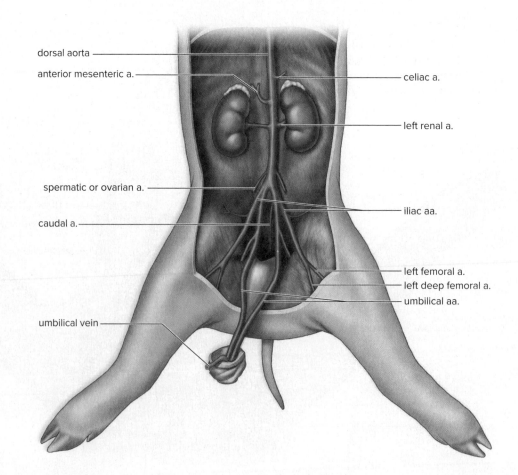

dorsal aorta
anterior mesenteric a.
celiac a.
left renal a.
spermatic or ovarian a.
iliac aa.
caudal a.
left femoral a.
left deep femoral a.
umbilical aa.
umbilical vein

**2.** Find the **iliac veins** alongside the iliac arteries, and follow these veins as they join the posterior vena cava (Fig. 22.8).

### Umbilical Arteries and Veins

**1.** Locate the **umbilical arteries** on either side of the bladder. Follow these arteries as they branch from the iliac arteries and as they pass into the **umbilical cord** (see Fig. 22.7).

**2.** When you were exposing the abdominal cavity, you cut the **umbilical vein.** Follow the umbilical vein from the umbilical cord to the liver. The umbilical vein is joined to the posterior vena cava by the **venous duct** (ductus venosus), which passes through the posterior portion of the liver.

### Posterior Vena Cava

**1.** Locate the **posterior vena cava,** which is easily seen as a large, blue vessel just ventral to the dorsal aorta (Fig. 22.8).

**2.** Note that this vessel seems to disappear in the region of the liver. Here the posterior vena cava receives the hepatic veins coming from the liver. Scrape away some of the liver tissue in order to see these veins.

**3.** Locate the posterior vena cava as it passes through the diaphragm into the thoracic cavity and as it enters the right atrium.

**4.** Follow the posterior vena cava from the iliac veins to the right atrium of the heart (see Fig. 22.4).

## Storage of Pigs

**1.** Before leaving the laboratory, place your pig in the plastic bag provided.

**2.** Expel excess air from the bag, and tie it shut.

**3.** Write your name and section on the tag provided, and attach it to the bag. Your instructor will indicate where the bags are to be stored until the next laboratory period.

**4.** Clean the dissecting tray and tools, and return them to their proper location.

**5.** Wipe off your goggles.

**6.** Wash your hands.

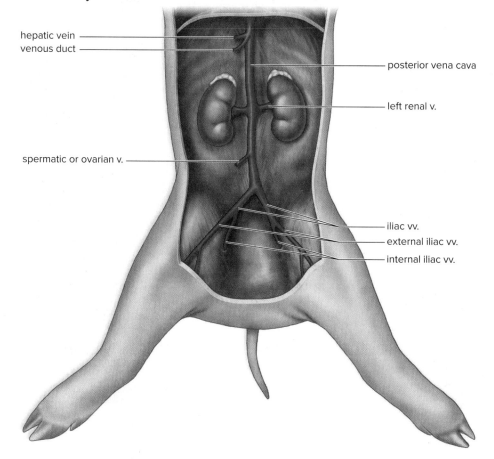

hepatic vein
venous duct
posterior vena cava
left renal v.
spermatic or ovarian v.
iliac vv.
external iliac vv.
internal iliac vv.

**Figure 22.8  Veins in the abdominal cavity of the fetal pig.**
The posterior vena cava divides into the common iliac veins. The common iliac veins branch into the external and internal iliac veins. (vv. = veins.) Use this diagram to identify blood vessels, but *do not remove any organs.*

# 22.4 Blood Vessel Comparison

## Pre-Lab

**7.** What is the purpose of valves within veins? _____

_____

**8.** How do arteries help regulate blood pressure? _____

_____

Blood pressure keeps the blood in arteries moving away from the heart. Skeletal muscle contraction pushing in on veins keeps the blood in veins moving toward the heart. Do you predict that arteries or veins are generally

more superficial in the body? _____

_____

## Wall of an Artery Compared with Wall of a Vein

Both arteries and veins have three distinct layers, or **tunicas,** that form a wall around the lumen, the space through which blood flows. The three tunicas are called the inner layer, the middle layer, and the outer layer. Figure 22.9 shows that arteries have a thicker wall than veins because the middle layer consisting of smooth muscle and elastic fibers is much thicker. The elastic fibers in the wall of an artery allows it to expand when the blood pours into it with each heartbeat. The smooth muscle allows the arterial wall to constrict when needed to keep blood pressure normal.

**Figure 22.9 Blood vessel comparison.**

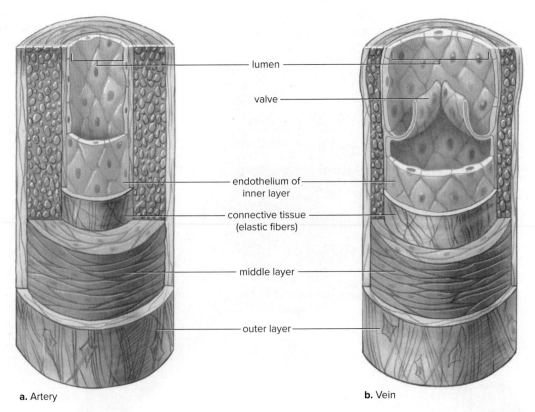

lumen

valve

endothelium of inner layer

connective tissue (elastic fibers)

middle layer

outer layer

**a.** Artery

**b.** Vein

1. Obtain a microscope slide that shows an artery and a vein in cross section.
2. View the slide, under both low and high power, and with the help of Figure 22.10, determine which is the artery and which is the vein.
3. Identify the **outer layer,** which contains many collagen and elastic fibers and often appears white in specimens.
4. Identify the **middle layer,** the thickest layer, which is composed of smooth muscle and elastic tissue.

   Does this layer appear thicker in arteries than in veins? _____

5. Identify the **inner layer,** a smooth lining of simple squamous epithelial cells called the endothelium. In veins, the endothelium forms valves that keep the blood moving toward the heart. Arteries do not have valves. Considering the relationship of arteries and veins to the heart, why do veins have valves, while

   arteries do not? _____

   _____

**Figure 22.10    Photomicrograph of an artery and a vein.**
Ed Reschke

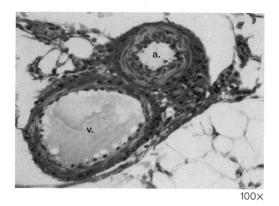

100×

## Conclusions

- Which type of blood vessel (artery or vein) has thicker walls? What makes the wall thicker? _____

  _____

  _____

- Which type of blood vessel has thinner walls? _____

  _____

- Which type of blood vessel is more apt to lose its elasticity, leading to a discoloration that can be

  externally observed? _____

  _____

  What is this condition called? _____

_____    1. What blood vessels connect arterioles to venules and exchange molecules with tissue fluid?

_____    2. Which blood vessels return blood to the right atrium?

_____    3. Which direction does the pulmonary artery carry blood?

_____    4. Is the blood entering the left ventricle $O_2$-poor or $O_2$-rich?

_____    5. What valve prevents the backflow of blood when the left ventricle relaxes?

_____    6. Does the pulmonary or systemic circuit carry blood to the spleen?

_____    7. What vessel transports blood to the liver from the digestive tract?

_____    8. Do pulmonary veins carry $O_2$-poor or $O_2$-rich blood?

_____    9. In addition to the esophagus and the posterior vena cava, what structure passes through the diaphragm?

_____    10. What are the names of the veins that serve the neck and head regions of the body?

_____    11. What organs are served by the anterior mesenteric artery?

_____    12. What arteries form from the branching of the posterior end of the aorta?

_____    13. What is the name of the vein that is joined to the posterior vena cava by the venous duct (ductus venosus)? (*Hint:* It was cut when the abdominal cavity was exposed.)

_____    14. Which type of blood vessel (artery or vein) has a thick wall with a lot of smooth muscle that helps regulate blood pressure?

## Thought Questions

15. Explain why narrowing or blockages of the coronary arteries can be life-threatening, and identify which arteries, when blocked, increase the risk of having a stroke.

16. Explain why an airline recently introduced an exercise video to combat deep vein thrombosis (clots that form in deep veins after prolonged inactivity). Consider the pressure of blood in veins and venous structure in your reply.

17. In figures illustrating arteries and veins, arteries usually appear red and veins usually appear blue. Explain why the pulmonary arteries seen in Figure 22.2 appear blue and the pulmonary veins in Figure 22.2 appear red.

# 23

# Basic Mammalian Anatomy II

## Learning Outcomes

**23.1 Urinary System**
- Identify the organs of the urinary system, and state their function.

**23.2 Male Reproductive System**
- Identify the organs of the male reproductive system, and state their function.

**23.3 Female Reproductive System**
- Identify the organs of the female reproductive system, and state their function.

**23.4 Review of the Respiratory, Digestive, and Cardiovascular Systems**
- Using preserved specimens, images, or charts, locate and identify the individual organs of the respiratory, digestive, and cardiovascular systems.
- Using preserved specimens, images, or charts, locate and identify the hepatic portal system.

## Introduction

### Pre-Lab

1. Which organ systems are the focus in this lab? _____

2. Male testes produce _____, and female ovaries produce _____.

The **urinary system** and the **reproductive system** are so closely associated in mammals that they are often considered together as the **urogenital system.** They are particularly associated in males, where certain structures have both urinary and reproductive functions. In this laboratory, we will focus first on dissecting the urinary and reproductive systems in the fetal pig.

The kidneys of the urinary system produce urine, which is stored in the bladder before being released to the exterior. As the kidneys produce urine, they also regulate the volume and the composition of the blood so that the water and salt balance and the acid-base balance of the blood stays within normal limits.

In mammalian reproductive systems, the testes are the male gonads, and the ovaries are the female gonads. The testes produce sperm, and the ovaries produce oocytes that become eggs. We will compare the anatomy of the reproductive systems in pigs with those in humans.

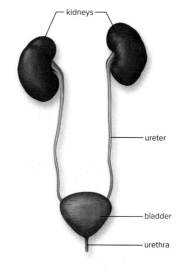

Urinary system

**Figure 23.1  Urinary system of the fetal pig.**
In (**a**) females and (**b**) males, urine is made by the kidneys, transported to the bladder by the ureters, stored in the bladder, and then excreted from the body through the urethra.

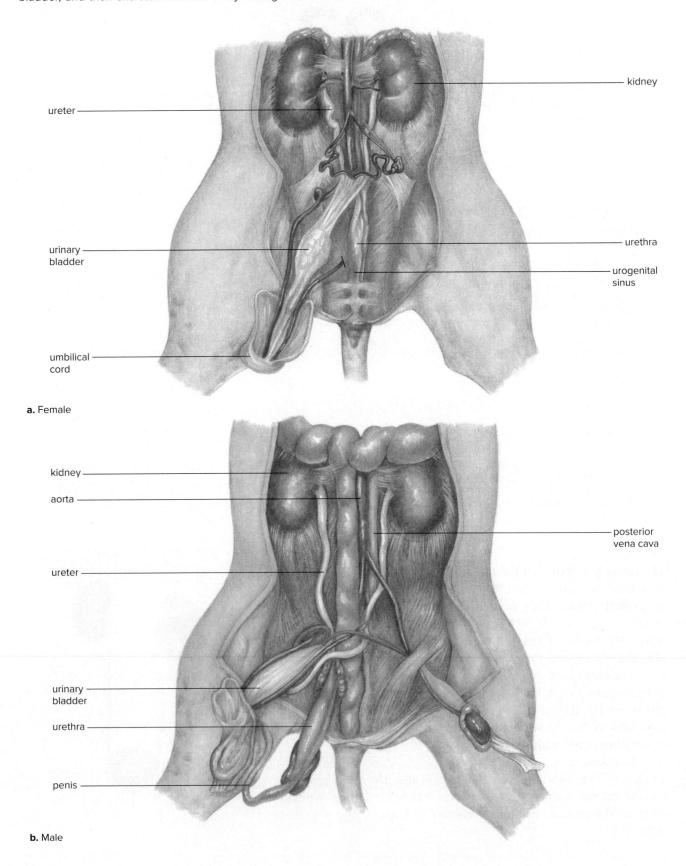

ureter

kidney

urinary bladder

urethra

urogenital sinus

umbilical cord

**a.** Female

kidney

aorta

posterior vena cava

ureter

urinary bladder

urethra

penis

**b.** Male

# 23.1 Urinary System

The urinary system consists of the **kidneys,** which produce urine; the **ureters,** which transport urine to the **urinary bladder,** where urine is stored; and the **urethra,** which transports urine to the outside of the body. In males, the urethra also transports sperm during ejaculation.

### Observation: Urinary System in Pigs

During this dissection, compare the urinary system structures of male and female fetal pigs. Later in this laboratory period, exchange specimens with a neighboring team for a more thorough inspection.

1. The large, paired kidneys (Fig. 23.1) are reddish organs covered by **peritoneum,** a membrane that anchors them to the dorsal wall of the abdominal cavity, sometimes called the **peritoneal cavity.** Clean the peritoneum away from one of the kidneys, and study it more closely.

> ⚠️ **Wear protective latex gloves and eyewear** when handling preserved animal organs. Exercise caution when using sharp instruments during this laboratory. Wash hands thoroughly upon completion of this laboratory.

2. Locate the ureters, which leave the kidneys and run posteriorly under the peritoneum (Fig. 23.1).
3. Clean the peritoneum away, and follow a ureter to the urinary bladder, which normally lies in the ventral portion of the abdominal cavity. The urinary bladder is on the inner surface of the flap of tissue to which the umbilical cord was attached.
4. The urethra arises from the bladder posteriorly and joins the urogenital sinus. Follow the urethra until it passes from view into the ring formed by the pelvic girdle.
5. Sequence the organs in the urinary system to trace the path of urine from its production to its exit from the body. _____
6. Using a scalpel, section one of the kidneys in place, cutting it lengthwise (Fig. 23.2). At the center of the medial portion of the kidney is an irregular, cavity-like reservoir, the **renal pelvis.** The outermost portion of the kidney (the **renal cortex**) shows many small striations perpendicular to the outer surface. This region and the more even-textured **renal medulla** contain **nephrons** (excretory tubules), microscopic organs that produce urine.

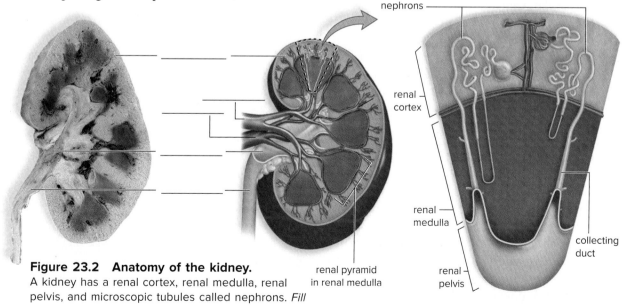

**Figure 23.2   Anatomy of the kidney.**
A kidney has a renal cortex, renal medulla, renal pelvis, and microscopic tubules called nephrons. *Fill in the missing labels with the help of your textbook.*

MedImage/Science Source

## 23.2 Male Reproductive System

The **male reproductive system** consists of the **testes** (sing., testis), which produce sperm, and the **epididymides** (sing., epididymis), which store sperm before they enter the **vasa deferentia** (sing., vas deferens). Just prior to ejaculation, sperm leave the vasa deferentia and enter the urethra, which eventually passes into the penis. The **penis** is the male organ of sexual intercourse. **Seminal vesicles,** the **prostate gland,** and the **bulbourethral glands** (Cowper's glands) add fluid to semen after sperm reach the urethra. Table 23.1 summarizes the male reproductive organs.

The testes begin their development in the abdominal cavity, just anterior and dorsal to the kidneys. Before birth, however, they

| Table 23.1 Male Reproductive Organs and Functions | |
| --- | --- |
| **Organ** | **Function** |
| Testis | Produces sperm and sex hormones |
| Epididymis | Stores sperm as they mature |
| Vas deferens | Conducts and stores sperm |
| Seminal vesicle | Contributes secretions to semen |
| Prostate gland | Contributes secretions to semen |
| Urethra | Conducts sperm |
| Bulbourethral glands | Contribute secretions to semen |
| Penis | Organ of copulation |

gradually descend into paired **scrotal sacs** within the scrotum, suspended anterior to the anus. Each scrotal sac is connected to the body cavity by an **inguinal canal,** the opening of which can be found in the pig. The passage of the testes from the body cavity into the scrotal sacs is called the descent of the testes. The testes in most of the male fetal pigs being dissected will probably be partially or fully descended.

### Observation: Male Reproductive System in Pigs

While doing this dissection, consult Table 23.1 for the function of the male reproductive organs.

#### Inguinal Canal, Testis, Epididymis, and Vas Deferens

1. Locate the opening of the left inguinal canal, which leads to the left scrotal sac (Fig. 23.3).
2. Expose the canal and sac by making an incision through the skin and muscle layers from a point over this opening back to the left scrotal sac.
3. Open the sac, and find the testis. Note the much-coiled tubule—the epididymis—that lies alongside the testis. An epididymis is continuous with a vas deferens, which runs toward the abdominal cavity.
4. Each vas deferens loops over an umbilical artery and ureter and unites with the urethra as it leaves the urinary bladder. We will dissect this juncture below.

#### Penis, Urethra, and Accessory Glands

1. Cut through the ventral skin surface just posterior to the umbilical cord. This will expose the rather undeveloped penis, which contains a long portion of the urethra.
2. Lay the penis to one side, and then cut down through the ventral midline, laying the legs wide apart in the process (Fig. 23.4). The cut will pass between muscles and through pelvic cartilage (bone has not developed yet). Do not cut any of the ducts or tracts in the region.
3. You will now see the urethra ventral to the rectum. It is somewhat heavier in the male due to certain accessory glands:
   a. Bulbourethral glands (Cowper's glands), about 1 cm in diameter, are further along the urethra and are more prominent than the other accessory glands.

## Figure 23.3  Male reproductive system of the fetal pig.

In males, the urinary system and the reproductive system are joined. The vasa deferentia (sing., vas deferens) enter the urethra, which also carries urine.

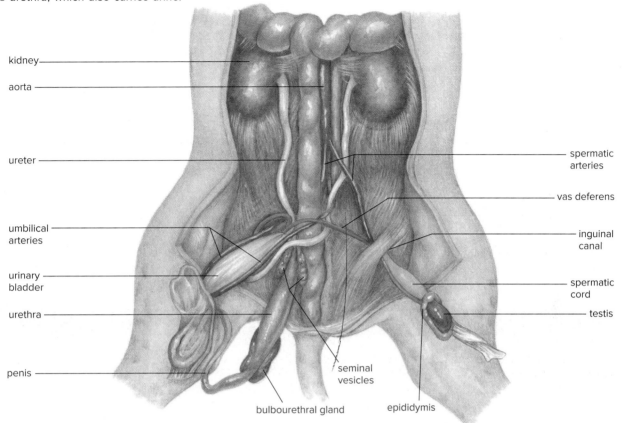

## Figure 23.4  Photograph of the male reproductive system of the fetal pig.

Compare Figure 23.3 to this photograph to help identify the structures of the male urogenital system.

Carlyn Iverson/McGraw Hill

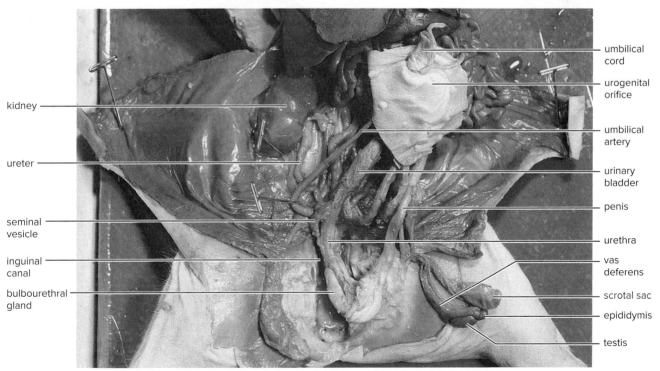

**b.** The prostate gland, about 4 mm across and 3 mm thick, is located on the dorsal surface of the urethra, just posterior to the juncture of the urinary bladder with the urethra. It is often difficult to locate and is not shown in Figures 23.3 and 23.4.

**c.** Small, paired seminal vesicles may be seen on either side of the prostate gland.

4. Trace the urethra as it leaves the bladder. When it nears the end of the abdominal cavity, it turns rather abruptly and runs anteriorly just under the skin where you have just dissected it. This portion of the urethra is within the penis.

5. Now you should be able to see the vasa deferentia enter the urethra. If necessary, free these structures from surrounding tissue to see them enter the urethra near the location of the prostate gland. In males, the urethra transports sperm and urine to the urogenital opening.

6. Sequence the organs in the male reproductive system to trace the path of sperm from the organ of

production to the penis. _____

_____

## Comparison of Male Fetal Pig and Human Male

Use Figure 23.5 to help compare the male pig reproductive system with the human male reproductive system. Complete Table 23.2, which compares the location of the penis in these two mammals.

| Table 23.2 Location of Penis in Male Fetal Pig and Human Male | |
|---|---|
| **Fetal Pig** | **Human** |
| Penis | |

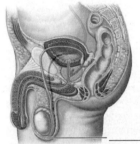

**Figure 23.5 Human male urogenital system.**
In the fetal pig, but not in the human male, the penis lies beneath the skin and exits at the urogenital opening. *Fill in the missing labels with the help of your textbook.*

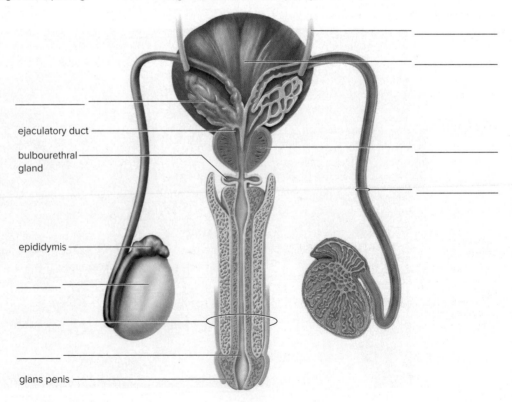

ejaculatory duct

bulbourethral gland

epididymis

glans penis

# 23.3  Female Reproductive System

**Pre-Lab**

5. *With the help of your textbook, complete the labeling for Figure 23.8.*

The **female reproductive system** (Table 23.3) consists of the **ovaries** (sing., ovary), which produce eggs, and the **oviducts,** which transport eggs to the **uterus,** where if the egg is fertilized, development will occur. In the fetal pig, the uterus does not form a single organ, as in humans, but is partially divided into structures called **uterine horns,** which connect with the oviduct. The **vagina** is the birth canal and the female organ of sexual intercourse.

| Table 23.3   Female Reproductive Organs and Functions | |
|---|---|
| **Organ** | **Function** |
| Ovary | Produces egg and sex hormones |
| Oviduct (fallopian tube) | Conducts egg toward uterus |
| Uterus | Houses developing fetus |
| Vagina | Receives penis during copulation and serves as birth canal |

## Observation: Female Reproductive System in Pigs

While doing this dissection, consult Table 23.3 for the function of the female reproductive organs.

### Ovaries and Oviducts

1. Locate the paired ovaries, small bodies suspended from the peritoneal wall in mesenteries, posterior to the kidneys (Figs. 23.6 and 23.7).
2. Closely examine one ovary. Note the small, short, coiled oviduct, sometimes called the **fallopian tube.** The oviduct does not attach directly to the ovary but ends in a funnel-shaped structure with fingerlike processes (fimbriae) that partially encloses the ovary. The egg produced by an ovary enters an oviduct, where it is fertilized by a sperm, if reproduction will occur. Any resulting embryo passes to the uterus.

### Uterine Horns

1. Locate the **uterine horns.** (Do not confuse the uterine horns with the oviducts; the latter are much smaller and are found very close to the ovaries.)
2. Find the body of the uterus, located where the uterine horns join.

### Vagina

1. Separate the hindlimbs of your specimen, and cut down along the midventral line. The cut will pass through muscle and the cartilaginous pelvic girdle. With your fingers, spread the cut edges apart, and use blunt dissecting instruments to separate connective tissue.
2. Now find the vagina, which passes from the uterus to the urogenital sinus. The vagina is dorsal to the urethra, which also enters the urogenital sinus, and ventral to the rectum, which exits at the anus. The urogenital sinus opens at the urogenital papilla (Figs. 23.6 and 23.7).
3. The vagina is the organ of copulation and is the birth canal. The receptacle for the vagina in a pig, the urogenital sinus, is absent in adult humans and several other adult female mammals in which both the urethra and vagina have their own openings.
4. The vagina plays a critical role in reproduction, even though development of the offspring occurs in the

   uterus. Explain. _____

   _____

**Figure 23.6  Female reproductive system of the fetal pig.**
In the fetal pig, both the vagina and the urethra enter the urogenital sinus, which opens at the urogenital papilla.

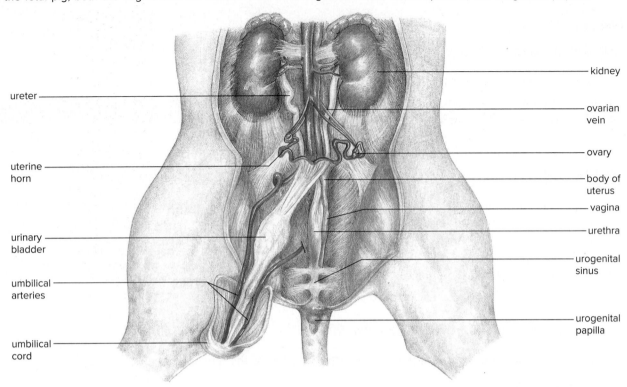

**Figure 23.7  Photograph of the female reproductive system of the fetal pig.**
Compare Figure 23.6 with this photograph to help identify the structures of the female urinary and reproductive systems.

Carlyn Iverson/McGraw Hill

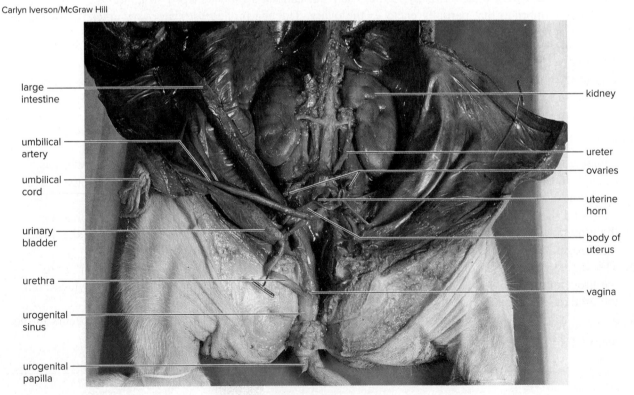

## Comparison of Female Fetal Pig with Human Female

Use Figure 23.8 to compare the female pig reproductive system with the human female reproductive system. Complete Table 23.4, which compares the appearance of the oviducts and the uterus, as well as the presence or absence of a urogenital sinus, in these two mammals.

**Figure 23.8  Human female reproductive system.**
Especially compare the anatomy of the oviducts in humans with that of the uterine horns in a pig. In a pig, the fetuses develop in the uterine horns; in a human female, the fetus develops in the body of the uterus. *Fill in the missing labels with the help of your textbook.*

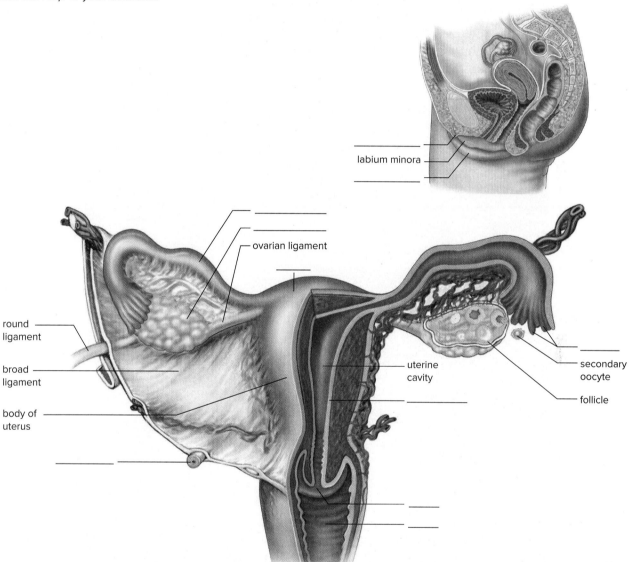

labium minora

ovarian ligament

round ligament

broad ligament

body of uterus

uterine cavity

secondary oocyte

follicle

| Table 23.4  Comparison of Female Fetal Pig with Human Female | | |
|---|---|---|
| | **Fetal Pig** | **Human** |
| Oviducts | | |
| Uterus | | |
| Urogenital sinus | | |

# Figure 23.9 Internal anatomy of the fetal pig.

*Label the figure with the help of your textbook.* Most of the major organs are shown in this photograph. The stomach has been removed. The spleen, gallbladder, and pancreas are not visible.

Ken Taylor/Wildlife Images

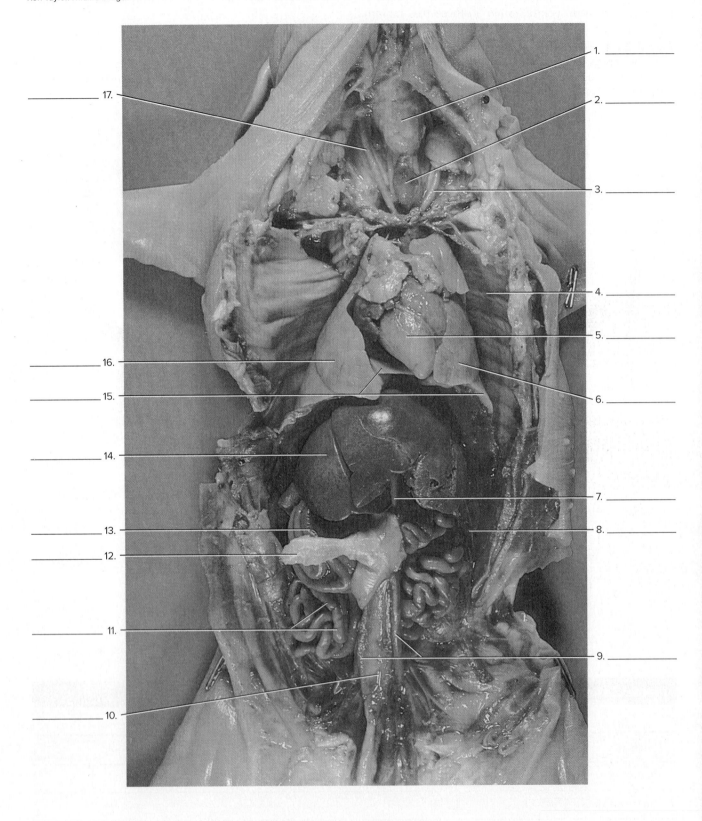

1. _____

2. _____

3. _____

4. _____

5. _____

6. _____

7. _____

8. _____

9. _____

10. _____

11. _____

12. _____

13. _____

14. _____

15. _____

16. _____

17. _____

# 23.4  Review of the Respiratory, Digestive, and Cardiovascular Systems

## Pre-Lab

6. List the functions of the respiratory, digestive, and cardiovascular systems. _____

_____

_____

_____

You previously dissected the respiratory, digestive, and cardiovascular systems of the fetal pig. Review your knowledge of these systems by reexamining your dissection of the fetal pig and *label Figure 23.9.* In this portion of today's lab, you will review each system and examine some organs in more detail. **Do not remove any organs** unless told to do so by your instructor.

### Observation: Respiratory System in Pigs

1. Using these terms (*bronchiole, bronchus, glottis, larynx, pharynx, trachea*), trace the path of air from the nasal passages to the lungs. List the first three organs in the left column and the last three organs in the right column.

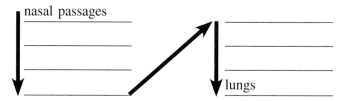

2. Make sure you have cut the corners of the mouth. In the **pharynx,** you should be able to locate the **glottis,** an opening to the _____.
3. If necessary, make a midventral incision in the neck to expose the **larynx.**
4. Clear away the "straplike" muscles covering the **trachea.** Now you should be able to feel the cartilaginous rings that hold the trachea open. Locate the esophagus, which lies below (dorsal to) the trachea.
5. If available, observe a slide on display showing a section through the trachea and esophagus. Notice in Figure 23.10 that the air and food pathways cross in the pharynx.
6. Open the pig's mouth, insert a blunt probe into the glottis, and carefully work the probe down through the larynx to the level of the **bronchi.**

**Figure 23.10  Air and food passages in the fetal pig.**
A probe can pass from the mouth to the larynx and to the esophagus.

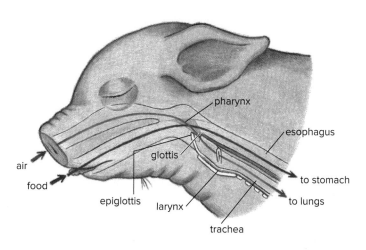

7. Observe the **lungs,** and if available, observe a prepared slide of lung tissue.
8. If so directed by your instructor, remove a portion of the trachea, the bronchi, and the lungs, keeping them all in one piece. Place this specimen in a small container of water. Holding the trachea with your forceps, gently but firmly stroke the lung repeatedly with the blunt wooden base of one of your probes. If you work carefully, the alveolar tissue will be fragmented and rubbed away, leaving the branching system of air tubes and blood vessels.

## Observation: Digestive System in Pigs

1. Using these terms (*esophagus, large intestine, small intestine, stomach*), trace the path of food from the mouth to the anus:

mouth

anus

2. Open the **mouth,** and insert a blunt probe into the esophagus (Fig. 23.10). Then trace the **esophagus** to the stomach.
3. Open one side of the **stomach,** and examine its interior surface. Does it appear smooth or rough? _____
4. Find the pyloric sphincter, the muscle that surrounds the entrance to the duodenum, the first part of the **small intestine.** Record the length of the small intestine. _____

   If you have not done so before, find the bile duct that empties into the duodenum. The bile duct comes

   from the _____.
5. Find the **cecum,** a projection where the small intestine enters the large intestine. How does the appearance

   of the pig's large intestine differ from that of a human? _____

   _____

6. Carefully cut the mesenteries holding the colon of the **large intestine** in place, and uncoil the large

   intestine. Record the length of the large intestine. _____ How does the length of the large intestine

   compare with that of the small intestine? _____

7. Locate again the liver, pancreas, and gallbladder, three accessory organs of digestion.

## Observation: Cardiovascular System in Pigs

### Heart

1. Trace the path of blood through the heart, starting with the vena cava and ending with the aorta. Mention all the chambers of the heart and the valves.

To the heart:     From the lungs:

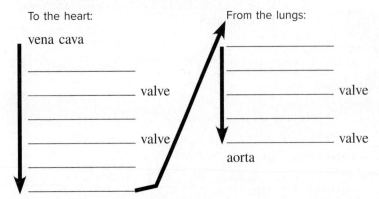

vena cava

_____ valve

_____ valve

_____ valve

_____ valve

aorta

2. Keeping the heart inside the pig, cut the pericardial sac (the tissue that surrounds the heart).
3. Look for and identify the vessels attached to the heart.
4. Section the heart, and look for its four chambers. Remnants of the atrioventricular valves can be seen as thin sheets of whitish tissue attached to fine, white, tendinous strands.
5. With your blunt probe, find the oval opening in the wall between the two atrial chambers. Recall that this is a shunt that allows blood to bypass lung circulation prior to birth.

### Blood Vessels

1. In general, arteries take blood _____ the heart, and veins take blood _____ the heart.
2. Locate the following blood vessels in your pig. State their origin and destination.

   a. coronary artery _____          f. subclavian vein _____
   b. cardiac vein _____             g. renal artery _____
   c. carotid artery _____           h. renal vein _____
   d. jugular vein _____             i. iliac artery _____
   e. subclavian artery _____        j. iliac vein _____

### Hepatic Portal System and Associated Vessels

A **portal system** goes from one capillary bed to another without passing through the heart. For example, the hepatic portal vein takes blood from the intestinal capillaries to capillaries in the liver. The liver plays an important role in processing and storing materials absorbed from the intestine. The hepatic veins take blood from the liver to the inferior (posterior) vena cava.

1. In your pig, the hepatic portal vein is dorsal to the bile duct may or may not be blue depending on if the latex entered it. To find the hepatic portal vein, break the mesenteries in the region of the bile duct (Fig. 23.11).
2. The **hepatic veins** consist of three or four vessels from the liver to the inferior vena cava. To see them, scrape away the liver with the blunt side of the scalpel until liver material has been removed and only a mass of cords remains.
3. Identify the **umbilical vein** leading into the liver and, on its posterior surface, the **venous duct,** which is the main channel through the liver in the fetus.

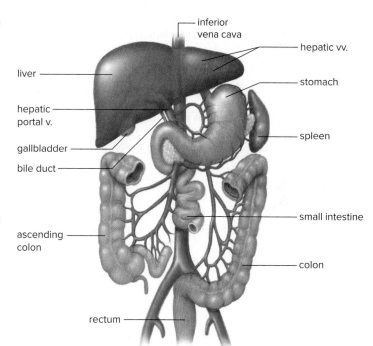

**Figure 23.11  Hepatic portal system.**
The hepatic portal vein lies between the digestive tract and the liver. The hepatic veins enter the inferior vena cava. (v. = vein; vv. = veins)

## Storage of Pigs

1. Before leaving the laboratory, place your pig in the plastic bag provided.
2. Expel excess air from the bag, and tie it shut.
3. Write your name and section on the tag provided, and attach it to the bag. Your instructor will indicate where the bags are to be stored until the next laboratory period.
4. Clean the dissecting tray and tools, and return them to their proper location.
5. Wipe off your goggles and wash your hands.

_____  **1.** What two individual systems make up the urogenital system?

_____  **2.** What is the cavity-like space in a kidney called?

_____  **3.** What organ stores urine?

_____  **4.** What urinary structure transports both urine and sperm?

_____  **5.** Where are sperm stored while they mature?

_____  **6.** When descending, what do the testes travel through to get to the scrotal sacs?

_____  **7.** What accessory gland for reproduction in human males is inferior (posterior) to the urinary bladder?

_____  **8.** What female reproductive structures transport eggs to the location where embryos/ fetuses develop?

_____  **9.** Where, in a pig, are the embryos/fetuses located during development?

_____  **10.** What reproductive structures produce sex hormones in the male and in the female?

_____  **11.** What respiratory structure has cartilaginous rings that keep the airway open?

_____  **12.** What does food travel through to enter the duodenum when leaving the stomach?

_____  **13.** What is the tissue surrounding the heart called?

_____  **14.** How does blood pass from the intestinal capillaries to the capillaries of the liver?

**Thought Questions**

**15.** Why do females experience more urinary tract infections than males? It may help to observe Figures 23.5 and 23.8 (the lateral view at top right) and think about where the bladder and urethra are in the drawings. You may also refer to figures of the human female reproductive system in your textbook.

**16.** What male reproductive structure is severed during a vasectomy? Do the testes still produce sperm following a vasectomy? Why does sterility result from a vasectomy?

**17.** The hormone insulin lowers blood glucose. How will insulin affect what happens to glucose once blood travels through the hepatic portal system?

# 24

# Nervous System and Senses

## Learning Outcomes

**24.1 Central Nervous System**
- Identify the parts of the brain studied, and state the functions of each part.
- Describe the anatomy of the spinal cord, and tell how the cord functions as a relay station.

**24.2 Peripheral Nervous System**
- Distinguish between cranial nerves and spinal nerves on the basis of location and function.
- Describe the anatomy and physiology of a spinal reflex arc.

**24.3 The Human Eye**
- Identify the parts of the eye, and state a function for each part.

**24.4 The Human Ear**
- Identify the parts of the ear, and state a function for each part.

**24.5 Sensory Receptors in Human Skin**
- Describe the anatomy of the human skin, and explain the distribution and function of sensory receptors.

**24.6 Human Chemoreceptors**
- Relate the ability to distinguish foods to the senses of smell and taste.

## Introduction

### Pre-Lab

1. What are the two divisions of the nervous system? _____

The nervous system has two major divisions. One is the central nervous system (CNS), which consists of the brain and spinal cord. The other is the peripheral nervous system (PNS), which contains cranial nerves and spinal nerves (Fig. 24.1). Sensory receptors detect changes in environmental stimuli, and nerve impulses move along sensory nerve fibers to the brain and the spinal cord. The brain and spinal cord sum up the data before sending impulses via motor nerve fibers to effectors (muscles and glands), so a response to stimuli is possible. The brain and spinal cord contain all parts of neurons—dendrites, cell body, axon, and axon terminal—but nerves contain only axons.

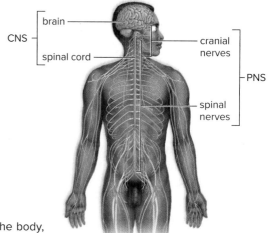

**Figure 24.1   The nervous system.**
The central nervous system (CNS) is in the midline of the body, and the peripheral nervous system (PNS) is outside the CNS.

## 24.1 Central Nervous System

The brain is the enlarged anterior end of the spinal cord. It contains centers that receive input from and can command other regions of the nervous system.

> ⚠️ **Latex gloves** Wear protective latex gloves when handling preserved animal organs. Use protective eyewear and exercise caution when using sharp instruments during this laboratory. Wash hands thoroughly upon completion of this laboratory.

### Preserved Sheep Brain

The sheep brain (Fig. 24.2) is often used to study the brain. It is easily available and large enough that individual parts can be identified.

### Observation: Preserved Sheep Brain

Examine the exterior and a midsagittal (longitudinal) section of a preserved sheep brain or a model of the human brain, and with the help of Figure 24.2, identify the following:

1. **Ventricles:** interconnecting spaces that produce and serve as a reservoir for cerebrospinal fluid, which cushions the brain. Toward the anterior, note the lateral ventricle on one longitudinal section and similarly a lateral ventricle on the other longitudinal section.
2. **Cerebrum:** most developed area of the brain; responsible for higher mental capabilities. The cerebrum is divided into the right and left **cerebral hemispheres,** joined by the **corpus callosum,** a broad sheet of white matter. The outer portion of the cerebrum is highly convoluted and divided into the following surface lobes (see Fig. 24.4):

   a. **Frontal lobe:** controls motor functions and permits voluntary muscle control; it is also responsible for abilities to think, problem solve, speak, and smell.

   b. **Parietal lobe:** receives information from sensory receptors located in the skin and also the taste receptors in the mouth. A groove called the **central sulcus** separates the frontal lobe from the parietal lobe.

   c. **Occipital lobe:** interprets visual input and combines visual images with other sensory experiences. The optic nerves split and enter opposite sides of the brain at the optic chiasma, located in the diencephalon.

   d. **Temporal lobe:** has sensory areas for hearing and smelling. The olfactory bulb contains nerve fibers that communicate with the olfactory cells in the nasal passages and take nerve impulses to the temporal lobe.
3. **Diencephalon:** portion of the brain where the third ventricle is located. The hypothalamus and thalamus are also located here.

   a. **Thalamus:** two connected lobes located in the roof of the third ventricle. The thalamus is the highest portion of the brain to receive sensory impulses before the cerebrum. It is believed to control which received impulses are passed on to the cerebrum. For this reason, the thalamus sometimes is called the "gatekeeper to the cerebrum."

   b. **Hypothalamus:** forms the floor of the third ventricle and contains control centers for appetite, body temperature, blood pressure, and water balance. Its primary function is homeostasis. The hypothalamus also has centers for pleasure, reproductive behavior, hostility, and pain.
4. **Cerebellum:** located just posterior to the cerebrum as you observe the brain dorsally, the cerebellum's two lobes make it appear rather like a butterfly. In cross section, the cerebellum has an internal pattern that looks like a tree. The cerebellum coordinates equilibrium and motor activity to produce smooth movements.

# Figure 24.2 The sheep brain.

(a–c) ©Dr. J. Timothy Cannon

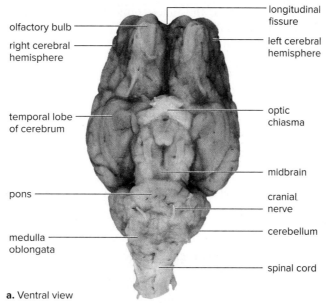

olfactory bulb

right cerebral hemisphere

temporal lobe of cerebrum

pons

medulla oblongata

longitudinal fissure

left cerebral hemisphere

optic chiasma

midbrain

cranial nerve

cerebellum

spinal cord

**a.** Ventral view

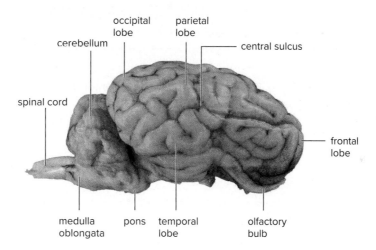

occipital lobe

parietal lobe

cerebellum

central sulcus

spinal cord

frontal lobe

medulla oblongata

pons

temporal lobe

olfactory bulb

**b.** Lateral view

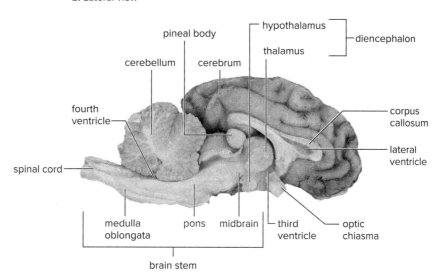

pineal body

hypothalamus

thalamus

diencephalon

cerebellum

cerebrum

fourth ventricle

corpus callosum

lateral ventricle

spinal cord

medulla oblongata

pons

midbrain

third ventricle

optic chiasma

brain stem

**c.** Longitudinal section

**5. Brain stem:** part of the brain that connects with the spinal cord. Because it includes the pons and medulla oblongata, it contains centers for the functioning of internal organs. Based on its location, it serves as a relay station for nerve impulses passing from the cord to the brain. Therefore, it helps keep the rest of the brain alert and functioning.

    **a. Midbrain:** anterior to the pons, the midbrain serves as a relay station for sensory input and motor output. It also contains a reflex center for eye muscles.

    **b. Pons:** the ventral, bulblike enlargement on the brain stem. It serves as a passageway for nerve impulses running between the medulla and the higher brain regions.

    **c. Medulla oblongata** (or simply **medulla**): the most posterior portion of the brain stem. It controls internal organs; for example, blood pressure, cardiac, and breathing control centers are present in the medulla. Nerve impulses pass from the spinal cord through the medulla to and from higher brain regions.

## The Human Brain

Based on your knowledge of the sheep brain, complete Table 24.1 by stating the major functions of each part of the brain listed.

| Table 24.1    Summary of Brain Functions | |
|---|---|
| **Part** | **Major Functions** |
| Cerebrum | |
| Cerebellum | |
| **Diencephalon** Thalamus | |
| Hypothalamus | |
| **Brain stem** Midbrain | |
| Pons | |
| Medulla oblongata | |

Which parts of the brain work together to achieve the following?

    **1.** Maintaining hand–eye coordination _____

    **2.** Concentrating on homework when TV is playing _____

    **3.** Avoiding dark alleys while walking home at night _____

    **4.** Keeping the blood pressure within the normal range _____

## Cerebral Lobes

As stated previously, the outer portion of the cerebrum is highly convoluted and divided into lobes as illustrated in Figure 24.4. The various sense organs send nerve impulses to a particular lobe where the nerve

## Figure 24.3 The human brain (longitudinal section).

The cerebrum is larger in humans than in sheep. Label where indicated using: *brain stem, cerebellum, cerebrum, diencephalon, hypothalamus, medulla oblongata, midbrain, pons, thalamus.* Use your textbook and corresponding diagrams of the sheep brain to finish labeling the figure.

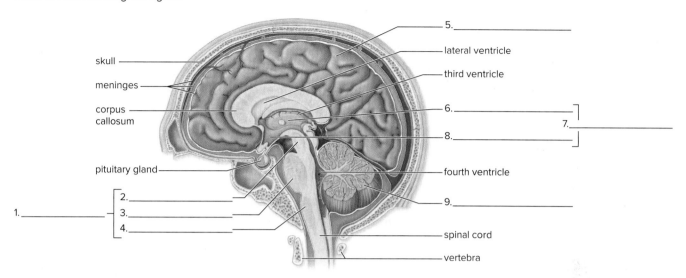

impulses are integrated to give us our senses of vision, hearing, smell, taste, and touch. Although not stated in Figure 24.4, the frontal lobe helps us remember smells of some significance to us. *Write the name of a sense next to the appropriate lobe in Figure 24.4.*

## The Spinal Cord

The spinal cord is a part of the central nervous system. It lies in the middorsal region of the body and is protected by vertebrae.

## Figure 24.4 The cerebral lobes.

Each lobe has centers for integrating nerve impulses received from a particular type of sense organ. Our five senses result from this activity of the brain.

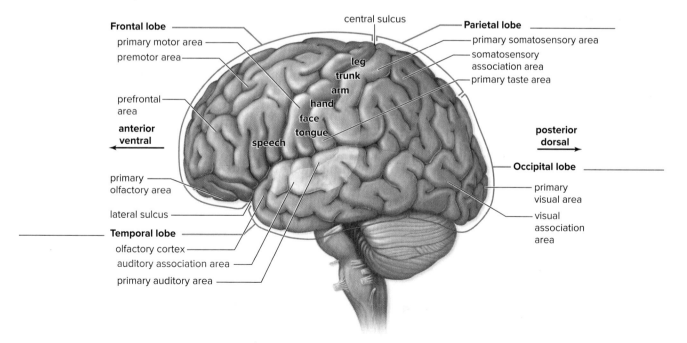

1. Examine a prepared slide of a cross section of the spinal cord under the lowest magnification possible. For example, some microscopes are equipped with a short scanning objective that enlarges about 3.5×, with a total magnification of 35×. If a scanning objective is not available, observe the slide against a white background with the naked eye.
2. Identify the following with the help of Figure 24.5:

   a. **Gray matter:** a central, butterfly-shaped area composed of axons of sensory neurons, interneurons, and motor neuron cell bodies.

   b. **White matter:** masses of long fibers that lie outside the gray matter and carry impulses up and down the spinal cord. In living animals, white matter appears white because an insulating myelin sheath surrounds long fibers.

**Figure 24.5   The spinal cord.**
Photomicrograph of spinal cord cross section.

Dr. Keith Wheeler/Science Source

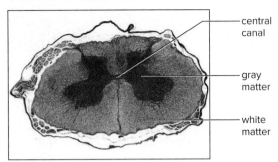

central canal

gray matter

white matter

## 24.2   Peripheral Nervous System

**Pre-Lab**

3. How is a reflex different from a regular nervous system response? _____
_____

The peripheral nervous system contains the cranial nerves and the spinal nerves. Twelve pairs of cranial nerves project from the inferior surface of the brain. The cranial nerves are largely concerned with nervous communication between the head, neck, and facial regions of the body and the brain. The 31 pairs of spinal nerves emerge from either side of the spinal cord (Fig. 24.6).

### Spinal Nerves

Each spinal nerve contains long fibers of sensory neurons and long fibers of motor neurons. In Figure 24.6, identify the following:

1. **Sensory neuron:** conducts nerve impulses from a sensory receptor to the spinal cord. The cell body of a sensory neuron is in the dorsal root ganglion.
2. **Interneuron:** lies completely within the spinal cord. Some interneurons have long fibers and conduct nerve impulses to and from the brain. The interneuron in Figure 24.6 transmits nerve impulses from the sensory neuron to the motor neuron.
3. **Motor neuron:** conducts nerve impulses from the spinal cord to an effector—in this case, a muscle. Muscle contraction is one type of response to stimuli.

Suppose you were walking barefoot and stepped on a locust thorn. Describe the pathway of information, starting with the pain receptor in your foot, that would allow you to both feel and respond to this unwelcome

stimulus. _____
_____
_____

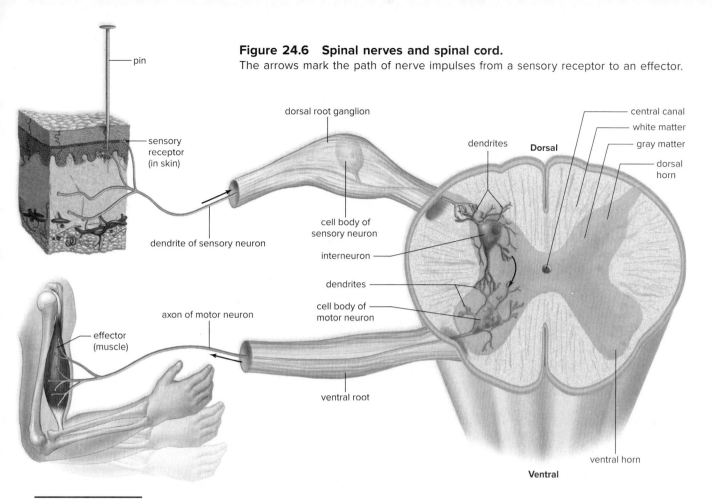

**Figure 24.6 Spinal nerves and spinal cord.**
The arrows mark the path of nerve impulses from a sensory receptor to an effector.

pin

sensory receptor (in skin)

dendrite of sensory neuron

dorsal root ganglion

cell body of sensory neuron

interneuron

dendrites

cell body of motor neuron

axon of motor neuron

effector (muscle)

ventral root

dendrites

**Dorsal**

central canal

white matter

gray matter

dorsal horn

ventral horn

**Ventral**

## Spinal Reflexes

A **reflex** is an involuntary and predictable response to a given stimulus that allows a quick response to environmental stimuli without communicating with the brain. When you touch a sharp tack, you immediately withdraw your hand (Fig. 24.6). When a spinal reflex occurs, a sensory receptor is stimulated and generates nerve impulses that pass along the three neurons mentioned earlier—the sensory neuron, interneuron, and motor neuron—until the effector responds. During the investigation of the knee-jerk reflex (Fig. 24.7), a receptor detects the tap, and sensory neurons conduct nerve impulses to interneurons in the spinal cord. The interneurons send a message via motor neurons to the effectors, muscles in the leg or foot. These reflexes are involuntary because the brain is not involved in formulating the response. *Consciousness* of the stimulus lags behind the response because information must be sent up the spinal cord to the brain before you can become aware of the tap.

### Experimental Procedure: Spinal Reflex

Although many reflexes occur in the body, only a tendon reflex is investigated in this Experimental Procedure. One easily tested tendon reflex involves

**Figure 24.7 Knee-jerk reflex.**
The quick response when the patellar tendon is stimulated by tapping with a reflex hammer indicates that a reflex has occurred.

Ph. Gerbier/Science Source

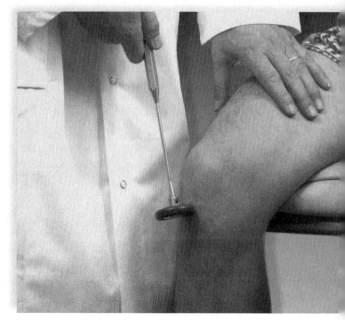

Knee-jerk (patellar) reflex

the **patellar tendon.** When this tendon is tapped with a reflex hammer (Fig. 24.7) or meter stick, the attached muscle is stretched. This causes a receptor to generate nerve impulses which are transmitted along sensory neurons to the spinal cord. Nerve impulses from the cord then pass along motor neurons and stimulate the muscle, causing it to contract. As the muscle contracts, it pulls on the tendon, causing movement of the bone opposite the joint. Receptors in other tendons, such as the Achilles tendon, respond similarly. Such reflexes help the body automatically maintain balance and posture.

### Knee-Jerk (Patellar) Reflex

1. Have the subject sit on a stool or table so that their legs hang freely.
2. Sharply tap one of the patellar tendons just below the patella (kneecap) with a meter stick.
3. In this relaxed state, does the leg flex (move toward the buttocks) or extend (move away from the buttocks)? _____

## 24.3 The Human Eye

### Pre-Lab

4. *With the help of your textbook, complete the labeling for Figure 24.8.*

The human eye is responsible for sight. Light rays enter the eye and strike the **rod cells** and **cone cells,** the photoreceptors for sight. The rods and cones generate nerve impulses that go to the brain via the optic nerve.

### Observation: The Human Eye

1. Examine a human eye model, and identify the structures listed in Table 24.2 and depicted in Figure 24.8.
2. List the structures light passes from outside the eye to the retina. _____
   _____
   _____

3. During **accommodation,** the lens rounds up to aid in viewing near objects or flattens to aid in viewing distant objects. Which structure holds the lens and is involved in accommodation? _____
   _____
   _____

4. **Refraction** is the bending of light rays so that they can be brought to a single focus. Which of the structures listed in Table 24.2 aid in refracting and focusing light rays? _____
   _____

5. Specifically, what are the sensory receptors for sight, and where are they located in the eye? _____
   _____

6. What structure takes nerve impulses to the brain from the rod cells and cone cells? _____
   _____

## Table 24.2 Parts of the Human Eye

| Part | Location | Function |
|---|---|---|
| Sclera | Outer layer of eye | Protects and supports eyeball |
| Cornea | Transparent portion of sclera | Refracts light rays |
| Choroid | Middle layer of eye | Absorbs stray light rays |
| Retina | Inner layer of eye | Contains receptors for sight |
| Rod cells | In retina | Make black-and-white vision possible |
| Cone cells | Concentrated in fovea centralis | Make color vision possible |
| Fovea centralis | Special region of retina | Makes acute vision possible |
| Lens | Interior of eye between cavities | Refracts and focuses light rays |
| Ciliary body | Extension from choroid | Holds lens in place; functions in accommodation |
| Iris | More anterior extension of choroid | Regulates light entrance |
| Pupil | Opening in middle of iris | Admits light |
| Humors (aqueous and vitreous) | Fluid media in anterior and posterior compartments, respectively, of eye | Transmit and refract light rays; support eyeball |
| Optic nerve | Extension from posterior of eye | Transmits impulses to occipital lobe of brain |

**Figure 24.8  Anatomy of the human eye.**
The sensory receptors for vision are the rod cells and cone cells present in the retina of the eye. *Fill in the missing labels with the help of your textbook.*

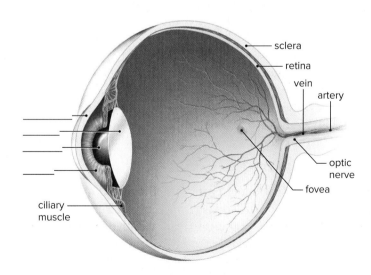

# The Blind Spot of the Eye

The **blind spot** occurs where the optic nerve fibers exit the retina. No vision is possible at this location because of the absence of rod cells and cone cells.

This Experimental Procedure requires a laboratory partner. Figure 24.9 shows a small circle and a cross several centimeters apart.

**Figure 24.9   Blind spot.**
This dark circle (or cross) will disappear at one location because there are no rod cells or cone cells at each eye's blind spot, where vision does not occur.

### Left Eye

1. Hold Figure 24.9 approximately 30 cm from your eyes. The cross should be directly in front of your left eye. If you wear glasses, keep them on.
2. Close your right eye.
3. Stare only at the cross with your left eye. You should also be able to see the circle in the same field of vision. Slowly move the paper toward you until the circle disappears.
4. Repeat the procedure as many times as needed to find the blind spot.
5. Then slowly move the paper closer to your eyes until the circle reappears. Because only your left eye is open, you have found the blind spot of your left eye.
6. With your partner's help, measure the distance from your eye to the paper when the circle first

   disappeared. Left eye: _____ cm

### Right Eye

1. Hold Figure 24.9 approximately 30 cm from your eyes. The circle should be directly in front of your right eye. If you wear glasses, keep them on.
2. Close your left eye.
3. Stare only at the circle with your right eye. You should also be able to see the cross in the same field of vision. Slowly move the paper toward you until the cross disappears.
4. Repeat the procedure as many times as needed to find the blind spot.
5. Then slowly move the paper closer to your eyes until the cross reappears. Because only your right eye is open, you have found the blind spot of your right eye.
6. With your partner's help, measure the distance from your eye to the paper when the cross first

   disappeared. Right eye: _____ cm

   Why are you unaware of a blind spot under normal conditions? Although the eye detects patterns of light and color, it is the brain that determines what we visually perceive. The brain interprets the visual input based in part on past experiences. In this exercise, you created an artificial situation in which you became aware of how your perception of the world is constrained by the eye's anatomy.

## Accommodation of the Eye

When the eye accommodates to see objects at different distances, the shape of the lens changes. The lens shape is controlled by the ciliary muscles attached to it. When you are looking at a distant object, the lens is in a flattened state. When you are looking at a closer object, the lens becomes more rounded. The elasticity of the lens determines how well the eye can accommodate. Lens elasticity decreases with increasing age, a condition called **presbyopia.** Presbyopia is the reason many older people need bifocals or reading glasses to see near objects.

### Experimental Procedure: Accommodation of the Eye

This Experimental Procedure requires a laboratory partner. It tests accommodation of either your left or your right eye.

1. Hold a pencil upright by the eraser and at arm's length in front of whichever of your eyes you are testing (Fig. 24.10).
2. Close the opposite eye.
3. Move the pencil from arm's length toward your eye.
4. Focus on the end of the pencil.
5. Move the pencil toward you until the end is out of focus. Measure the distance (in centimeters)

   between the pencil and your eye: _____ cm. This is the distance at which your eye can no longer accommodate for distance.

6. If you wear glasses, repeat this experiment without your glasses, and note the accommodation distance of your eye without glasses:

   _____ cm. (Contact lens wearers need not make these determinations, and they *should write the words* contact lens *in this blank.*)

7. The "younger" lens can easily accommodate for closer distances. The nearest point at which the end of the pencil can be clearly seen is called the **near point.** The more elastic the lens, the "younger" the eye

   (Table 24.3). How "old" is the eye you tested? _____

**Figure 24.10  Accommodation.**
When testing the ability of your eyes to accommodate to see a near object, always keep the pencil in this position.

| Table 24.3  Near Point and Age Correlation | | | | | | |
|---|---|---|---|---|---|---|
| Age (years) | 10 | 20 | 30 | 40 | 50 | 60 |
| Near point (cm) | 9 | 10 | 13 | 18 | 50 | 83 |

# 24.4  The Human Ear

### Pre-Lab

  5. *With the help of your textbook, complete the labeling for Figure 24.11.*

The human ear, whose parts are listed and depicted in Table 24.4 and Figure 24.11, serves two functions: hearing and balance.

## Table 24.4 Parts of the Human Ear

| Part | Medium | Function | Mechanoreceptor |
|---|---|---|---|
| *Outer ear* | Air | | |
|    Pinna | | Collects sound waves | — |
|    Auditory canal | | Filters air | — |
| *Middle ear* | Air | | |
|    Tympanic membrane and ossicles | | Amplify sound waves | — |
|    Auditory tube | | Equalizes air pressure | — |
| *Inner ear* | Fluid | | |
|    Semicircular canals | | Rotational equilibrium | Stereocilia embedded in cupula |
|    Vestibule (contains utricle and saccule) | | Gravitational equilibrium | Stereocilia embedded in otolithic membrane |
|    Cochlea (spiral organ) | | Hearing | Stereocilia embedded in tectorial membrane |

**Figure 24.11 Anatomy of the human ear.**

The outer ear extends from the pinna to the tympanic membrane. The middle ear extends from the tympanic membrane to the oval window. The inner ear encompasses the semicircular canals, the vestibule, and the cochlea. *Fill in the missing labels with the help of your textbook.*

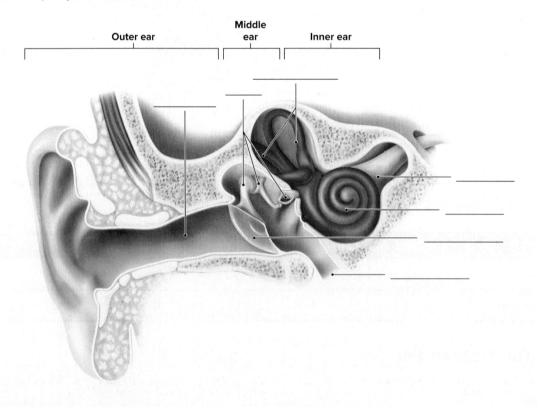

## Observation: The Human Ear

Examine a human ear model, and find the structures depicted in Figure 24.11 based on the information given in Table 24.4.

When you hear, sound waves are picked up by the tympanic membrane and amplified by the malleus, incus, and stapes. This creates pressure waves in the canals of the cochlea that lead to stimulation of hair cells, the receptors for hearing. Hair cells in the utricle and saccule of the vestibule and in semicircular canals are receptors for equilibrium (i.e., balance). Nerve impulses from the cochlea travel by way of the cochlear nerve and the vestibular nerve to the brain and eventually are interpreted by the brain as sound.

## Experimental Procedure: Locating Sound

Humans locate the direction of sound according to how fast it is detected by either or both ears. A difference in the hearing ability of the two ears can lead to a mistaken judgment about the direction of sound. You and a laboratory partner should perform this Experimental Procedure on each other. Enter the data for *your* ears, not your partner's ears, in the spaces provided.

1. Ask the subject to be seated, with eyes closed. Then strike a tuning fork or rap two spoons together at the five locations listed in number 2. *Use a random order.*
2. Ask the subject to give the exact location of the sound in relation to their head. Record the subject's perceptions when the sound is
   a. directly below and behind the head. _____
   b. directly behind the head. _____
   c. directly above the head. _____
   d. directly in front of the face. _____
   e. to the side of the head. _____

3. Is there an apparent difference in hearing between your two ears? _____

# 24.5   Sensory Receptors in Human Skin

## Pre-Lab

6. *With the help of your textbook, complete the labeling for Figure 24.12.*

The sensory receptors in human skin respond to touch, pain, temperature, and pressure (Fig. 24.12). There are individual sensory receptors for each of these stimuli, as well as free nerve endings able to respond to pressure, pain, and temperature.

## Sense of Touch

The dermis of the skin contains touch receptors, whose concentration differs in various parts of the body.

## Experimental Procedure: Sense of Touch

You will need a laboratory partner to perform this Experimental Procedure. Enter *your* data, not the data of your partner, in the spaces provided.

1. Ask the subject to be seated, with eyes closed.
2. Then test the subject's ability to discriminate between the two points of a hairpin, a pair of scissors, or calipers at the four locations listed in number 7.
3. Gently touch the given skin area with the points of the hairpin, scissors, or calipers closed (making one point).

**Figure 24.12**
**Sensory receptors in the skin.**
Each type of receptor shown responds primarily to a particular stimulus.

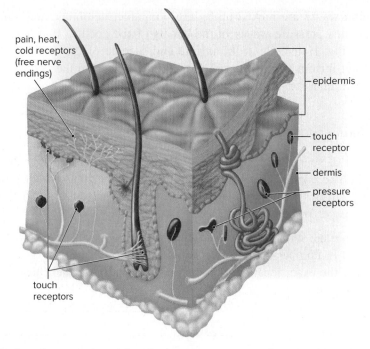

pain, heat, cold receptors (free nerve endings)

epidermis

touch receptor

dermis

pressure receptors

touch receptors

4. Ask the subject whether the experience involves one or two touch sensations.
5. If the subject reports one touch sensation, separate the points of the test item slightly. Gently touch the given skin area again and repeat number 4.
6. Repeat number 5 until the subject reports feeling two touch sensations.
7. Record the shortest distance between the hairpin, scissors, or caliper points for a two-point discrimination.

   **a.** Forearm: _____ mm          **c.** Index finger: _____ mm

   **b.** Back of the neck: _____ mm          **d.** Back of the hand: _____ mm

8. Which of these areas apparently contains the greatest density of touch receptors? _____

   Why is this useful? _____

9. Do you have a sense of touch at every point in your skin? _____ Explain. _____

   _____

## Sense of Heat and Cold

Temperature receptors respond to a change in temperature. If the temperature remains constant, the temperature receptors will experience accommodation to that temperature.

### Experimental Procedure: Sense of Heat and Cold

1. Obtain three 1,000-ml beakers and add 500 ml *cold* (tap with *a bit* of ice added to it) water to one, *room temperature* water to the second, and *warm water* (45°–50°C) to the third.
2. Immerse your left hand in the cold-water beaker and your right hand in the warm-water beaker for 30 seconds.
3. Then place both hands in the beaker with room-temperature tap water.
4. Record the sensation in the right and left hands.

   **a.** Right hand: _____

   **b.** Left hand: _____

5. Explain your results: _____

   _____

## 24.6 Human Chemoreceptors

### Pre-Lab

**7.** Which portions of the brain interpret taste and smell? _____

The taste receptors, called *taste buds*, located in the mouth, and the smell receptors, called *olfactory cells*, located in the nasal cavities, are the chemoreceptors that respond to molecules in the air and water.

### Experimental Procedure: Sense of Taste and Smell

You will need a laboratory partner to perform the following procedures. It will not be necessary for all tests to be performed on both partners. You should take turns being either the subject or the experimenter.

#### Taste and Smell

1. Students work in groups. Each group has one experimenter and several subjects.
2. The experimenter should obtain a LifeSavers candy from the various flavors available, without letting the subject know what flavor it is.
3. The subject closes both eyes and holds their nose.
4. The experimenter gives the LifeSavers candy to the subject, who places it on their tongue.
5. The subject, while still holding their nose, guesses the flavor of the candy. The experimenter records the guess in Table 24.5.
6. The subject releases their nose and guesses the flavor again. The experimenter records the guess and the actual flavor in Table 24.5.

### Table 24.5  Taste and Smell Experiment

| Subject | Actual Flavor | Flavor While Holding Nose | Flavor After Releasing Nose |
|---------|---------------|---------------------------|-----------------------------|
| 1 | | | |
| 2 | | | |
| 3 | | | |
| 4 | | | |
| 5 | | | |

### Conclusions: Sense of Taste and Smell

• From your results, how would you say smell affects the taste of LifeSavers candy? _____

_____

• What do you conclude about the effect of smell on your sense of taste? _____

_____

_____

_____

_____  1. Is the sciatic nerve part of the central nervous system or the peripheral nervous system?

_____  2. What part of the brain is divided into right and left hemispheres?

_____  3. What portion of the brain looks like a tree in cross section and what is its function?

_____  4. What is the most posterior portion of the brain stem?

_____  5. What cerebral lobe is associated with the sense of vision?

_____  6. What structures protect the spinal cord?

_____  7. What makes the white matter of the spinal cord appear white?

_____  8. What type of neuron is found completely within the central nervous system?

_____  9. What type of neuron is responsible for transmitting nerve impulses from the spinal cord to an effector?

_____  10. What part of the eye refracts and focuses light rays?

_____  11. What part of the ear contains the sensory receptors for hearing?

_____  12. What is the anatomical name for the ear structure that detects sound waves?

_____  13. What layer of the skin contains Meissner and Pacinian corpuscles?

_____  14. What are the names of the receptors that respond to molecules in air and water?

## Thought Questions

15. Explain why people who develop cataracts (cloudy eye lenses) require surgery to implant new lenses.

16. Explain why an inner ear infection can influence a person's sense of equilibrium.

17. Describe why a cut to the epidermis will produce minimal amounts of pain, if any at all.

# 25

## Effects of Pollution
## on Ecosystems

---

### Learning Outcomes

**25.1 Studying the Effects of Pollutants**
- Determine the effects of pollution on species composition and diversity in ecosystems.
- Using data from a seed germination experiment, predict the effect of acid rain on crop yield.
- Using *Gammarus* as the subject, predict the effect of acid rain on food chains involving animals.

**25.2 Studying the Effects of Cultural Eutrophication**
- Predict the effect of cultural eutrophication on food chains.

---

## Introduction

### Pre-Lab

**1.** What are the causes of thermal pollution, acid pollution, and cultural eutrophication?

_____

_____

_____

_____

This laboratory will consider three causes of aquatic pollution: thermal pollution, acid pollution, and cultural eutrophication. **Thermal pollution** occurs when water temperature rises above normal for a given environment. As water temperature rises, the amount of oxygen dissolved in water decreases, possibly depriving organisms of an adequate supply of oxygen. Deforestation, soil erosion, and climate change contribute to thermal pollution, but the chief cause is use of water from a lake or the ocean as a coolant for the waste heat of a power plant.

When sulfur dioxide and nitrogen oxides enter the atmosphere, usually from the burning of fossil fuels, they are converted to acids, which return to Earth as **acid deposition** (acid rain or snow). Acid deposition kills plants, aquatic invertebrates, and also decomposers, threatening the entire ecosystem.

**Figure 25.1 Cultural eutrophication.**
Eutrophic lakes tend to have large populations of algae and rooted plants.

Pat Watson/McGraw Hill

**Cultural eutrophication,** or overenrichment, is due to runoff from agricultural fields, wastewater from sewage treatment plants, and even excess detergents. These sources of excess nutrients cause an algal bloom seen as a green scum on a lake (Fig. 25.1). When algae overgrow and die, decomposition robs the lake of oxygen, causing a fish die-off.

# 25.1 Studying the Effects of Pollutants

## Pre-Lab

2. How do you predict species composition and diversity will be affected by pollution? _____

_____

3. What will happen to enzymes within seeds if they are exposed to low pH conditions in the Effect of Acid Rain on Seed Germination Experimental Procedure? _____

_____

We are going to study the effects of pollution by observing its effects on hay infusion organisms, on seed germination, and on an animal called *Gammarus*.

---

## Study of Hay Infusion Culture

A hay infusion culture (hay soaked in water) contains various microscopic organisms, but we will be concentrating on how the pollutants in our study affect the protozoan populations in the culture. We will consider both of these aspects:

**Species composition:** number of different types of microorganisms.
**Species diversity:** composition and the abundance of each type of microorganism.

### Experimental Procedure: Effect of Pollutants on a Hay Infusion Culture

During this Experimental Procedure, you will examine, by preparing a wet mount, hay infusion cultures that have been treated in the following manner.

1. **Control culture:** This culture simulates the species composition and diversity of an untreated culture. Prepare a wet mount and answer the following questions:

    With the assistance of Figure 25.2 and Figure 3.9 on page 27, identify as many different types of microorganisms as possible in the hay infusion culture. State whether species composition is high, medium, or low. Record your estimation in the second column of Table 25.1. Do you judge species diversity to be high, medium, or low? Record your estimation in the third column of Table 25.1.

2. **Oxygen-deprived culture:** Thermal pollution causes water to be oxygen-deprived; therefore, when we study the effects of low oxygen on a hay infusion culture, we are studying an effect of thermal pollution. Prepare a

# Figure 25.2  Microorganisms in hay infusion cultures.

Organisms are not to size.

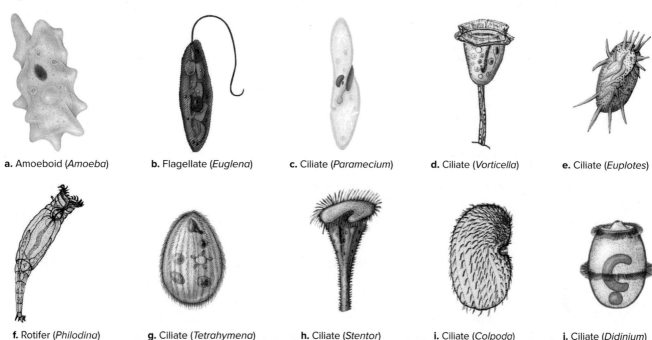

**a.** Amoeboid (*Amoeba*)  **b.** Flagellate (*Euglena*)  **c.** Ciliate (*Paramecium*)  **d.** Ciliate (*Vorticella*)  **e.** Ciliate (*Euplotes*)

**f.** Rotifer (*Philodina*)  **g.** Ciliate (*Tetrahymena*)  **h.** Ciliate (*Stentor*)  **i.** Ciliate (*Colpoda*)  **j.** Ciliate (*Didinium*)

wet mount of this culture and determine if there is a difference in species composition and diversity. Record the species composition and species diversity as high, medium, or low in Table 25.1.

3. **Acidic culture:** In this culture, the pH has been adjusted to 4 with sulfuric acid ($H_2SO_4$). This simulates the effect of acid rain on a hay infusion culture. Prepare a wet mount of this culture and determine if there is a difference in species composition and diversity. Record the species composition and species diversity as high, medium, or low in Table 25.1.

4. **Enriched culture:** More organic nutrients have been added to this culture. These nutrients will cause the algae population, which is food for most protozoans, to increase. In the short term, their species composition should increase. Eventually, as the algae die off, decomposition will rob the water of oxygen and the protozoans may start to die off. Prepare a wet mount of this culture and determine if there is a difference in species composition and diversity. Record the species composition and species diversity as high, medium, or low in Table 25.1.

| Table 25.1  Effect of Pollution on a Hay Infusion Culture | | | |
|---|---|---|---|
| **Type of Culture** | **Species Composition (High, Medium, or Low)** | **Species Diversity (High, Medium, or Low)** | **Explanation** |
| Control | | | |
| Oxygen-deprived | | | |
| Acidic | | | |
| Enriched | | | |

## Conclusions

Complete the fourth column in Table 25.1 by explaining the physiological or environmental reasons why pollution affected the culture.

## Effect of Acid Rain on Seed Germination

Seeds depend on favorable environmental conditions of temperature, light, and moisture to germinate, grow, and reproduce. Like any other biological process, germination requires enzymatic reactions that can be adversely affected by an unfavorable pH.

### Experimental Procedure: Effect of Acid Rain on Seed Germination

In this Experimental Procedure, we will test whether there is a relationship between acid concentration and germination.

Your instructor has placed 20 sunflower seeds in each of five containers with water of increasing acidity: 0% vinegar (tap water), 1% vinegar, 5% vinegar, 20% vinegar, and 100% vinegar.

1. Test and record the pH of solutions having the vinegar concentrations noted above. Record the pH of each solution in Table 25.2.
2. Count the number of germinated sunflower seeds in each container, and complete Table 25.2.

### Table 25.2 Effect of Increasing Acidity on Germination of Sunflower Seeds

| Concentration of Vinegar | pH | Number of Seeds That Germinated | Percent Germination |
|---|---|---|---|
| 0% | | | |
| 1% | | | |
| 5% | | | |
| 20% | | | |
| 100% | | | |

### Conclusion: Effect of Increasing Acidity on Germination of Sunflower Seeds

- As you know, each enzyme has an optimum pH. Explain why acid rain is expected to inhibit metabolism, and, therefore, seedling development. _____

## Study of *Gammarus*

A small crustacean called *Gammarus* lives in ponds and streams (Fig. 25.3) where it feeds on debris, algae, or anything smaller than itself, such as some of the microorganisms in Figure 25.2. In turn, fish like to feed on *Gammarus*.

**Figure 25.3   *Gammarus*.**
*Gammarus* is a type of crustacean classified in a subphylum that also includes shrimp.

William Amos/Photoshot

- Add 25 ml of spring water to a beaker and record the pH of the water. _____ pH

- Add four *Gammarus* to the container. Do they all use their legs in swimming? _____

- Which legs are used in jumping and climbing? _____

- What do *Gammarus* do when they "bump" into each other? _____

### Control Sample

After observing *Gammarus,* decide what behaviors are most often observed. During a 5-minute time span, total the amount of time spent doing each of these behaviors.

| Behaviors | Amount of Time | Total Time |
|---|---|---|
| 1. _____ | _____ | _____ |
| 2. _____ | _____ | _____ |
| 3. _____ | _____ | _____ |

### Test Sample

If so directed by your instructor, put a *Gammarus* in a beaker of spring water adjusted to pH 4 by adding vinegar. During a 5-minute time span, total the amount of time spent doing each of these behaviors.

| Behaviors | Amount of Time | Total Time |
|---|---|---|
| 1. _____ | _____ | _____ |
| 2. _____ | _____ | _____ |
| 3. _____ | _____ | _____ |

### Conclusion

- Draw a conclusion from this study: _____
_____

## Conclusions: Studying the Effects of Pollutants

- Give an example to show that the hay infusion study pertains to real ecosystems. _____
_____

- What are the potential consequences of acid rain on crops that reproduce by seeds? _____
_____

  On the food chains of the ocean? _____

- How does the addition of nutrients affect species composition and species diversity of an ecosystem over time? _____

## 25.2 Studying the Effects of Cultural Eutrophication

### Pre-Lab

4. Would an increase in *Daphnia* in a body of water increase or decrease the algae population? _____

5. What device will be used to measure the algal population in the *Daphnia* Feeding on *Chlorella* Experimental Procedure? What does it do? _____

_____

*Chlorella,* the green alga used in this study, is considered to be representative of algae in bodies of fresh water. The crustacean *Daphnia* feeds on green algae such as *Chlorella* (Fig. 25.4). First, you will observe how *Daphnia* feeds, and then you will determine the extent to which *Daphnia* could keep the effects of cultural eutrophication from occurring in a hypothetical example.

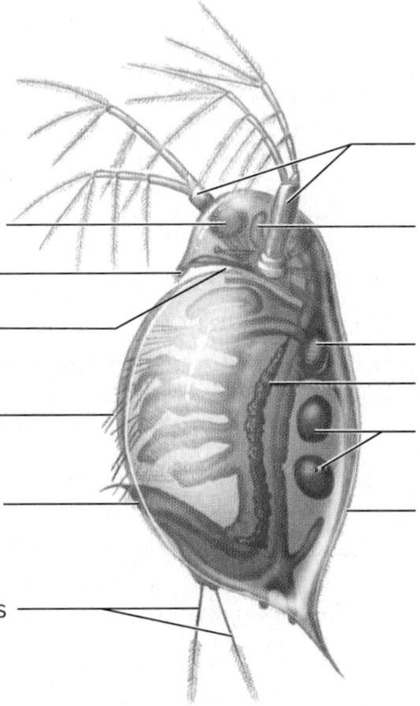

### Observation: Daphnia *Feeding*

1. Place a small pool of petroleum jelly in the center of a small petri dish.
2. Use a dropper to take a *Daphnia* from the stock culture, place it on its back (covered by water) in the petroleum jelly, and observe it under the stereomicroscope (Fig. 25.4).
3. Add a drop of *carmine solution,* and observe how the *Daphnia* filters the "food" from the water and passes it through the gut. The gut is more visible if you push the animal onto its side. In this position, you may also observe the heart beating in the region above the gut and just behind the head.
4. Allow the *Daphnia* to filter-feed for up to 30 minutes, and observe the progress of the carmine particles through the gut.

**Figure 25.4   Anatomy of *Daphnia*.**

### Experimental Procedure: Daphnia *Feeding on* Chlorella

This exercise requires the use of a spectrophotometer. Absorbance will be a measure of the algal population level; the greater the number of algal cells, the greater the absorbance.

1. Obtain two spectrophotometer tubes (cuvettes) and a Pasteur pipet.
2. Fill one of the cuvettes with distilled water, and use it to zero the spectrophotometer. Save this tube for step 6.
3. Use the Pasteur pipet to fill the second cuvette with *Chlorella.* Gently aspirate and expel the sample several times (without creating bubbles) to give a uniform dispersion of the algae.
4. Add 10 hungry *Daphnia,* and following your instructor's directions, immediately measure the absorbance with the spectrophotometer. Record your reading in the first column of Table 25.3.
5. Remove the cuvette with the *Daphnia* to a safe place in a test tube rack. Allow the *Daphnia* to feed for 30 minutes.
6. Rezero the spectrophotometer with the distilled water cuvette.
7. Measure the absorbance of the experimental cuvette again. Record your data in the second column of Table 25.3, and explain your results in the third column.

## Table 25.3  Spectrophotometer Data/*Daphnia* Feeding on *Chlorella*

| Absorbance Before Feeding | Absorbance After Feeding | Explanation |
|---|---|---|
|  |  |  |

### Experimental Procedure: Case Study in Cultural Eutrophication

The following problem will test your understanding of the ecological value of a single species—in this case, *Daphnia*.

1. Assume that developers want to build condominium units on the shores of Silver Lake. Homeowners in the area have asked the regional council to determine how many units can be built without altering the nature of the lake. As a member of the council, you have been given the following information:

    The current population of *Daphnia,* 10 animals/liter, presently filters 24% of the lake per day, meaning that it removes this percentage of the algal population per day. This is sufficient to keep the lake essentially clear. Predation—the eating of the algae—will allow the *Daphnia* population to increase to no more than 50 animals/liter. Therefore, 50 *Daphnia*/liter will be available for feeding on the increased number of algae that would result from building the condominiums.

    Using this information, complete Table 25.4.

## Table 25.4  *Daphnia* Filtering

| Number of *Daphnia*/Liter | Percent of Lake Filtered |
|---|---|
| 10 | 24% |
| 50 |  |

2. The sewage system of the condominiums will add nutrients to the lake. Phosphorus output will be 1 kg per day for every 10 condominiums. This will cause a 30% increase in the algal population. Using this information, complete Table 25.5.

## Table 25.5  Cultural Eutrophication

| Number of Condominiums | Phosphorus Added | Increase in Algal Population |
|---|---|---|
| 10 | 1 kg | 30% |
| 20 |  |  |
| 30 |  |  |
| 40 |  |  |
| 50 |  |  |

### Conclusion: Cultural Eutrophication

- Assume that phosphorus is the only nutrient that will cause an increase in the algal population and that *Daphnia* is the only type of zooplankton available to feed on the algae. How many condominiums would you allow the developer to build? _____

_____

_____  1. What is produced when sulfur dioxide and nitrogen oxides, formed by burning fossil fuels, enter the atmosphere?

_____  2. What kind of pollution is produced when water from a lake or the ocean is used to disperse waste heat?

_____  3. What happens to the oxygen level in water when the temperature increases?

_____  4. What typically occurs when there is a large runoff from an agricultural field that enters a body of water?

_____  5. Which population experiences growth referred to as a "bloom" when excess nutrients enter a body of water?

_____  6. Explain what can happen to the fish population of a pond if there is excessive algae growth.

_____  7. What do we call the number of different microorganisms observed and their relative abundance in the hay infusion culture?

_____  8. What chemical was used to simulate the effect of acid rain on a hay infusion culture?

_____  9. What does an unfavorable pH change that will inhibit seed germination?

_____  10. In the Experimental Procedure on the effect of acid rain on seed germination, if 18 of the sunflower seeds germinated in the 1% vinegar solution, what percent germinated?

_____  11. What kind of animal is *Gammarus*?

_____  12. What will *Daphnia* do that might prevent cultural eutrophication from occurring?

_____  13. Are there more or fewer algal cells present in a water sample as the absorption of light decreases?

_____  14. What element from the Silver Lake condominiums' sewage output contributes to the increased algal population?

## Thought Questions

15. A "dead zone" forms where the Mississippi River empties into the Gulf of Mexico due to nutrient runoff into the river that is carried to the Gulf. Explain the cause of the dead zone and the impact it has upon humans.

16. Pollutants often affect the producers in an ecosystem first and then have far-ranging effects. Use acid rain and a food chain to illustrate your understanding of the impact pollutants have after they enter an ecosystem.

17. Describe how the cultural eutrophication case study illustrates the need to have a balance of population sizes in an ecosystem.

# A

# Preparing a Laboratory Report/ Laboratory Report Form

A laboratory report has the sections noted in the outline that follows. Use this outline and a copy of the Laboratory Report Form on page A–3 to help you write a report assigned by your instructor. In general, do not use the words *we, my, our, your, us,* or *I* in the report. Use scientific measurements and their proper abbreviations. (For example, cm is the proper notation for centimeter and sec is correct for seconds.)

1. **Introduction:** Tell the reader what the experiment was about.
   a. **Background information:** Begin by giving an overview of the topic. Review the Introduction section of the laboratory (and/or the introduction to the section for which you are writing the report). Do not copy the information, but use it to get an idea about what background information to include.

   > For example, suppose you are doing a laboratory report titled "Solar Energy" in Laboratory 6 (Photosynthesis). You might give a definition of photosynthesis and explain the composition of white light.

   b. **Purpose:** Think about the steps of the experiment and state what the experiment was about. Include the independent and dependent variable.

   > For example, you might state that the purpose of the photosynthesis experiment was to determine the effect of white light versus green light on the photosynthetic rate. The independent variable was the color of light and the dependent variable was the rate of photosynthesis.

   c. **Hypothesis:** Consider the expected results of the experiment in order to state the hypothesis. It's possible that the Introduction section of the laboratory might hint at the expected results. State this in the form of a hypothesis.

   > For example, you might state: It was hypothesized that white light would be more effective than green light for photosynthesis.

2. **Method:** Tell the reader how you did the experiment.
   a. **Equipment and sample used:** Use any illustrations in the laboratory manual that show the experimental setup to describe the equipment and the sample (subject) used.

   > For example, for the photosynthesis experiment look at Figure 6.4 and describe what you see. You might state that a 150-watt lamp was the source of white light directed at *Elodea*, an aquatic plant, placed in a test tube filled with a solution of sodium bicarbonate ($NaHCO_3$). A beaker of water placed between the lamp and the test tube was a heat absorber.

   b. **Collection of data:** Think about what you did during the experiment, such as what you observed or what you measured. Look at any tables you filled out in order to recall how the data were collected and what control(s) were used.

   > For example, for the photosynthesis experiment, you might state that the rate of photosynthesis was determined by the amount of oxygen released and was measured by how far water moved in a side arm placed in a stopper of the test that held *Elodea*. A control was the same experimental setup, except the test tube lacked *Elodea*.

3. **Results:** Present the data in a clear manner.
   a. **Graph or table:** If at all possible, show your data in table or graph form. You could reproduce a table you filled in or a graph you drew to show the results of the experiment. Be sure to include the title of the table; do not include any interpretation of the data column in the table.

   For example, for the photosynthesis experiment you might reproduce Table 6.3.

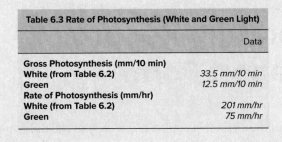

| Table 6.3 Rate of Photosynthesis (White and Green Light) | |
| --- | --- |
| | Data |
| **Gross Photosynthesis (mm/10 min)** | |
| **White (from Table 6.2)** | *33.5 mm/10 min* |
| **Green** | *12.5 mm/10 min* |
| **Rate of Photosynthesis (mm/hr)** | |
| **White (from Table 6.2)** | *201 mm/hr* |
| **Green** | *75 mm/hr* |

   Or for Figure 5.3 Effect of Temperature on Enzyme Activity, you might show this graph as your results.

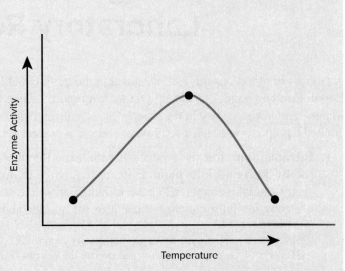

   b. **Description of data:** Examine your data, and decide what they tell you. Then, below any table or graph, add a description to help the reader understand what the table or graph is showing. Define any terms in the table that are not readily understandable.

   For example, below Table 6.3 you might state that these data indicate that the rate of photosynthesis with white light is faster than with green light. Also, you should define gross photosynthesis. Or below the graph that shows the effect of temperature on enzyme activity, you might state that these data show that the rate of enzymatic activity speeds up until boiling occurs and then it drops off.

4. **Conclusion:** Tell if the data support or do not support the hypothesis.
   a. **Compare the hypothesis with the data:** Do your data agree or disagree with the hypothesis?

   For example, for the photosynthesis experiment you might state: These results support the hypothesis that white light is more effective for photosynthesis than green light.

   b. **Explanation:** Explain why you think you obtained these results. Look at any questions you answered while in the laboratory, and use them to help you decide on an appropriate explanation.

   For example, the answers to the questions in 6.2 Solar Energy might help you state that white light gives a higher rate of photosynthesis because it contains all the visible light rays. Green light gives a lower rate because green plants such as *Elodea* do not absorb green light.

   If your results do not support the hypothesis, explain why you think this occurred.

   In this instance you might state that while white light contains all visible light rays and green light is not absorbed by a green plant, the experiment did not support the hypothesis because of failure to use a heat absorber when doing the green light experiment.

**Laboratory Report for** _____

_____

1. **Introduction**
   **a.** Background information




   **b.** Purpose




   **c.** Hypothesis




2. **Method**
   **a.** Equipment used




   **b.** Collection of data

# 3. Results

**a.** Graph or table
(Place these on attached sheets.)

**b.** Description of data

# 4. Conclusion

**a.** Compare the hypothesis with the data

**b.** Explanation

**c.** Conclusion

# B

# Metric System

| Unit and Abbreviation | Metric Equivalent | Approximate English-to-Metric Equivalents | Units of Temperature |
|---|---|---|---|
| **Length** | | | |
| nanometer (nm) | $= 10^{-9}$ m $(10^{-3}$ μm) | | |
| micrometer (μm) | $= 10^{-6}$ m $(10^{-3}$ mm) | | |
| millimeter (mm) | $= 0.001$ $(10^{-3})$ m | | |
| centimeter (cm) | $= 0.01$ $(10^{-2})$ m | 1 inch $= 2.54$ cm | |
| | | 1 foot $= 30.5$ cm | |
| meter (m) | $= 100$ $(10^{2})$ cm | 1 foot $= 0.30$ m | |
| | $= 1,000$ mm | 1 yard $= 0.91$ m | |
| kilometer (km) | $= 1,000$ $(10^{3})$ m | 1 mi $= 1.6$ km | |
| **Weight (mass)** | | | |
| nanogram (ng) | $= 10^{-9}$ g | | |
| microgram (μg) | $= 10^{-6}$ g | | |
| milligram (mg) | $= 10^{-3}$ g | | |
| gram (g) | $= 1,000$ mg | 1 ounce $= 28.3$ g | |
| | | 1 pound $= 454$ g | |
| kilogram (kg) | $= 1,000$ $(10^{3})$ g | $= 0.45$ kg | |
| metric ton (t) | $= 1,000$ kg | 1 ton $= 0.91$ t | |
| **Volume** | | | |
| microliter (μl) | $= 10^{-6}$ l $(10^{-3}$ ml) | | |
| milliliter (ml) | $= 10^{-3}$ l | 1 tsp $= 5$ ml | |
| | $= 1$ cm$^3$ (cc) | 1 fl oz $= 30$ ml | |
| | $= 1,000$ mm$^3$ | | |
| liter (l) | $= 1,000$ ml | 1 pint $= 0.47$ l | |
| | | 1 quart $= 0.95$ l | |
| | | 1 gallon $= 3.79$ l | |
| kiloliter (kl) | $= 1,000$ l | | |

**Units of Temperature**

| °F | | °C |
|---|---|---|
| | 230 | 110 |
| | 220 | |
| 212° — 210 | | 100 — 100° |
| | 200 | |
| | 190 | 90 |
| | 180 | 80 |
| | 170 | |
| 160° — 160 | | 70 — 71° |
| | 150 | |
| | 140 | 60 |
| 134° — 130 | | 57° |
| | 120 | 50 |
| 105.8° — 110 | | 41° |
| 98.6° — 100 | | 40 |
| | | 37° |
| | 90 | 30 |
| | 80 | |
| | 70 | 20 |
| 56.66° — 60 | | 13.7° |
| | 50 | 10 |
| | 40 | |
| 32° — 30 | | 0 — 0° |
| | 20 | |
| | 10 | -10 |
| | 0 | -20 |
| | -10 | |
| | -20 | -30 |
| | -30 | |
| | -40 | -40 |

**Common Temperatures**

| °C | °F | |
|---|---|---|
| 100 | 212 | Water boils at standard temperature and pressure. |
| 71 | 160 | Flash pasteurization of milk. |
| 57 | 134 | Highest recorded temperature in the United States, Death Valley, July 10, 1913. |
| 41 | 105.8 | Average body temperature of a marathon runner in hot weather. |
| 37 | 98.6 | Human body temperature. |
| 13.7 | 56.66 | Human survival is still possible at this temperature. |
| 0 | 32.0 | Water freezes at standard temperature and pressure. |

*See temperature scale

To convert temperature scales:

$$°C = \frac{5}{9}(°F - 32)$$

$$°F = \frac{9}{5}(°C + 32)$$

## Practical Examination Answer Sheet

1. _____
2. _____
3. _____
4. _____
5. _____
6. _____
7. _____
8. _____
9. _____
10. _____
11. _____
12. _____
13. _____
14. _____
15. _____
16. _____
17. _____
18. _____
19. _____
20. _____
21. _____
22. _____
23. _____
24. _____
25. _____

26. _____
27. _____
28. _____
29. _____
30. _____
31. _____
32. _____
33. _____
34. _____
35. _____
36. _____
37. _____
38. _____
39. _____
40. _____
41. _____
42. _____
43. _____
44. _____
45. _____
46. _____
47. _____
48. _____
49. _____
50. _____

# Index

Note: Page numbers followed by *f* refer to figures; page numbers followed by *t* refer to tables.

Corpus callosum, 334, 335*f*, 337*f*
Cortex
    flowering plant roots, 202, 202*f*
    flowering plant stems, 205, 205*f*,
      209, 209*f*
Cotyledons, 199*f*, 200
Coverslips, 25, 25*f*
Crabs, 229*f*
Cranial nerves, 333, 333*f*, 335*f*, 338
Cranium, 141, 142*t*
Crayfish, 230–231, 230*f*
Crenation, 39, 39*f*
Cretaceous period, 131*t*
Cro-Magnons, 141
Crop
    earthworms, 226, 227*f*
    grasshoppers, 232*f*
    pigeon, 255*f*
Crossing-over, 81, 85*f*
Crustaceans, 229, 229*f*, 230, 352–355, 354*f*
Cultural eutrophication, 350, 354–355
Cuticles, leaf, 210*f*
Cyanobacteria, 151–153
Cycads, 183, 183*f*
Cyprus, 184
Cystic fibrosis, 119–120
Cytochrome *c*, 143, 143*f*
Cytokinesis
    in animal cells, 68, 70, 70*f*
    defined, 65, 65*f*
    in plant cells, 70, 73, 73*f*
Cytoplasm
    buffering properties, 42–43
    defined, 29, 30
    division during mitosis, 65, 65*f*, 70, 70*f*
    in plant cells, 34*f*
    in protists, 160*f*
Cytosine, 103, 103*t*

### D

*Daphnia*, 354–355, 354*f*
Data, defined, 3
Daughter cells
    meiosis, 77, 78, 79*f*, 84, 85*f*
    mitosis, 65, 67, 84, 85*f*
Daughter chromosomes, 67, 85*f*
Daughter colonies, *Volvox*, 155, 155*f*
Decomposers, 148
Degradation reactions, 46, 46*f*
Denaturation, 48
Deoxyribonucleic acid (DNA)
    genetic disorders and, 111–113
    isolating, 110–111
    replication, 66–67, 102, 103–104, 104*f*
    structure, 101, 102–103, 102*f*
    triplet code, 108
Depth of field (microscope), 24, 24*f*

Dermal tissue, flowering plants,
    197, 197*t*, 198
Descent of testes, 322
Detritus, 226
Deuterostomes, 216, 217*f*
Devonian period, 131*t*
Dialysis tubing, 36–37
Diameter measurement, 10
Diameter of field (microscope), 23
Diaphragm
    biological benefit, 259
    comparative anatomy, 254
    fetal pig, 266, 267*f*, 269*f*, 271*f*
    microscope, 21, 22
    rat, 256*f*
Diaries, 285, 286, 289
Diatoms, 154*f*, 158, 158*f*
*Didinium*, 27*f*, 351*f*
Diencephalon, 252*f*, 334, 335*f*
Diet, recording, 286–288
*Difflugia*, 27*f*
Diffusion, 34–41
Digestion, 276–283
Digestive glands, 231, 237*f*, 238, 354*f*
Digestive systems
    comparative anatomy, 257
    crayfish, 231
    fetal pig, 330
    frog, 247
    human, 276*f*
    sea star, 237, 237*f*
Digits, fetal pig, 263*f*, 267*f*
Dihybrid crosses, 92, 93*f*
*Dileptus*, 27*f*
Dinoflagellates, 158, 159*f*
Dinosaurs, 131*t*, 134*f*
Diploid cells, 78, 83
Diploid life cycles, 169
Discrimination, 346
Diseases
    bacterial, 148–150
    fungal, 165, 165*f*
    genetic, 111–113, 119–121
    protozoan, 160*f*
Dissecting microscopes, 16, 17–19, 18*f*
Dissection instructions, fetal pig, 261,
    266, 267*f*, 331
DNA. *See* Deoxyribonucleic acid (DNA)
Dominant alleles, 87, 115
Dominant disorders, 119–120
Dorsal, defined, 262
Dorsal aorta, 247, 254*f*, 312*f*, 314*f*
Dorsal blood vessel, earthworms,
    226, 227*f*, 228*f*
Dorsal fins, 243, 244*f*, 254*f*
Dorsal horn, 339*f*
Dorsal root ganglia, 339*f*
Dorsal spinal cord, 137*f*

Dorsal tubular nerve cord, 241, 241*f*, 244*f*
Double fertilization, 189
Double helix structure, 103
Down syndrome, 124
*Drosophila*, 95–99, 95*f*
Duckbill dinosaur, 134*f*
Ductus arteriosus, 310*t*
Ductus venosus, 310*t*
Duodenum, 254*f*, 255*f*, 270
Dyes, 16

### E

Earlobes, 115*f*, 116*f*, 344*f*
Ears
    comparative anatomy, 343–345
    fetal pig, 263*f*, 267*f*
    vertebrate, 244
Earthworms
    anatomy, 223*f*, 225–228, 225*f*,
      227*f*, 228*f*
    locomotion, 223*f*
    mating, 224*f*
Ecdysozoa, 217*f*
*Echinarachnius parma*, 236*f*
Echinoderms, 133*f*, 217*f*, 235–238
Effectors, motor neuron, 333, 338, 339
Eggs
    *Daphnia*, 354*f*
    in flowering plants, 192, 192*f*
    fusion with sperm, 77, 77*f*
    gymnosperm, 187
    in nonvascular plants, 170, 171, 171*f*
    of reptiles, 243
    seedless vascular plants, 176*f*
Ejaculatory ducts, 324*f*
Elbows, fetal pig, 263*f*, 267*f*
Electron microscopes, 15, 16–17, 17*f*
Electrophoresis, 110–111, 111*f*
*Elodea*
    cell structure, 33, 34*f*
    photosynthesis demonstration,
      58–59, 61, 62
    plasmolysis demonstration, 40
Embryo sac, 192, 192*f*
Embryos
    flowering plants, 189, 190*f*
    in gymnosperms, 184*f*, 185
    vertebrate similarities, 137–138, 137*f*
Emulsification, 279*f*
Endodermis, 202, 202*f*
Endoskeletons, 235, 237*f*
Endosperm, 189
Endospores, 151
Energy
    average daily intake, 286–288
    average daily needs, 289–293
    in selected foods, 287*t*

Left internal cervical artery, 312*f*
Left internal jugular vein, 313*f*
Left internal thoracic vein, 313*f*
Left pleural cavity, 268
Left pulmonary artery, 305*f*, 307*f*
Left pulmonary veins, 305*f*, 307*f*
Left renal artery, 312*f*, 314*f*
Left renal vein, 315*f*
Left subclavian artery, 305*f*, 307*f*, 311, 312*f*
Left subclavian vein, 313*f*
Left subscapular artery, 313*f*
Left subscapular vein, 313*f*
Left ventricle, 269*f*, 304, 305*f*, 307*f*, 309*f*
Legs, 230*f*, 231, 232, 232*f*. *See also* Hindlimbs
Length measurement, 10–11
Lens of eye
    accommodation, 340, 343
    anatomy in human, 341*t*
Lenses, microscope, 16, 17, 18, 20–21
Lenticels, 208, 208*f*
*Lepas anatifera*, 229*f*
Letter *e* slides, 22
Life cycles, 83
LifeSavers candy, 347
Ligaments, 219*f*, 327*f*, 341*f*
Light reactions, 55*f*
Lignin, 198–199
Lily anther, 78–79
*Limulus polyphemus*, 229*f*
Liquid measures, 12, 12*f*
Liquids, speed of diffusion in, 35
Liter, 12
Liver
    fetal pig, 269*f*, 270, 271*f*
    frog, 247, 248*f*, 251*f*
    main functions in human, 331
    role in digestion, 279*f*
    squid, 222*f*
Liverworts, 168*f*, 170, 172–173, 173*f*
Lobe-finned fishes, 242, 242*f*
Locomotion in earthworms, 223, 223*f*
Longitudinal fissure, 335*f*
Longitudinal muscles, 223*f*
Lophotrochozoa, 217*f*
Lower epidermis of leaf, 210
Low-power diameter of field, 23
Low-power objective, 21
*Lumbricus*, 224*f. See also* Earthworms
Lungs
    alveoli, 329
    fetal pig, 268, 269*f*, 271*f*
    frog, 247, 248*f*, 250*f*, 251*f*
    rat, 256*f*

Lycophytes, 168*f*, 173, 174, 174*f*
*Lycopodium*, 174, 174*f*
Lymphocytes, 17*f*
*Lysmata grabhami*, 229*f*

# M

M stage, 67
*Macrocystis*, 157*f*
Macronuclei, 160*f*
Madreporite, 237
Magnification, 22–23
Magnification changing knobs, 19
Malaria, 159
Male gametophytes
    in conifers, 184*f*, 185
    in flowering plants, 188, 190*f*, 191
    in seed plants, 181, 182, 182*f*
Male reproductive system, mammalian, 322–324, 323*f*, 324*f*
Malleus, 345
Malpighian tubules, 232, 232*f*
Mammals
    basic features, 245*f*
    cardiovascular systems, 304–308, 307*f*, 309*f*
    evolution, 242*f*
    external anatomy, 262–263, 263*f*
    fishes and reptiles compared, 252–259
    fossils, 134, 134*f*
    on geologic timescale, 131*t*
    oral cavity and pharynx, 264–265, 264*f*–265*f*
    rat anatomy, 256*f*
    reproductive systems, 322–327, 323*f*, 324*f*, 326*f*, 327*f*–328*f*
    respiratory and digestive systems, 329–331
    thoracic and abdominal cavities, 266, 267*f*, 268–272, 269*f*, 271*f*
    urinary systems, 320*f*, 321, 321*f*
Mammary glands, 262
Mantle, 218, 219, 220*f*, 222*f*
Maple leaf, 135*f*
*Marchantia*, 172, 173*f*
Mass extinctions, 131*t*, 132, 132*f*
Maxillary teeth, 246, 246*f*
Mayfly, 133*f*
Meats, 287*t*
Mechanical stage, 20*f*, 21
Medulla, frog, 252*f*
Medulla oblongata, 335*f*, 336
Megaphylls, 177
Megasporangia, 184*f*
Megaspores, 184*f*, 185, 187
Meiosis
    in gametogenesis, 65
    in human life cycle, 82–83, 83*f*

mitosis versus, 84, 85*f*
    overview, 77
    phases, 78–79
    variation produced by, 80–82
Meiosis I, 77, 78–79, 78*f*, 80–81, 84
Meiosis II, 77, 79*f*, 82
Meissner corpuscles, 346*f*
Mendel, Gregor, 87, 88, 90
Mendelian genetics, 87
Meninges, 337*f*
Menisci, 12, 12*f*
Meristematic tissues, 197, 198*f*, 201
Merkel disks, 346*f*
Mesenteries, 270
Mesonephric ducts, frog, 249, 250*f*, 251*f*
Mesophyll, 198*f*, 210
Mesozoic era, 131*t*
Messenger RNA, 101, 106, 107, 107*f*
Metacarpals, 136*f*, 137
Metamorphosis, 233–234
Metaphase, 69*f*, 71*f*
Metaphase I, 78*f*, 81
Metaphase II, 79*f*, 82
Metaphase plates, 67*t*, 85*f*
Meter, 10, 10*t*
Meter sticks, 10–11
Methylene blue, 25, 26
Metric system, 9–13
Micrometer, 10, 10*t*
Micronuclei, 160*f*
Microphylls, 174
Microscopes
    main types, 15–17
    using compound light microscopes, 19–24
    using stereomicroscopes, 17–19
    wet mount observations, 25–27
Microsporangia, 184*f*
Microspores, 184*f*, 185, 187*f*
Midbrain, 335*f*, 336
Middle ear, 344*t*
Middle layer of blood vessels, 316*f*, 317
Milk, 287*t*
Milligram, 11
Milliliter, 12
Millimeter, 10, 10*t*
Millipedes, 229, 229*f*
Miocene epoch, 131*t*
Mitochondria, 160*f*
Mitosis
    in animal cells, 67–70
    associated structures, 67*t*
    defined, 65
    in human life cycle, 83
    meiosis versus, 84, 85*f*
    in plant cells, 70–73
Mitral valve, 306
Molecular evidence for evolution, 143–145

Molecular genetics, 101, 143–145
Molluscs
    anatomy, 217–223, 218, 219*f*
    evolution, 217*f*
    fossils, 133*f*
Molting, 1
Monocots
    eudicots versus, 189, 189*t*,
        199–200, 199*f*
    herbaceous stems, 206, 206*f*
    root systems, 203, 203*f*
Monohybrid crosses, 88, 89*f*
Morels, 148*f*, 162*f*
Mosses, 168*f*, 170, 170*f*, 171–172
Motor neurons, 338, 339*f*
Mouth
    bivalve, 220*f*
    crayfish, 230*f*, 231
    *Daphnia*, 354*f*
    earthworms, 225*f*, 227*f*
    frog, 246, 246*f*
    grasshoppers, 232*f*
Mouthparts, grasshoppers, 232, 232*f*
Multiple-allele traits, 122–123
Muscles
    annelids, 223*f*
    bivalves, 219, 220*f*
    earthworms, 228*f*
Muscular foot, 218
Mushrooms, 164–165, 164*f*
Mycelia, 162, 163*f*

## N

Nanometer, 10, 10*t*
Nares, frog, 246, 246*f*
Nasopharynx, 264, 264*f*
Nautilus, 218*f*
Near point, 343
Neck, fetal pig, 268
Needles, pine, 185, 185*f*
Nephridia, 226, 227*f*, 228*f*
Nephrons, 321, 321*f*
*Nereis*, 224*f*
*Nereocystis*, 157*f*
Nerve cord, 241, 241*f*
Nerve ganglia, 232*f*
Nervous system
    animal ears, 344*f*
    central nervous system anatomy,
        334–338, 335*f*, 337*f*
    frog, 252, 252*f*
    genetic disorders, 120
    overview, 333
    peripheral nervous system anatomy,
        338–340, 339*f*
    skin receptors, 345–346
    tissues, 333

Net patterns, 199
Net photosynthesis, 58
Neurofibromatosis, 119
Neurons, 338
Nictitating membrane, frog, 246
Nipples, 262, 263*f*
Nitrogen oxides, 349
Nodes, 196, 208, 208*f*
Nonvascular plants, 170–173
Northern leopard frog, 245*f*
Nose, fetal pig, 263*f*, 267*f*
Nosepiece of microscope, 20
Nostrils, frog, 246
Notochord, 241, 241*f*, 242*f*
Nuclei
    division in mitosis, 67*t*
    *Euglena*, 161*f*
Nucleoids, 30*f*, 31, 149, 149*f*
Nucleoli, 67*t*, 78*f*
Nucleotides
    DNA, 101, 102–103, 102*f*
    RNA, 105, 105*f*
Nutrients in selected foods, 287*t*
Nymphs, grasshopper, 233, 234*f*

## O

Obesity, 295
Objectives (microscope), 20–21
Observations, 2, 3–4
Occipital lobe, 334, 335*f*
*Odontaster validus*, 236*f*
Oil immersion objective, 21
Olfactory bulb, 252*f*, 334, 335*f*
Olfactory nerve, 252*f*
Oligocene epoch, 131*t*
Oligochaetes, 224
Omnivores, 140
One-trait crosses, 88–92
Onion epidermal cells, 25–26, 25*f*
Onion root tip cells, 71–72
Oogenesis, 83
Open circulatory systems, 221, 230
*Ophiopholis aculeata*, 236*f*
Opisthokonts, 154*f*
Opposing leaves, 211
Optic chiasma, 334, 335*f*
Optic lobe, 252*f*
Optic nerve, 341*f*, 341*t*, 342
Oral cavity, fetal pig, 264, 264*f*
Oral groove, *Paramecium*, 160*f*
Oral hemal ring, 237*f*
Oral hood of lancelet, 244, 244*f*
Oral thrush, 165
Ordovician period, 131*t*
Organ systems, 276*f*
Organelles, 30
*Oscillatoria*, 152, 152*f*

Osmosis, 37–41, 38*f*
Ossicles, 344*t*
Ostium, 251*f*
Outer ear, 344*t*
Outer layer of blood vessels, 316*f*, 317
Oval opening, 310*t*, 331
Ovarian arteries, 314*f*
Ovarian ligaments, 327*f*
Ovarian vein, 315*f*, 326*f*
Ovaries
    earthworms, 226
    fetal pig, 326*f*
    flowering plants, 188, 188*f*, 189, 190*f*
    frog, 247, 248*f*, 251*f*
    grasshoppers, 232*f*
    mammalian, 319, 325
Overweight, 295
Oviducts
    frog, 249, 251*f*
    grasshoppers, 232*f*
    human, 327*f*
    mammalian, 325
Ovipositors, 232, 233*f*
Ovules
    archegonia versus, 181
    flowering plants, 187, 188*f*, 190*f*
    of gymnosperms, 184*f*, 185, 186
Oxygen, from photosynthesis, 57.
    *See also* Photosynthesis
Oxygen deprivation, 350–351
*Oxymetra erinacea*, 236*f*
*Oxytricha*, 27*f*

## P

Pacinian corpuscles, 346*f*
Pain receptors, 346*f*
Paleocene epoch, 131*t*
Paleozoic era, 130, 131*t*
Palisade mesophyll, 210
Palmately compound leaves, 211
Pancreas
    fetal pig, 269*f*, 272
    frog, 248
    human, 276*f*
    perch, 254*f*
    pH in, 50
    pigeon, 255*f*
    role in digestion, 279–282, 279*f*
Pancreatic amylase, 281–283
Pancreatic duct, 279*f*
Pancreatic juice, 279*f*
Pancreatic lipase, 279–280
*Pandorina*, 27*f*
Papillary muscles, 306, 307*f*
*Paramecium*, 27*f*, 148*f*, 160*f*, 351*f*
Parenchyma cells, 198
Parent cells, 67, 68, 77

P-T extinction, 132
Pteridophytes, 173, 174–175, 175f
Pulmonary arteries and veins
    fetal pig, 269f, 310
    human, 304, 305f, 307f, 309f
Pulmonary circuit, 304, 308, 309f, 310
Pulmonary semilunar valve, 306, 307f
Pulmonary systems, 257, 257f
Pulmonary trunk
    fetal pig, 310, 312f, 313f
    human, 304, 305f, 307f
Punnett squares, 88, 88f, 118, 118f
Pupa stage, 233, 234f
Pupil, 341f, 341t
Pure-breeding plants, 92
Pyloric stomach, sea star, 237f, 238
Pyrenoids
    *Euglena*, 27f, 161, 161f
    *Spirogyra*, 153, 155, 155f

## Q

Quaternary period, 131t

## R

Radial canals, 237f, 238
Radial symmetry, 216, 235
Radiolarians, 154f
Radius, 136f, 137
Radulas, 221
Rat anatomy, 256f
Rate of photosynthesis, 59, 59t, 61, 61t
Ray-finned fishes, 242
Rays, 209, 209f
Reactants, 45
Receptacles, 156, 188, 188f
Receptors. *See* Sensory receptors
Recessive alleles, 87, 115
Recessive disorders, 120
Rectum
    bivalve, 220f
    fetal pig, 272
    functions in human, 276f
    grasshoppers, 233f
    pigeon, 255f
Red algae, 156, 157f
Red blood cells. *See* Erythrocytes
Red perception, 121
Red pine needles, 185f
Red tides, 158
Reflexes, spinal, 339–340
Refraction, 340
Renal arteries
    fetal pig, 314, 314f, 321f
    human, 249f
Renal cortex, 321, 321f
Renal medulla, 321, 321f

Renal pelvis, 321, 321f
Renal pyramids, 321f
Renal veins, 249f, 321f
Replication of DNA, 65, 66, 67, 67f,
    101, 102, 103–104, 104f
Reproduction. *See also* Asexual
    reproduction; Sexual reproduction
    earthworms, 224f, 225, 226
    in flowering plants, 189–193, 190f
    fungi, 162–165
    in gymnosperms, 184f, 185
    land plant life cycle, 169
    nonvascular plants, 170–173
    protists, 153, 154f, 155
    seed plant life cycle, 182, 182f
    seedless vascular plants, 173–179
Reproductive systems
    frog, 248–249
    mammalian females, 325–327, 326f, 327f
    mammalian males, 322–324, 323f, 324f
Reptiles
    adaptations to land, 242, 242f
    basic features, 245f
    birds as, 134, 134f, 254
    embryos, 137f
    fishes and mammals compared,
        252–259
    fossils, 134f
    on geologic timescale, 131t
Resolution (microscope), 16
Respiratory systems
    comparative anatomy, 258–259
    fetal pig, 329–330, 329f
    frog, 247
Results, reporting, 358, 360
Retina, 341f, 341t
Rh factor, 122
*Rhapidostreptus virgato*, 229f
Rhizaria, 154f
Rhizoids, 172, 173f, 175f, 176f, 179f
Rhizomes, 174, 176f, 177, 207, 207f
*Rhizopus*, 164
Rib cage, 259
Ribonucleic acid (RNA), 101,
    105–109, 105f
Ribose, 105
Ribosomal RNA, 101, 106
Ribosomes
    animal and plant, 31
    in bacteria, 149, 149f
    prokaryotic, 30, 30f
    protein synthesis in, 108, 109, 109f
Right atrioventricular valve, 306, 307f
Right atrium
    fetal pig, 269f
    frog, 247
    human, 304, 305f, 307f, 309f
    rat, 256f

Right brachial artery, 313f
Right common carotid artery, 271f,
    311, 313f
Right coronary artery, 305f
Right external jugular vein, 312f
Right iliac vein, 312f
Right internal jugular vein, 312f
Right internal thoracic vein, 312f
Right ovarian vein, 312f
Right pleural cavity, 268
Right pulmonary artery, 304, 307f
Right pulmonary veins, 304, 307f
Right spermatic vein, 311, 312f
Right subclavian artery, 311, 313f
Right subclavian vein, 311, 312f
Right ventricle, 269f, 304, 305f, 307f,
    308, 309f
Ring canal, 237f, 238
Ringworm, 165, 165f
RNA polymerase, 107, 107f
Rockweed, 156, 157f
Rod cells, 340, 341f, 341t
Root cap, 201, 201f
Root hair plexuses, 346f
Root hairs, 196f, 201, 201f, 202f
Root systems, flowering plants, 196,
    196f, 200–204, 201f
Root tips, 198f, 200–201, 201f
Roots
    basic functions, 195
    eudicot, 202, 202f
    monocot, 203, 203f
    tissues, 197–200, 200–203, 201f
    types, 205
Rotating lenses (microscope), 18
Rotifers, 217f, 351f
Rough ER, 31
Round ligament, 327f
Round window, 344f
Ruffini endings, 346f

## S

S stage, 66–67
*Sabella*, 224f
Saccule, 344t, 345
Sagittal crest, 140, 140f
Salivary glands, 232f, 276f
*Salmonella*, 148f
Sand dollars, 235, 236f
SAR supergroup, 154f
Sassafras leaf, 135f
Scallops, 217, 218f
Scanning electron microscopes, 16, 17f
Scanning objective, 20
Scars, on woody stems, 208, 208f
Scavengers, 225, 230
Scientific method, 2–3, 2f